Bayesian Statistics and Marketing

Bayesian Statistics and Marketing

Peter E. Rossi
University of Chicago, USA

Greg M. Allenby
Ohio State University, USA

Robert McCulloch
University of Chicago, USA

John Wiley & Sons, Ltd

John Wiley & Sons Ltd, The Atrium, Southern Gate, Chichester,
West Sussex PO19 8SQ, England

Telephone (+44) 1243 779777

Email (for orders and customer service enquiries): cs-books@wiley.co.uk
Visit our Home Page on www.wiley.com

Reprinted with corrections August 2006

Other Wiley Editorial Offices

John Wiley & Sons Inc., 111 River Street, Hoboken, NJ 07030, USA

Jossey-Bass, 989 Market Street, San Francisco, CA 94103-1741, USA

Wiley-VCH Verlag GmbH, Boschstr. 12, D-69469 Weinheim, Germany

John Wiley & Sons Australia Ltd, 42 McDougall Street, Milton, Queensland 4064, Australia

John Wiley & Sons (Asia) Pte Ltd, 2 Clementi Loop #02-01, Jin Xing Distripark, Singapore 129809

John Wiley & Sons Canada Ltd, 22 Worcester Road, Etobicoke, Ontario, Canada M9W 1L1

Wiley also publishes its books in a variety of electronic formats. Some content that appears
in print may not be available in electronic books.

Library of Congress Cataloging-in-Publication Data:

Rossi, Peter E. (Peter Eric), 1955-
Bayesian, statistics and marketing / Peter Rossi and Greg Allenby, Rob McCulloch.
p. cm.
Includes bibliographical references and index.
ISBN-13: 978-0-470863-67-1 (HB)
ISBN-10: 0-470863-67-6 (HB)
1. Marketing research – Mathematical models. 2. Marketing – Mathematical models. 3. Bayesian statistical decision theory. I. Allenby, Greg M. (Greg Martin), 1956- II. McCulloch, Robert E. (Robert Edward) III. Title.
HF5415.2.R675 2005
658.8′3′015118 – dc22

2005016418

British Library Cataloguing in Publication Data

A catalogue record for this book is available from the British Library

ISBN-13: 978-0-470863-67-1 (HB)
ISBN-10: 0-470863-67-6 (HB)

Typeset in 10/12pt Galliard by Laserwords Private Limited, Chennai, India

To our wives –
Laurie, Tricia, and Nancy

for our children –
Ben and Emily, Kate and Mark, Kate and Stephen

and with thanks to our parents –
Alice and Peter, Marilyn and Stan, Ona and Ernest

Contents

1 Introduction

The past ten years have seen a dramatic increase in the use of Bayesian methods in marketing. Bayesian analyses have been conducted over a wide range of marketing problems from new product introduction to pricing, and with a wide variety of data sources. While the conceptual appeal of Bayesian methods has long been recognized, the recent popularity stems from computational and modeling breakthroughs that have made Bayesian methods attractive for many marketing problems. This book aims to provide a self-contained and comprehensive treatment of Bayesian methods and the marketing problems for which these methods are especially appropriate. There are unique aspects of important problems in marketing that make particular models and specific Bayesian methods attractive. We, therefore, do not attempt to provide a generic treatment of Bayesian methods. We refer the interested reader to classic treatments by Robert and Casella (2004), Gelman *et al.* (2004), and Berger (1985) for more general-purpose discussion of Bayesian methods. Instead, we provide a treatment of Bayesian methods that emphasizes the unique aspects of their application to marketing problems.

Until the mid-1980s, Bayesian methods appeared to be impractical since the class of models for which the posterior inference could be computed was no larger than the class of models for which exact sampling results were available. Moreover, the Bayes approach does require assessment of a prior, which some feel to be an extra cost. Simulation methods, in particular Markov chain Monte Carlo (MCMC) methods, have freed us from computational constraints for a very wide class of models. MCMC methods are ideally suited for models built from a sequence of conditional distributions, often called hierarchical models. Bayesian hierarchical models offer tremendous flexibility and modularity and are particularly useful for marketing problems.

There is an important interaction between the availability of inference methods and the development of statistical models. Nowhere has this been more evident than in the application of hierarchical models to marketing problems. Hierarchical models match closely the various levels at which marketing decisions are made – from individual consumers to the marketplace. Bayesian researchers in marketing have expanded on the standard set of hierarchical models to provide models useful for marketing problems. Throughout this book, we will emphasize the unique aspects of the modeling problem in marketing and the modifications of method and models that researchers in marketing have devised. We hope to provide the requisite methodological knowledge and an appreciation of how these methods can be used to allow the reader to devise and analyze

Bayesian Statistics and Marketing P. E. Rossi, G. M. Allenby and R. McCulloch

new models. This departs, to some extent, from the standard model of a treatise in statistics in which one writes down a set of models and catalogues the set of methods appropriate for analysis of these models.

1.1 A BASIC PARADIGM FOR MARKETING PROBLEMS

Ultimately, marketing data results from customers taking actions in a particular context and facing a particular environment. The marketing manager can influence some aspects of this environment. Our goal is to provide models of these decision processes and then make optimal decisions conditional on these models. Fundamental to this prospective is that customers are different in their needs and wants for marketplace offerings, thus expanding the set of actions that can be taken. At the extreme, actions can be directed at specific individuals. Even if one-on-one interaction is not possible, the models and system of inference must be flexible enough to admit nonuniform actions.

Once the researcher acknowledges the existence of differences between customers, the modeling task expands to include a model of these differences. Throughout this book, we will take a stand on customer differences by modeling differences via a probability distribution. Those familiar with standard econometric methods will recognize this as related to a random coefficients approach. The primary difference is that we regard the customer-level parameters not as nuisance parameters but as the goal of inference. Inferences about customer differences are required for any marketing action, from strategic decisions associated with formulating offerings to tactical decisions of customizing prices. Individuals who are most likely to respond to these variables are those who find highest value in the offering's attributes and those who are most price-sensitive, neither of whom are well described by parameters such as the mean of the random coefficients distribution.

Statistical modeling of marketing problems consists of three components: (i) within-unit behavior; (ii) across-unit behavior; (iii) action. 'Unit' refers the particular level of aggregation dictated by the problem and data availability. In many instances, the unit is the consumer. However, it is possible to consider both less and more aggregate levels of analyses. For example, one might consider a particular consumption occasion or survey instance as the 'unit' and consider changes in preferences across occasions or over time as part of the model (an example of this is in Yang *et al.* 2002). In marketing practice, decisions are often made at a much higher level of aggregation such as the 'key account' or sales territory. In all cases, we consider the 'unit' as the lowest level of aggregation considered explicitly in the model.

The first component of the problem is the conditional likelihood for the unit-level behavior. We condition on unit-specific parameters which are regarded as the sole source of between-unit differences. The second component is a distribution of these unit-specific parameters over the population of units. Finally, the decision problem is the ultimate goal of the modeling exercise. We typically postulate a profit function and ask what is the optimal action conditional on the model and the information in the data. Given this view of marketing problems, it is natural to consider the Bayesian approach to inference which provides a unified treatment of all three components.

1.2 A SIMPLE EXAMPLE

As an example of the components outlined in Section 1.1, consider the case of consumers observed making choices between different products. Products are characterized by some vector of choice attribute variables which might include product characteristics, prices and advertising. Consumers could be observed to make choices either in the marketplace or in a survey/experimental setting. We want to predict how consumers will react to a change in the marketing mix variables or in the product characteristics. Our ultimate goal is to design products or vary the marketing mix so as to optimize profitability.

We start with the 'within-unit' model of choice conditional on the observed attributes for each of the choice alternatives. A standard model for this situation is the multinomial logit model,

$$\Pr[i|x_1, \ldots, x_p, \beta] = \frac{\exp(x_i'\beta)}{\sum_{j=1}^{p} \exp(x_j'\beta)}. \quad (1.2.1)$$

If we observe more than one observation per consumer, it is natural to consider a model which accommodates differences between consumers. That is, we have some information about each consumer's preferences and we can start to tease out these differences. However, we must recognize that in many situations we have only a small amount information about each consumer. To allow for the possibility that each consumer has different preferences for attributes, we index the β vectors by c for consumer c. Given the small amount of information for each consumer, it is impractical to estimate separate and independent logits for each of the C consumers. For this reason, it is useful to think about a distribution of coefficient vectors across the populations of consumers. One simple model would be to assume that the βs are distributed normally over consumers:

$$\beta_c \sim N(\mu, V_\beta). \quad (1.2.2)$$

One common use of logit models is to compute the implication of changes in marketing actions for aggregate market shares. If we want to evaluate the effect on market share for a change in x for alternative i, then we need to integrate over the distribution in (1.2.2). For a market with a large number of consumers, we might view the expected probability as market share and compute the derivative of market share with respect to an element of x:[1]

$$\frac{\partial \mathrm{MS}(i)}{\partial x_{i,j}} = \frac{\partial}{\partial x_{i,j}} \int \Pr[i|x_1, \ldots, x_p, \beta]\varphi(\beta|\mu, V_\beta)\, d\beta. \quad (1.2.3)$$

Here $\varphi(\cdot)$ is the multivariate normal density.

The derivatives given in (1.2.3) are necessary to evaluate uniform marketing actions such as changing price in a situation in which all consumers face the same price. However, many marketing actions are aimed at a subset of customers or, in some cases, individual customers. In this situation, it is desirable to have a way of estimating not only the

[1] Some might object to this formulation of the problem as the aggregate market shares are deterministic functions of x. It is a simple matter to add an additional source of randomness to the shares. We are purposely simplifying matters for expositional purposes.

common parameters that drive the distribution of βs across consumers but also the individual βs.

Thus, our objective is to provide a way of inferring about $\{\beta_1, \ldots, \beta_C\}$ as well as μ, V_β. We also want to use our estimates to derive optimal marketing policies. This will mean to maximize expected profits over the range of possible marketing actions:

$$\max_a \mathrm{E}[\pi(a|\Omega)]. \tag{1.2.4}$$

Ω represents the information available about the distribution of the outcomes resulting from marketing actions. Clearly, information about the distribution of choice given the model parameters as well as information about the parameters will be relevant to selecting the optimal action. Our goal, then, is to adopt a system of inference and decision-making that will make it possible to solve (1.2.4). In addition, we will require that there be practical ways of implementing this system of inference. By 'practical' we mean computable for problems of the size which practitioners in marketing encounter.

Throughout this book, we will consider models similar to the simple case considered here and develop these inference and computational tools. We hope to convince the reader that the Bayesian alternative is the right choice.

1.3 BENEFITS AND COSTS OF THE BAYESIAN APPROACH

At the beginning of Chapter 2, we outline the basics of the Bayesian approach to inference and decision-making. There are really no other approaches which can provide a unified treatment of inference and decision as well as properly accounting for parameter and model uncertainty. However compelling the logic behind the Bayesian approach, it has not been universally adopted. The reason for this is that there are nontrivial costs of adopting the Bayesian perspective. We will argue that some of these 'costs' have been dramatically reduced and that some 'costs' are not really costs but are actually benefits.

The traditional view is that Bayesian inference provides the benefits of exact sample results, integration of decision-making, 'estimation', 'testing', and model selection, and a full accounting of uncertainty. Somewhat more controversial is the view that the Bayesian approach delivers the answer to the right question in the sense that Bayesian inference provides answers conditional on the observed data and not based on the distribution of estimators or test statistics over imaginary samples not observed. Balanced against these benefits are three costs: (i) formulation of a prior; (ii) requirement of a likelihood function; and (iii) computation of various integrals required in Bayesian paradigm. Development of various simulation-based methods in recent years has drastically lowered the computational costs of the Bayesian approach. In fact, for many of the models considered in this book, non-Bayesian computations would be substantially more difficult or, in some cases, virtually impossible. Lowering of the computational barrier has resulted in a huge increase in the amount of Bayesian applied work.

In spite of increased computational feasibility or, indeed, even computational superiority of the Bayesian approach, some are still reluctant to use Bayesian methods because of the requirement of a prior distribution. From a purely practical point of view, the prior is yet another requirement that the investigator must meet and this imposes a cost on the use of Bayesian approaches. Others are reluctant to utilize prior information based on concerns of scientific 'objectivity'. Our answer to those with concerns about 'objectivity' is twofold. First, to our minds, scientific standards require that replication is possible. Bayesian inference with explicit priors meets this standard. Secondly, marketing is an applied field which means that the investigator is facing a practical problem often in situations with little information and should not neglect sources of information outside of the current data set.

For problems with substantial data-based information, priors in a fairly broad range will result in much the same a posteriori inferences. However, in any problem in which the data-based information to 'parameters' ratio is low, priors will matter. In models with unit-level parameters, there is often relatively little data-based information, so that it is vital that the system of inference incorporate even small amounts of prior information. Moreover, many problems in marketing explicitly involve multiple information sets so that the distinction between the sample information and prior information is blurred.

High-dimensional parameter spaces arise due to either large numbers of units or the desire to incorporate flexibility in the form of the model specification. Successful solution of problems with high-dimensional parameter spaces requires additional structure. Our view is that prior information is one exceptionally useful way to impose structure on high-dimensional problems. The real barrier is not the philosophical concern over the use of prior information but the assessment of priors in high-dimensional spaces. We need devices for inducing priors on high-dimensional spaces that incorporate the desired structure with a minimum of effort in assessment. Hierarchical models are one particularly useful method for assessing and constructing priors over parameter spaces of the sort which routinely arise in marketing problems.

Finally, some have argued that any system of likelihood-based inference is problematic due to concerns regarding misspecification of the likelihood. Tightly parameterized likelihoods can be misspecified, although the Bayesian is not required to believe that there is a 'true' model underlying the data. In practice, a Bayesian can experiment with a variety of parametric models as a way of guarding against misspecification. Modern Bayesian computations and modeling methods make the use of a wide variety of models much easier than in the past. Alternatively, more flexible non- or semi-parametric models can be used. All nonparametric models are just high-dimensional models to the Bayesian and this simply underscores the need for prior information and Bayesian methods in general. However, there is a school of thought prominent in econometrics that proposes estimators which are consistent for the set of models outside one parametric class (method of moments procedures are the most common of this type). However, in marketing problems, parameter estimates without a probability model are of little use. In order to solve the decision problem, we require the distribution of outcome measures conditional on our actions. This distribution requires not only point estimates of parameters but also a specification of their distribution. If we regard the relevant distribution as part of the parameter space, then this statement is equivalent to the need for estimates of all rather than a subset of model parameters.

In a world with full and perfect information, revealed preference should be the ultimate test of the value of a particular approach to inference. The increased adoption of Bayesian methods in marketing shows that the benefits do outweigh the costs for many problems of interest. However, we do feel that the value of Bayesian methods for marketing problem is underappreciated due to lack of information. We also feel that many of the benefits are as yet unrealized since the models and methods are still to be developed. We hope that this book provides a platform for future work on Bayesian methods in marketing.

1.4 AN OVERVIEW OF METHODOLOGICAL MATERIAL AND CASE STUDIES

Chapters 2 and 3 provide a self-contained introduction to the basic principles of Bayesian inference and computation. A background in basic probability and statistics at the level of Casella and Berger (2002) is required to understand this material. We assume a familiarity with matrix notation and basic matrix operations, including the Cholesky root. Those who need a refresher or a concise summary of the relevant material might examine Appendices A and B of Koop (2003). We will develop some of the key ideas regarding joint, conditional, and marginal densities at the beginning of Chapter 2 as we have found that this is an area not emphasized sufficiently in standard mathematical statistics or econometrics courses.

We recognize that a good deal of the material in Chapters 2 and 3 is available in many other scattered sources, but we have not found a reference which puts it together in a way that is useful for those interested in marketing problems. We also will include some of the insights that we have obtained from the application of these methods.

Chapters 4 and 5 develop models for within-unit and across-unit analysis. We pay much attention to models for discrete data as much disaggregate marketing data involves aspects of discreteness. We also develop the basic hierarchical approach to modeling differences across units and illustrate this approach with a variety of different hierarchical structures and priors.

The problem of model selection and decision theory is developed in Chapter 6. We consider the use of the decision-based metric in valuing information sources and show the importance of loss functions in marketing applications.

Chapter 7 treats the important problem of simultaneity. In models with simultaneity, the distinction between dependent and independent variables is lost as the models are often specified as a system of equations which jointly or simultaneously determine the distribution of a vector of random variables conditional on some set of explanatory or exogeneous variables. In marketing applications, the marketing mix variables and sales are jointly determined given a set of exogeneous demand or cost shifters.

These core chapters are followed by five case studies from our research agenda. These case studies illustrate the usefulness of the Bayesian approach by tackling important problems which involve extensions or elaborations of the material covered in the first seven chapters. Each of the case studies has been rewritten from their original journal form to use a common notation and emphasize the key points of differentiation for each article. Data and code are available for each of the case studies.

1.5 COMPUTING AND THIS BOOK

It is our belief that no book on practical statistical methods can be credible unless the authors have computed all the methods and models contained therein. For this reason, we have imposed the discipline on ourselves that nothing will be included that we have not computed. It is impossible to assess the practical value of a method without applying it in a realistic setting. Far too often, treatises on statistical methodology gloss over the implementation. This is particularly important with modern Bayesian methods applied to marketing problems. The complexity of the models and the dimensionality of the data can render some methods impractical. MCMC methods can be theoretically valid but of little practical value. Computations lasting more than a day can be required for adequate inference due to high autocorrelation and slow computation of an iteration of the chain.

If a method takes more than 3 or 4 hours of computing time on standard equipment, we deem it impractical in the sense that most investigators are unwilling to wait much longer than this for results. However, what is practical depends not only on the speed of computing equipment but also on the quality of the implementation. The good news is that in 2005 even the most pedestrian computing equipment is capable of truly impressive computations, unthinkable at the beginning of the MCMC revolution in the late 1980s and early 1990s. Achieving the theoretical capabilities of the latest CPU chip may require much specialized programming, use of optimized BLAS libraries and the use of a low-level language such as C or FORTRAN. Most investigators are not willing to make this investment unless their primary focus is on the development of methodology. Thus, we view a method as 'practical' if it can be computed in a relatively high-level computing environment which mimics as closely as possible the mathematical formulas for the methods and models. For even wider dissemination of our methods, some sort of prepackaged set of methods and models is also required.

For these reasons, we decided to program the models and methods of this book in the R language. In addition, we provide a website for the book which provides further data and code for models discussed in the case studies. R is free, widely accepted in the statistical community, and offers much of the basic functionality needed and support for optimized matrix operations. We have taken the philosophy advocated by Chambers and others that one should code in R, profile the code, and, if necessary, rewrite in a lower-level language such as C. This philosophy has produced over 4000 lines of R code to implement the models and methods in this book and less than 500 lines of C and C++ code. We have been impressed by the speed of R running on standard computing equipment. This is a testimony to the speed of modern CPUs as well as the hard work of the R development team and contributors. In many instances, we can achieve more than adequate speed without any low-level code. We do not claim to have come anywhere near the theoretical speed possible for these applications. The gap between what is theoretically possible and what is achievable in R is only important if you are faced with a computing bottleneck.

CPU speed is not the only resource that is important in computing. Memory is another resource which can be a bottleneck. Our view is that memory is so cheap that we do not want to modify our code to deal with memory constraints. All of our programs are designed to work entirely in memory. All of our applications use less than 1 GB of memory. Given that 512 MB is now relatively standard for Windows machines, we think this is reasonable as much as it might raise eyebrows among those who were brought up in a memory-poor world.

Our experiences coding and profiling the applications in this book have changed our views on statistical computing. We were raised to respect minor changes in the speed of computations via various tricks and optimization of basic linear algebra operations. When we started to profile our code, we realized that, to a first approximation, linear algebra is free. The mainstay of Bayesian computations is the Cholesky root. These are virtually free on modern equipment (for example, one can compute the Cholesky root of 1000×1000 matrices at the rate of at least 200 per minute on standard-issue laptop computers). We found conversions from vectors to matrices and other 'minor' operations to be more computationally demanding. Minimizing the number of matrix decompositions or taking advantage of the special structure of the matrices involved often has only minor impact. Optimization frequently involves little more than avoiding loops over the observations in the data set.

Computing also has an important impact on those who wish to learn from this book. We recognized, from the start, that our audience may be quite diverse. It is easy to impose a relatively minimal requirement regarding the level of knowledge of Bayesian statistics. It is harder to craft a set of programs which can be useful to readers with differing computing expertise and time to invest in computing. We decided that a two-prong attack was necessary: First, for those who want to use models pretty much 'off-the-shelf', we have developed an R package to implement most of the models developed in the book; and second, for those who want to learn via programming and who wish to extend the methods and models, we provide detailed code and examples for each of the chapters of the book and for each of the case studies.

R Our R package, *bayesm*, is available on the Comprehensive R Archive Network (CRAN, google 'R language' for the URL). *bayesm* implements all of the models and methods discussed in Chapters 1–7 (see Appendix B for more information on *bayesm* and Appendix A for an introduction to R). The book's website, www.wiley.com/go/bsm, provides documented code, data sets and additional information for those who wish to adapt our models and methods. Throughout this book, a boldface "R" in the left margin indicates a reference to *bayesm*.

We provide this code and examples with some trepidation. In some sense, those who really want to learn this material intimately will want to write their own code from scratch, using only some of our basic functions. We hope that providing the 'answers' to the problem will not discourage study. Rather, we hope many of our readers will take our code as a base to improve on. We expect to see much innovation and improvement on what we think is a solid base.

ACKNOWLEDGEMENTS

We owe a tremendous debt to our teachers, Seymour Geisser, Dennis Lindley and Arnold Zellner, who impressed upon us the value and power of the Bayesian approach. Their enthusiasm is infectious. Our students (including Andrew Ainslie, Neeraj Arora, Peter Boatwright, Yancy Edwards, Tim Gilbride, Lichung Jen, Ling-Jing Kao, Jaehwan Kim, Alan Montgomery, Sandeep Rao, and Sha Yang) have been great collaborators as well as patient listeners as we have struggled to articulate this program of research. Junhong Chu and Maria Ana Vitorino read the manuscript very carefully and rooted out numerous errors and expositional problems. Greg would like to thank Vijay Bhargava for his encouragement in both personal and professional life.

McCulloch and Rossi gratefully acknowledge support from the Graduate School of Business, University of Chicago. Rossi also thanks the James M. Kilts Center for Marketing, Graduate School of Business, University of Chicago for support. Allenby gratefully acknowledges support from the Fisher College of Business, Ohio State University.

2

Bayesian Essentials

Using this Chapter

This chapter provides a self-contained introduction to Bayesian inference. For those who need a refresher in distribution theory, Section 2.0 provides an introduction to marginal, joint, and conditional distributions and their associated densities. We then develop the basics of Bayesian inference, discuss the role of subjective probability and priors and provide some of the most compelling arguments for adopting the Bayesian point of view. Regression models (both univariate and multivariate) are considered, along with their associated natural conjugate priors. Asymptotic approximations and importance sampling are introduced as methods for nonconjugate models. Finally, a simulation primer for the basic distributions/models in Bayesian inference is provided. Those who want an introduction to Bayesian inference without many details should concentrate on Sections 2.1–2.5 and Section 2.8.1.

2.0 ESSENTIAL CONCEPTS FROM DISTRIBUTION THEORY

Bayesian inference relies heavily on probability theory and, in particular, distributional theory. This section provides a review of basic distributional theory with examples designed to be relevant to Bayesian applications.

A basic starting point for probability theory is a *discrete* random variable, X. X can take on a countable number of values, each with some probability. The classic example would be a Bernoulli random variable, where X takes the value 1 with probability p and 0 with probability $1 - p$. X denotes some event such as whether a company will sell a product tomorrow. p represents the probability of a sale. For now, let us set aside the question of whether this probability can represent a long-run frequency or whether it represents a subjective probability (note: it is hard to understand the long-run frequency argument for this example since it requires us to imagine an infinite number of 'other worlds' for the event of a sale tomorrow). We can easily extend this example to the number of units sold tomorrow. Then X is still discrete but can take on the values $0, 1, 2, \ldots, m$ with probabilities, $p_0, p_1, \ldots, p_m$. X now has a nontrivial probability distribution. With knowledge of this distribution, we can answer any question such as the probability that there will be at least one sale tomorrow, the probability that there

Bayesian Statistics and Marketing P. E. Rossi, G. M. Allenby and R. McCulloch

will be between 1 and 10 sales, etc. In general, we can compute the probability that sales will be in any set simply by summing over the probabilities of the elements in the set:

$$\Pr(X \in A) = \sum_{x \in A} p_x. \tag{2.0.1}$$

We can also compute the *expectation* of the number of units sold tomorrow as the average over the probability distribution.

$$E[X] = \sum_{i=0}^{m} i p_i. \tag{2.0.2}$$

If we are looking at aggregate sales of a popular consumer product, we might approximate sales as a *continuous* random variable which can take on any nonnegative real number. For this situation, we must summarize the probability distribution of X by a probability density. A density function is a *rate* function which tells us the probability per volume or unit of X. X has a density function, $p_X(x)$; p_X is a positive-valued function which integrates to one. To find the probability that X takes on any set of values we must integrate $p_X(\bullet)$ over this set:

$$\Pr(X \in A) = \int_A p_X(x|\theta)\, \mathrm{d}x. \tag{2.0.3}$$

This is very much the analog of the discrete sum in (2.0.1). The sense in which p is a rate function is that the probability that $X \in (x_0, x_0 + \mathrm{d}x)$ is approximately $p_X(x_0)\,\mathrm{d}x$. Thus, the probability density function, $p_X(\bullet)$, plays the same role as the discrete probability (sometimes called the probability mass function) in the discrete case. We can easily find the expectation of any function of X by computing the appropriate integral:

$$E[f(X)] = \int f(x) p(x|\theta)\, \mathrm{d}x. \tag{2.0.4}$$

In many situations, we will want to consider the *joint* distribution of two or more random variables, both of which are continuous. For example, we might consider the joint distribution of sales tomorrow in two different markets. Let X denote the sales in market A and Y denote the sales in market B. For this situation, there is a bivariate density function, $p_{X,Y}(x, y)$. This density gives the probability rate per unit of area in the plane. That is, the probability that both $X \in (x_0, x_0 + \mathrm{d}x)$ and $Y \in (y_0, y_0 + \mathrm{d}y)$ is approximately $p_{X,Y}(x_0, y_0)\,\mathrm{d}x\,\mathrm{d}y$. With the joint density, we compute the probability of any set of (X,Y) values. For example, we can compute the probability that both X and Y are positive. This is the area of under the density for the positive orthant:

$$\Pr(X > 0 \text{ and } Y > 0) = \int_0^\infty \int_0^\infty p_{X,Y}(x, y)\, \mathrm{d}x\, \mathrm{d}y. \tag{2.0.5}$$

For example, the multinomial probit model, considered in Chapter 4, has choice probabilities defined by the integrals of a multivariate normal density over various cones. If $p_{X,Y}(\bullet, \bullet)$ is a bivariate normal density, then (2.0.5) is one such equation.

Given the joint density, we can also compute the *marginal* densities of each of the variables X and Y. That is to say, if we know everything thing about the joint distribution, we certainly know everything about the marginal distribution. The way to think of this is via simulation. Suppose we were able to simulate from the joint distribution. If we look at the simulated distribution of either X or Y alone, we have simulated the marginal distribution.

To find the marginal density of X, we must average the *joint* density over all possible values of Y:

$$p_X(x) = \int p_{X,Y}(x, y)\,\mathrm{d}y. \tag{2.0.6}$$

A simple example will help make this idea clear. Suppose X, Y are uniformly distributed over the triangle $\{X, Y : 0 < Y < 1 \text{ and } Y < X < 1\}$, depicted in Figure 2.1. A uniform distribution means that the density is constant over the shaded triangle. The area of this triangle is $\frac{1}{2}$, so this means that the density must be 2 in order to ensure that the joint density integrates to 1:

$$\begin{aligned}\int_0^1 \int_y^1 p_{X,Y}(x, y)\,\mathrm{d}x\,\mathrm{d}y &= \int_0^1 \int_y^1 2\,\mathrm{d}x\,\mathrm{d}y = \int_0^1 (2x|_y^1)\,\mathrm{d}y \\ &= \int_0^1 (2 - 2y)\,\mathrm{d}y = (2y - y^2)|_0^1 = 1\end{aligned}$$

This means that the joint density is a surface over the triangle with height 2.

We can use (2.0.6) to find the marginal distribution of X by integrating out Y:

$$p_X(x) = \int p_{X,Y}(x, y)\,\mathrm{d}y = \int_0^x 2\,\mathrm{d}y = 2y|_0^x = 2x.$$

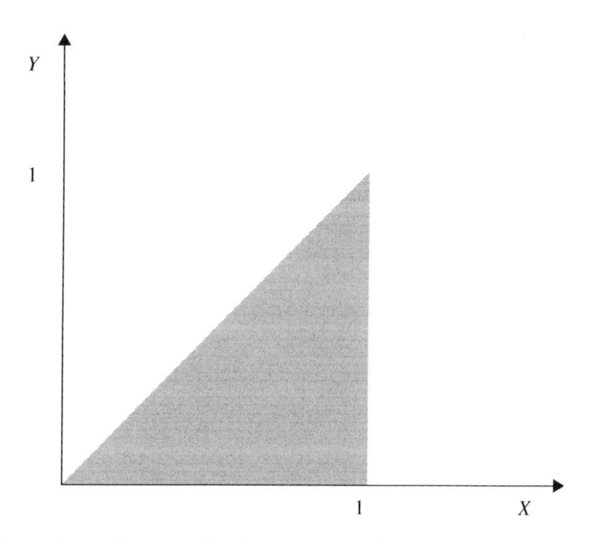

Figure 2.1 Support for the example of a bivariate distribution

Thus, the marginal distribution of X is not uniform! The density increases as x increases toward 1. The marginal density of Y can easily be found to be of the 'reverse' shape, $p_Y(y) = 2 - 2y$. This makes intuitive sense as the joint density is defined over the 'widest' area with X near 1 and with Y near 0.

We can also define the concept of a *conditional* distribution and conditional density. If X, Y have a joint distribution, we can ask what is the conditional distribution of Y given $X = x$. If X, Y are continuous random variables, then the conditional distribution of Y given $X = x$ is also a continuous random variable. The conditional density of $Y|X$ can be derived from the marginal and joint densities:[1]

$$p_{Y|X}(y|x) = \frac{p_{X,Y}(x, y)}{p_X(x)}. \tag{2.0.7}$$

The argument of the conditional density on the left-hand side of (2.0.7) is written $y|x$ to emphasize that there is a different density for every value of the conditioning argument x. We note that the conditional density is proportional to the joint! The marginal only serves to get the right normalization.

Let us return to our simple example. The conditional distribution of $Y|X = x$ is simply a slice of the joint density along a vertical line at the point x. This is clearly uniform but only extends from 0 to x. We can use (2.0.7) to get the right normalization.

$$p_{Y|X}(y|x) = \frac{2}{2x}, \qquad y \in (0, x).$$

Thus, if $x = 1$, then the density is uniform over $(0, 1)$ with height 1. The dependence between X and Y is only evidenced by the fact that the range of Y is restricted by the value of x.

In many statistics courses, we are taught that correlation is a measure of the dependence between two random variables. This stems from the bivariate normal distribution which uses correlation to drive the shape of the joint density. Let us start with two independent standard normal random variables, Z and W. This means that their joint density factorizes:

$$p_{Z,W}(z, w) = p_Z(z)p_W(w) \tag{2.0.8}$$

(this is because of the product rule for independent events). Each of the standard normal densities is given by

$$p_Z(z) = \frac{1}{\sqrt{2\pi}} \exp\left(-\frac{1}{2}z^2\right) \tag{2.0.9}$$

If we create X and Y by an appropriate linear combination of Z and W, we can create correlated or dependent random variables.

$$X = Z, \qquad Y = \rho Z + \sqrt{(1 - \rho^2)}\,W.$$

[1] The Borel paradox not withstanding.

X and Y have a correlated bivariate normal density with correlation coefficient ρ:

$$p_{X,Y}(x,y) = \frac{1}{2\pi\sqrt{(1-\rho^2)}} \exp\left\{-\frac{1}{2(1-\rho^2)}\left[x^2 - 2\rho xy + y^2\right]\right\}. \tag{2.0.10}$$

It is possible to show that

$$\operatorname{cov}(X,Y) = E[XY] = \iint xy p_{X,Y}(x,y)\,\mathrm{d}x\,\mathrm{d}y = \rho.$$

Both X, Y have marginal distributions which are standard normal and conditional distributions which are also normal but with a mean that depends on the conditioning argument:

$$X \sim N(0,1), Y \sim N(0,1), \qquad Y|X = x \sim N(\rho x, (1-\rho^2)).$$

We will return to this example when we consider methods of simulation from the bivariate and multivariate normal distributions. We also will consider this situation when introducing the Gibbs sampler in Chapter 3.

2.1 THE GOAL OF INFERENCE AND BAYES' THEOREM

The goal of statistical inference is to use information to make inferences about unknown quantities. One important source of information is data, but there is an undeniable role for non-data-based information. Information can also come from theories of behavior (such as the information that, properly defined, demand curves slope downward). Information can also come from 'subjective' views that there is a structure underlying the unknowns. For example, in situations with large numbers of different sets of parameters, an assumption that the parameter sets 'cluster' or that they are drawn from some common distribution is often used in modeling. Less controversial might be the statement that we expect key quantities to be finite or even in some range (for example, a price elasticity is not expected to be less than -50). Information can also be derived from prior analyses of other data, including data which is only loosely related to the data set under investigation.

Unknown quantity is a generic term referring to any value not known to the investigator. Certainly, parameters can be considered unknown since these are purely abstractions that index a class of models. In situations in which decisions are made, the unknown quantities can include the (as yet unrealized) outcomes of marketing actions. Even in a passive environment, predictions of 'future' outcomes are properly regarded as unknowns. There should be no distinction between a parameter and an unknown such as an unrealized outcome in the sense that the system of inference should treat each symmetrically.

Our goal, then, is to make inferences regarding unknown quantities *given* the information available. We have concluded that the information available can be partitioned into information obtained from the data as well as other information obtained independently or *prior* to the data. Bayesian inference utilizes probability statements as the basis for inference. What this means is that our goal is to make probability statements about unknown quantities *conditional* on the sample and prior information.

In order to utilize the elegant apparatus of conditional probability, we must encode the prior information as a probability distribution. This requires the view that probability can represent subjective beliefs and is not some sort of long-run frequency. There is much discussion in the statistics and probability theory literature as to whether or not this is a reasonable thing to do. We take a somewhat more practical view – there are many kinds of non-data-based information to be incorporated into our analysis. A subjective interpretation of probability is a practical necessity rather than a philosophical curiosity.

It should be noted that there are several paths which lead to the conclusion that Bayesian inference is a sensible system of inference. Some start with the view that decision-makers are expected utility maximizers. In this world, decision-makers must be 'coherent' or act in accordance with Bayes' theorem in order to avoid exposing themselves to sure losses. Others start with the view that the fundamental primitive is not utility but subjective probability. Still others adhere to the view that the likelihood principle (Section 2.2 below) more or less forces you to adopt the Bayesian form of inference. We are more of the subjectivist stripe but we hope to convince the reader, by example, that there is tremendous practical value to the Bayesian approach.

Bayes' Theorem

Denote the set of unknowns as θ. Our prior beliefs are expressed as a probability distribution, $p(\theta)$. $p(\bullet)$ is a generic notation for the appropriate density. In most cases, this represents a density with respect to standard Lebesgue measure, but it can also represent a probability mass function for discrete parameter spaces or a mixed continuous–discrete measure. The information provided by the data is introduced via the probability distribution for the data, $p(D|\theta)$, where D denotes the observable data. In some classical approaches, modeling is the art of choosing appropriate probability models for the data. In the Bayesian paradigm, the model for prior information is also important. Much of the work in Bayesian statistics is focused on developing a rich class of models to express prior information and devices to induce priors on high-dimensional spaces. In our view, the prior is very important and often receives insufficient attention.

To deliver on the goal of inference, we must combine the prior and likelihood to produce the distribution of the observables conditional on the data and the prior. Bayes' theorem is nothing more than an application of standard conditional probability to this problem:

$$p(\theta|D) = \frac{p(D,\theta)}{p(D)} = \frac{p(D|\theta)p(\theta)}{p(D)}. \tag{2.1.1}$$

$p(\theta|D)$ is called the *posterior* distribution and reflects the combined data and prior information. Equation (2.1.1) is often expressed using the likelihood function. Given D, any function which is proportional to $p(D|\theta)$ is called the *likelihood*, $\ell(\theta)$. The shape of the posterior is determined entirely by the likelihood and prior in the numerator of (2.1.1) and this is often emphasized by rewriting the equation:

$$p(\theta|D) \propto \ell(\theta)p(\theta). \tag{2.1.2}$$

If $\ell(\theta) = p(D|\theta)$, then the constant of proportionality is the marginal distribution of the data, $p(D) = \int p(D,\theta)\,d\theta = \int p(D|\theta)p(\theta)\,d\theta$. Of course, we are assuming here that this normalizing constant exists. If $p(\theta)$ represents a proper distribution (that is, it integrates to one), then this integral exists. With improper priors, it will be necessary to show that the integral exists, which will involve the tail behavior of the likelihood function.

2.2 CONDITIONING AND THE LIKELIHOOD PRINCIPLE

The likelihood principle states that the likelihood function, $\ell(\theta)$, contains all relevant information from the data. Two samples (not necessarily even from the same 'experiment' or method of sampling/observation) have equivalent information regarding θ if their likelihoods are proportional (for extensive discussion and derivation of the likelihood principle from conditioning and sufficiency principles, see Berger and Wolpert 1984). The likelihood principle, by itself, is not sufficient to build a method of inference but should be regarded as a minimum requirement of any viable form of inference. This is a controversial point of view for anyone familiar with the modern econometrics literature. Much of this literature is devoted to methods that do not obey the likelihood principle. For example, the phenomenal success of estimators based on the generalized method of moments procedure is driven by the ease of implementing these estimators even though, in most instances, generalized method of moments estimators violate the likelihood principle.

Adherence to the likelihood principle means that inferences are *conditional* on the observed data as the likelihood function is parameterized by the data. This is worth contrasting to any sampling-based approach to inference. In the sampling literature, inference is conducted by examining the sampling distribution of some estimator of θ, $\hat{\theta} = f(D)$. Some sort of sampling experiment[2] results in a distribution of D and, therefore, the estimator is viewed as a random variable. The sampling distribution of the estimator summarizes the properties of the estimator *prior* to observing the data. As such, it is irrelevant to making inferences given the data we actually observe. For any finite sample, this distinction is extremely important. One must conclude that, given our goal for inference, sampling distributions are simply not useful.

While sampling theory does not seem to deliver on the inference problem, it is possible to argue that it is relevant to the choice of estimating procedures. Bayesian inference procedures are simply one among many possible methods of deriving estimators for a given problem. Sampling properties are relevant to choice of procedures before the data is observed. As we will see in Section 2.5 below, there is an important sense in which one need never look farther than Bayes estimators even if the sole criterion is the sampling properties of the estimator.

2.3 PREDICTION AND BAYES

One of the appeals of the Bayesian approach is that all unknowns are treated the same. Prediction is defined as making probability statements about the distribution of as yet

---

[2] In the standard treatment, the sampling experiment consists of draws from the probability model for the data used in the likelihood. However, many other experiments are possible, including samples drawn from some other model for the data or various asymptotic experiments which involve additional assumptions regarding the data generation process.

unobserved data, denoted by D_f. The only real distinction between 'parameters' and unobserved data is that D_f is potentially observable:

$$p(D_f|D) = \int p(D_f, \theta|D)\, d\theta = \int p(D_f|\theta, D)p(\theta|D)\, d\theta. \tag{2.3.1}$$

Equation (2.3.1) defines the 'predictive' distribution of D_f given the observed data. In many cases, we assume that D and D_f are independent, conditional on θ. In this case, the predictive distribution simplifies to

$$p(D_f|D) = \int p(D_f|\theta)p(\theta|D)\, d\theta. \tag{2.3.2}$$

In (2.3.2), we average the likelihood for the unobserved data over the posterior of θ. This averaging properly accounts for uncertainty in θ when forming predictive statements about D_f.

2.4 SUMMARIZING THE POSTERIOR

For any problem of practical interest, the posterior distribution is a high-dimensional object. Therefore, summaries of the posterior play an important role in Bayesian statistics. Most schooled in classical statistical approaches are accustomed to reporting parameter estimates and standard errors. The Bayesian analog of this practice is to report moments of the marginal distributions of parameters such as the posterior mean and posterior standard deviation. It is far more useful and informative to produce the marginal distributions of parameters or relevant functions of parameters as the output of the analysis. Simulation methods are ideally suited for this. If we can simulate from the posterior distribution of the parameters and other unknowns, then we can simply construct the marginal of any function of interest. Typically, we describe these marginals graphically. As these distributions are often very nonnormal, the mean and standard deviations are not particularly useful. One major purpose of this book is to introduce a set of tools to achieve this goal of simulating from the posterior distribution.

Prior to the advent of powerful simulation methods, attention focused on the evaluation of specific integrals of the posterior distribution as a way of summarizing this high-dimensional object. The general problem can be written as finding the posterior expectation of a function of θ. (We note that marginal posteriors, moments, quantiles and probability of intervals are further examples of expectations of functions, as in (2.4.1) below with suitably defined h). For any interesting problem, only the unnormalized posterior, $\ell(\theta)p(\theta)$, is available so that two integrals must be performed to obtain the posterior expectation of $h(\theta)$:

$$E_{\theta|D}[h(\theta)] = \int h(\theta)p(\theta|D)\, d\theta = \frac{\int h(\theta)\ell(\theta)p(\theta)\, d\theta}{\int \ell(\theta)p(\theta)\, d\theta}. \tag{2.4.1}$$

For many years, only problems for which the integrals in (2.4.1) could be performed analytically were analyzed by Bayesians. Obviously, this restricts the set of priors and likelihoods to a very small set that produces posteriors of known distributional form and for which these integrals can be evaluated analytically. One approach would be to take various asymptotic approximations to these integrals. We will discuss the Laplace approximation method in Section 2.9 below. Unless these asymptotic approximations can be shown to be accurate, we should be very cautious about using them. In contrast, much of the econometrics and statistics literature uses asymptotic approximations to the sampling distributions of estimators and test statistics without investigating accuracy. In marketing problems, the combination of small amounts of sample information per parameter and the discrete nature of the data make it very risky to use asymptotic approximations. Fortunately, we do not have to rely on asymptotic approximations in modern Bayesian inference.

2.5 DECISION THEORY, RISK, AND THE SAMPLING PROPERTIES OF BAYES ESTIMATORS

We started our discussion by posing the problem of obtaining a system of inference appropriate for marketing problems. We could just as well have started on the most general level – finding an appropriate framework for making decisions of any kind. Parameter estimation is only one of many such decisions which occur under uncertainty.

The general problem considered in decision theory is to search among possible actions for the action which minimizes expected loss. The loss function, $L(a, \theta)$, associates a loss with a state of nature, θ, and an action, a. In Chapter 6, loss functions are derived for marketing actions from the profit function of the firm. We choose a decision which performs well, on average, where the averaging is taken across the posterior distribution of states of nature:

$$\min_a \left\{ \overline{L}(a) = E_{\theta|D}[L(a,\theta)] = \int L(a,\theta)p(\theta|D)\,d\theta \right\}. \tag{2.5.1}$$

In Chapter 6 we will explore the implications of decision theory for optimal marketing decisions and valuing of information sets. At this point it is important to note that (2.5.1) involves the entire posterior distribution and not just the posterior mean. With nonlinear loss functions, uncertainty or spread is just as important as location.

A special case of (2.5.1) is the estimation problem. If the action is the estimator and the state of nature is the unknowns to be estimated, then Bayesian decision theory produces a Bayes estimator. Typically, a symmetric function such as squared error or absolute error is used for loss. This defines the estimation problem as

$$\min_{\hat{\theta}} \{ L(\hat{\theta}) = E_{\theta|D}[L(\hat{\theta}, \theta)] \}. \tag{2.5.2}$$

For squared error loss, the optimal choice of estimator is the posterior mean:

$$\hat{\theta}_{\text{Bayes}} = E[\theta|D] = f(D|\tau). \tag{2.5.3}$$

Here τ is the prior hyperparameter vector (if any).

What are the sampling properties of the Bayes estimator and how do these compare to those of other competing general-purpose estimation procedures such as maximum likelihood? Recall that the sampling properties are derived from the fact that the estimator is a function of the data and therefore is a random variable whose distribution is inherited from the sampling distribution of the data. We can use the same loss function to define the 'risk' associated with an estimator, $\hat{\theta}$, as

$$r_{\hat{\theta}}(\theta) = E_{D|\theta}[L(\hat{\theta}, \theta)] = \int L(\hat{\theta}(D), \theta)p(D|\theta)\,\mathrm{d}D. \tag{2.5.4}$$

Note that the risk function for an estimator is a function of θ. That is, we have a different 'risk' at every point in the parameter space.

An estimator is said to be ***admissible*** if there exists no other estimator with a risk function that is less than or equal to the risk of the estimator in question. That is, we cannot find another estimator that does better (or at least as well), as measured by risk, for every point in the parameter space.[3] Define the expected risk

$$E[r(\theta)] = E_\theta[E_{D|\theta}[L(\hat{\theta}, \theta)]].$$

The outer expectation on the right-hand side is taken with respect to the prior distribution of θ. With a proper prior that has support over the entire parameter space, we can apply Fubini's theorem and interchange the order of integration and show that Bayes estimators have the property of minimizing expected risk and, therefore, are admissible:

$$E[r(\theta)] = E_\theta[E_{D|\theta}[L(\hat{\theta}, \theta)]] = \iint L(\hat{\theta}(D), \theta)p(D|\theta)p(\theta)\,\mathrm{d}D\,\mathrm{d}\theta$$

$$= E_D[E_{\theta|D}[L(\hat{\theta}, \theta)]] = \iint L(\hat{\theta}(D), \theta)p(\theta|D)p(D)\,\mathrm{d}D\,\mathrm{d}\theta. \tag{2.5.5}$$

The complete class theorem (see Berger 1985, Chapter 8) says even more – all admissible estimators are Bayes estimators. This provides a certain level of comfort and moral superiority, but little practical guidance. There can be estimators that outperform Bayes estimators in certain regions of (but not all of) the parameter space. Bayes estimators perform very well if you are in the region of the parameter space you expect to be in as defined by your prior. These results on admissibility also do not provide any guidance as to how to choose among an infinite number of Bayes estimators which are equivalent from the point of view of admissibility.

Another useful question to ask is what is the relationship between standard classical estimators such as the maximum likelihood estimator (MLE) and Bayes estimator. At least the MLE obeys the likelihood principle. In general, the MLE is not admissible so there can be no exact sample relationship. However, Bayes estimators are consistent,

[3] Obviously, if we have a continuous parameter space, we have to be a little more careful, but we leave such niceties for those more mathematically inclined.

asymptotically normal and efficient as long as mild regularity conditions[4] hold and the prior is nondogmatic in the sense of giving support to the entire parameter space. The asymptotic 'duality' between Bayes estimators and the MLE stems from the asymptotic behavior of the posterior distribution. As n increases, the posterior concentrates more and more mass in the vicinity of the 'true' value of θ. The likelihood term dominates the prior and the prior becomes more and more uniform in appearance in the region in which the likelihood is concentrating. Thus, the prior has no asymptotic influence and the posterior starts to look more and more normal:

$$p(\theta|D)\dot{\sim}N(\hat{\theta}_{\text{MLE}}, [-H_{\theta=\hat{\theta}_{\text{MLE}}}]^{-1}). \tag{2.5.6}$$

H_θ is the Hessian of the log-likelihood function. The very fact that, for asymptotics, the prior does not matter (other than its support) should be reason enough to abandon this method of analysis in favor of more powerful techniques.

2.6 IDENTIFICATION AND BAYESIAN INFERENCE

The set of models is limited only by the imagination of the investigator and the computational demands of the model and inference method. In marketing problems, we can easily write down a model that is very complex and may make extraordinary demands of data. A problem of identification is defined as the situation in which there is a set of different parameter values that give rise to the same distribution for the data. This set of parameter values are said to observationally equivalent in that the distribution of the data is the same for any member of this set.

Lack of identification is a property of the model and holds over all possible values of the data rather than just the data observed. Lack of identification implies that there will be regions over which the likelihood function is constant for any given data set. Typically, these can be flats or ridges in the likelihood.

From a purely technical point of view, identification is not a problem for a Bayesian analysis. First of all, the posterior may not have a 'flat' or region of constancy as the prior can modify the shape of the likelihood. Even if there are regions for which the posterior is constant, the Bayesian will simply report, correctly, that the posterior is uniform over these regions.

Lack of identification is often regarded as a serious problem which must be dealt with by imposing some sort of restriction on the parameter space. Methods of inference, such as maximum likelihood, that rely on maximization will encounter severe problems with unidentified models. A maximizer may climb up and shut down anywhere in the flat region of the likelihood created by lack of identification. Bayesian computational methods which use simulation as the basis for exploring the posterior are not as susceptible to these computational problems.

Rather than imposing some sort of constraint on the parameter space, the Bayesian can deal with lack of identification through an informative prior. With the proper

[4] It should be noted that, as the MLE is based on the maximum of a function while the Bayes estimator is based on an average, the conditions for asymptotic normality are different for the MLE than for the Bayes estimator. But both from a practical (that is, computational) and a theoretical perspective, averages behave more regularly than maxima.

informative prior, the posterior may not have any region of constancy. However, it should be remembered that lack of identification means that there are certain functions of the parameters for which the posterior is entirely driven by the prior. We can define a transformation function, $\tau = f(\theta)$, and a partition, $\tau' = (\tau_1', \tau_2')$, where dim $(\tau_1') = r$, such that $p(\tau_1|D) = p(\tau_1)$. This means that, for certain transformed coordinates, the posterior is the same as the prior and only prior information matters in a posteriori inference. r is the 'dimension' of the redundancies in the θ parameterization. The 'solution' to the identification problem can then be to report the marginal posterior for the parameters which are 'identified', $p(\tau_2|D)$. We will see this idea is useful for analysis of certain nonidentified models. It is important, however, to examine the marginal prior, $p_{\tau_2}(\bullet)$, which is induced by the prior on θ.

2.7 CONJUGACY, SUFFICIENCY, AND EXPONENTIAL FAMILIES

Prior to modern simulation methods, a premium was placed on models that would allow analytical expressions for various posterior summaries. Typically, this means that we choose models for which we can compute posterior moments analytically. What we want is for the posterior to be of a distributional form for which the posterior moments are available in analytical expressions. This requirement imposes constraints on both the choice of likelihood and prior distributions. One approach is to require that the prior distribution be conjugate to the likelihood. A prior is said to be *conjugate* to the likelihood if the posterior derived from this prior and likelihood is in the same class of distributions as the prior. For example, normal distributions have simple expressions for moments. We can get normal posteriors by combing normal priors and likelihoods based on normal sampling models, as we will see in Section 2.8. However, there can be conjugate priors for which no analytic expressions exist for posterior moments.

The key to conjugacy is the form of the likelihood since we can always pick priors with convenient analytic features. Likelihoods in the exponential family of distributions have conjugate priors (see Bernardo and Smith 1994, Section 5.2). The exponential family is a family of distributions with a minimal sufficient statistic of fixed dimension equal to the dimension of the parameter space. The duality of sufficient statistics and the parameter vector is what drives this result. Moreover, the exponential form means that combining an exponential family likelihood and prior will result in an exponential family posterior that is tighter than either the likelihood or the prior. To see this, recall the form of the regular parameterization of the exponential family. If we have a random sample $y' = (y_1, \ldots, y_n)$ from the regular exponential family, the likelihood is given by

$$p(y|\theta) \propto g(\theta)^n \exp\left\{\sum_{j=1}^{k} c_j \varphi_j(\theta) \bar{h}_j(y)\right\}, \qquad \bar{h}_j(y) = \sum_{i=1}^{n} h_j(y_i). \tag{2.7.1}$$

$\{\bar{h}_1, \ldots, \bar{h}_k\}$ are the set of minimal sufficient statistics for θ. A prior in the same form would be given by

$$p(\theta|\tau) \propto g(\theta)^{\tau_0} \exp\left\{\sum_{j=1}^{k} c_j \varphi_j(\theta) \tau_j\right\}, \tag{2.7.2}$$

where $\{\tau_0, \tau_1, \ldots, \tau_k\}$ are the prior hyperparameters. Clearly, the posterior is also in the exponential form with parameters $\tau_0^* = n + \tau_0, \tau_1^* = \tau_1 + \bar{h}_1, \ldots, \tau_k^* = \tau_k + \bar{h}_k$:

$$p(\theta|y) \propto g(\theta)^{\tau_0+n} \exp\left\{\sum_{j=1}^{k} c_j \varphi_j(\theta)(\bar{h}_j + \tau_j)\right\}. \tag{2.7.3}$$

Since the conjugate prior in (2.7.2) is of the same form as the likelihood, we can interpret the prior as the posterior from some other sample of data with τ_0 observations.

A simple example will illustrate the functioning of natural conjugate priors. Consider the Bernoulli probability model for the data, $y_i \sim \text{iid } B(\theta)$, where iid denotes 'independent and identically distributed'. θ is the probability of one of two possible outcomes for each $y_i = (0, 1)$. The joint density of the data is given by

$$p(y|\theta) = \theta^{\Sigma_i y_i}(1-\theta)^{n-\Sigma_i y_i}. \tag{2.7.4}$$

The choice of prior for this simple Bernoulli problem has been the subject of numerous articles. Some advocate the choice of a 'reference' prior or a prior that meets some sort of criteria for default scientific applications. Others worry about the appropriate choice of a 'non-informative' prior. Given that this problem is one-dimensional, one might want to choose a fairly flexible family of priors. The conjugate prior for this family is the beta prior:

$$p(\theta) \propto \theta^{\alpha-1}(1-\theta)^{\beta-1} \sim \text{Beta}(\alpha, \beta). \tag{2.7.5}$$

This prior is reasonably flexible with regard to location ($E[\theta] = \alpha/(\alpha+\beta)$), spread, and shape (either unimodal or 'U-shaped'). We can interpret this prior as the posterior from another sample of $\alpha + \beta - 2$ observations and $\alpha - 1$ values of 1. The posterior is also of the beta form

$$p(\theta|y) \propto \theta^{\alpha+\Sigma_i y_i - 1}(1-\theta)^{\beta+n-\Sigma_i y_i - 1} \sim \text{Beta}(\alpha', \beta'), \tag{2.7.6}$$

with $\alpha' = \alpha + \sum_i y_i$ and $\beta' = \beta + n - \sum_i y_i$. Thus, we can find the posterior moments from the beta distribution.

Those readers who are familiar with numerical integration methods might regard this example as trivial and not very interesting since one could simply compute whatever posterior integrals are required by univariate numerical integration. This would allow for the use of any reasonable prior. However, the ideas of natural conjugate priors are most powerful when applied to vectors of regression parameters and covariance matrices which we develop in the next section.

2.8 REGRESSION AND MULTIVARIATE ANALYSIS EXAMPLES

2.8.1 Multiple Regression

The regression problem has received a lot of attention in the Bayesian statistics literature and provides a very good and useful example of nontrivial natural conjugate priors. The standard linear regression model is a model for the conditional distribution of y given a vector of predictor variables in x:

$$y_i = x_i'\beta + \varepsilon_i, \qquad \varepsilon_i \sim \text{iid } N(0, \sigma^2) \tag{2.8.1}$$

or

$$y \sim N(X\beta, \sigma^2 I_n). \tag{2.8.2}$$

Here $N(\mu, \Sigma)$ is the standard notation for a multivariate distribution with mean μ and variance–covariance matrix Σ. X is the matrix whose ith row is x_i

We have only modeled the conditional distribution of y given x rather than the joint distribution. In nonexperimental situations, it can be argued that we need to choose a model for the joint distribution of both x and y. In order to complete the model, we would need a model for the marginal distribution of x:

$$p(x, y) = p(x|\psi)p(y|x, \beta, \sigma^2). \tag{2.8.3}$$

If ψ is a priori independent[5] of (β, σ^2), then the posterior factors into two terms, the posterior of the x marginal parameters and the posterior for the regression parameters, and we can simply focus on the rightmost term of (2.8.4),

$$p(\psi, \beta, \sigma^2|y, X) \propto [p(\psi)p(X|\psi)][p(\beta, \sigma^2)p(y|X, \beta, \sigma^2)]. \tag{2.8.4}$$

What sort of prior should be used for the regression model parameters? There are many possible choices for family of prior distributions and, given the choice of family, there is also the problem of assessing the prior parameters. Given that the regression likelihood is a member of the exponential family, a reasonable starting place would be to consider natural conjugate priors. The form of the natural conjugate prior for the regression model can be seen by examining the likelihood function.

To start, let us review the likelihood function for the normal linear regression model. We start from the distribution of the error terms which are multivariate normal. If $x \sim N(\mu, \Sigma)$, then the density of x is given by

$$p(x|\mu, \Sigma) = (2\pi)^{-k/2}|\Sigma|^{-1/2} \exp\left(-\frac{1}{2}(x-\mu)'\Sigma^{-1}(x-\mu)\right), \qquad \varepsilon \sim N(0, \sigma^2 I_n). \tag{2.8.5}$$

The density of ε is then

$$p(\varepsilon|\sigma^2) = (2\pi)^{-n/2}(\sigma^2)^{-n/2} \exp\left(-\frac{1}{2\sigma^2}\varepsilon'\varepsilon\right). \tag{2.8.6}$$

Given that the Jacobian from ε to y is 1 and we are conditioning on X, we can write the density of y given X, β and σ^2 as

$$p(y|X, \beta, \sigma^2) \propto (\sigma^2)^{-n/2} \exp\left(-\frac{1}{2\sigma^2}(y - X\beta)'(y - X\beta)\right). \tag{2.8.7}$$

[5] This rules out deterministic relationships between ψ and (β, σ^2) as well as stochastic dependence. In situations in which the X variables are set strategically by the marketer, this assumption can be violated. We explore this further in Chapter 7.

The natural conjugate prior is a joint density for β and σ^2 which is proportional to the likelihood. Given that the likelihood has a quadratic form in β in the exponent, a reasonable guess is that the conjugate prior for β is normal. We can rewrite the exponent so that the quadratic form is of the usual normal form. This can be done either by expanding out the existing expression and completing the square in β, or by recalling the usual trick of projecting y on the space spanned by the columns of X, $y = \hat{y} + e = X\hat{\beta} + (y - X\hat{\beta})$ and $\hat{y}'e = 0$. We have

$$\begin{aligned}(y - X\beta)'(y - X\beta) &= (y - X\hat{\beta})'(y - X\hat{\beta}) + (\beta - \hat{\beta})'X'X(\beta - \hat{\beta}) \\ &= \nu s^2 + (\beta - \hat{\beta})'X'X(\beta - \hat{\beta}),\end{aligned}$$

with $\nu s^2 = (y - X\hat{\beta})'(y - X\hat{\beta})$, $\nu = n - k$, and $\hat{\beta} = (X'X)^{-1}X'y$. If we substitute this expression in the exponent, we can factor the likelihood into two components, keeping the terms needed for a proper normal density for β and then the balance for σ:

$$p(y|X, \beta, \sigma^2) \propto (\sigma^2)^{-\nu/2} \exp\left\{-\frac{\nu s^2}{2\sigma^2}\right\} (\sigma^2)^{-(n-\nu)/2} \exp\left\{-\frac{1}{2\sigma^2}(\beta - \hat{\beta})'(X'X)(\beta - \hat{\beta})\right\}. \tag{2.8.8}$$

The first term in (2.8.8) suggests a form for the density[6] of σ^2 and the second for the density of β. We note that the normal density for β involves σ^2. This means that the conjugate prior will be specified as

$$p(\beta, \sigma^2) = p(\sigma^2)p(\beta|\sigma^2). \tag{2.8.9}$$

The first term in (2.8.8) suggests that the marginal prior on σ^2 has a density of the form $p(\theta) \propto \theta^{-\lambda} \exp(-\delta/\theta)$. This turns out to be in the form of the inverse gamma distribution. The standard form of the inverse gamma is

$$\theta \sim \mathrm{IG}(\alpha, \beta), \qquad p(\theta) = \frac{\beta^\alpha}{\Gamma(\alpha)}\theta^{-(\alpha+1)} \exp\left(-\frac{\beta}{\theta}\right). \tag{2.8.10}$$

The natural conjugate prior for σ^2 is of the form

$$p(\sigma^2) \propto (\sigma^2)^{-(\nu_0/2+1)} \exp\left\{-\frac{\nu_0 s_0^2}{2\sigma^2}\right\}, \tag{2.8.11}$$

the standard inverse gamma form with $\alpha = \nu_0/2$ and $\beta = \nu_0 s_0^2/2$. We note that the inverse gamma density requires a slightly different power for σ^2 than found in the likelihood. There is an extra term, $(\sigma^2)^{-1}$, in (2.8.11) which is not suggested by the form of the likelihood in (2.8.8). This term can be rationalized by viewing the conjugate prior as arising from the posterior of a sample of size ν_0 with sufficient statistics, $s_0^2, \bar{\beta}$, formed with the noninformative prior, $p(\beta, \sigma^2) \propto \frac{1}{\sigma^2}$.

[6] Note that we can speak interchangeably about σ or $\theta = \sigma^2$. However, we must remember the Jacobian of the transformation from σ to θ when we rewrite the density expressions converting from σ to θ (the Jacobian is 2σ).

The conjugate normal prior on β is given by:

$$p(\beta|\sigma^2) \propto (\sigma^2)^{-k/2} \exp\left\{-\frac{1}{2\sigma^2}(\beta - \bar{\beta})'A(\beta - \bar{\beta})\right\} \tag{2.8.12}$$

Priors (2.8.11) and (2.8.12) can be expressed in terms of common distributions using the relationship between the inverse gamma and the inverse of a chi-squared random variable:

$$\sigma^2 \sim \frac{\nu_0 s_0^2}{\chi^2_{\nu_0}} \quad \text{and} \quad \beta|\sigma^2 \sim N(\bar{\beta}, \sigma^2 A^{-1}). \tag{2.8.13}$$

The notation ν_0, s_0^2 is suggestive of the interpretation of the natural conjugate prior as based on another sample with ν_0 degrees of freedom and sum of squared errors $\nu_0 s_0^2$.

Given the natural conjugate priors for the regression problem (2.8.13), the posterior is in the same form:

$$\begin{aligned} p(\beta, \sigma^2|y, X) &\propto p(y|X, \beta, \sigma^2)p(\beta|\sigma^2)p(\sigma^2) \\ &\propto (\sigma^2)^{-n/2} \exp\left\{-\frac{1}{2\sigma^2}(y - X\beta)'(y - X\beta)\right\} \\ &\quad\times (\sigma^2)^{-k/2} \exp\left\{-\frac{1}{2\sigma^2}(\beta - \bar{\beta})'A(\beta - \bar{\beta})\right\} \\ &\quad\times (\sigma^2)^{-(\nu_0/2+1)} \exp\left\{-\frac{\nu_0 s_0^2}{2\sigma^2}\right\}. \end{aligned} \tag{2.8.14}$$

The quadratic forms in the first two exponents on the right-hand side can be combined by 'completing the square'. To remove some of the mystery from this operation, which appears to require that you know the answer prior to obtaining the answer, we rewrite the problem as the standard problem of decomposing a vector into its projection on a subspace spanned by some column vectors and an orthogonal 'residual'. Since A is positive definite, we can find its upper triangular Cholesky root, $U(A = U'U$, sometimes called the LU decomposition):

$$(y - X\beta)'(y - X\beta) + (\beta - \bar{\beta})'U'U(\beta - \bar{\beta}) = (\nu - W\beta)'(\nu - W\beta),$$

where

$$\nu = \begin{bmatrix} y \\ U\bar{\beta} \end{bmatrix}, \qquad W = \begin{bmatrix} X \\ U \end{bmatrix}.$$

We now project ν onto the space spanned by the W columns using standard least squares:

$$(\nu - W\beta)'(\nu - W\beta) = ns^2 + (\beta - \tilde{\beta})'W'W(\beta - \tilde{\beta}),$$

where

$$\tilde{\beta} = (W'W)^{-1}W'\nu = (X'X + A)^{-1}(X'X\hat{\beta} + A\bar{\beta}) \tag{2.8.15}$$

and

$$ns^2 = (\nu - W\tilde{\beta})'(\nu - W\tilde{\beta}) = (y - X\tilde{\beta})'(y - X\tilde{\beta}) + (\tilde{\beta} - \bar{\beta})'A(\tilde{\beta} - \bar{\beta}). \quad (2.8.16)$$

Using these results, we can rewrite (2.8.14):

$$\begin{aligned} p(\beta, \sigma^2 | y, X) &\propto (\sigma^2)^{-k/2} \exp\left\{-\frac{1}{2\sigma^2}(\beta - \tilde{\beta})'(X'X + A)(\beta - \tilde{\beta})\right\} \\ &\times (\sigma^2)^{-((n+\nu_0)/2+1)} \exp\left\{-\frac{(\nu_0 s_0^2 + ns^2)}{2\sigma^2}\right\} \end{aligned} \quad (2.8.17)$$

or

$$\begin{aligned} \beta | \sigma^2, y, X &\sim N(\tilde{\beta}, \sigma^2 (X'X + A)^{-1}), \\ \sigma^2 | y, X &\sim \frac{\nu_1 s_1^2}{\chi^2_{\nu_1}}, \qquad \text{with } \nu_1 = \nu_0 + n, \; s_1^2 = \frac{\nu_0 s_0^2 + ns^2}{\nu_0 + n}. \end{aligned} \quad (2.8.18)$$

The Bayes estimator corresponding to the posterior mean is $\tilde{\beta}$:

$$\mathrm{E}[\beta | y] = \mathrm{E}_{\sigma^2|y}\left[\mathrm{E}_{\beta|\sigma^2,y}[\beta]\right] = \mathrm{E}_{\sigma^2|y}[\tilde{\beta}] = \tilde{\beta}.$$

$\tilde{\beta}$ is a weighted average of the prior mean and the least squares estimator, $\hat{\beta}$. The weights depend on the prior precision and the sample information (recall that the information matrix for all n observations in the regression model is proportional to $X'X$). There are two important practical aspects of this estimator:

(i) The Bayes estimator is a 'shrinkage' estimator in the sense that the least squares estimator is 'shrunk' toward the prior mean. Similarly, the posterior distribution of σ^2 is 'centered' over s_1^2 which is a weighted average of the prior parameter and a sample quantity (however, the 'sample' sum of squares includes a term (see (2.8.16)) which represents the degree to which the prior mean differs from the least squares estimator).

(ii) As we acquire more sample information in the sense of total X variation, the Bayes estimator converges to the least squares estimator (this ensures consistency which we said is true, in general, for Bayes estimators).

These results for the posterior are special to the linear regression model and the natural conjugate prior set-up, but if the likelihood function is approximately normal with mean equal to the MLE then these results suggest that the Bayes estimator will be a shrinkage estimator which is a weighted average of the MLE and prior mean.

While the posterior mean of β is available by direct inspection of the conditional distribution of β given σ^2, the marginal distribution of β must be computed. It turns out that the marginal distribution of β is in the multivariate t form. The multivariate t distribution is in the elliptical class of distributions[7] and the marginal distribution of β

[7] Note that linear combinations of independent t distributions are not in the multivariate t form.

will have the same density contours as the conditional distribution but will have fatter, algebraic tails rather than the thin exponential tapering normal tails. Moments of the multivariate t distribution are available.[8] However, the results in (2.8.18) can be used to devise a simple simulation strategy for making iid draws from the posterior of β. Simply draw a value of σ^2 from its marginal posterior, insert this value into the expression for the covariance matrix of the conditional normal distribution of $\beta|\sigma^2, y$ and draw from this multivariate normal.[9]

Expression (2.8.18) illustrates one of the distinguishing characteristics of natural conjugate priors. The posterior is centered between the prior and the likelihood and is more concentrated than either. Thus, it appears that you always gain by combining prior and sample information. It should be emphasized that this is a special property of natural conjugate priors. The property comes, in essence, from the interpretation of the prior as the posterior from another sample of data. The posterior we calculate can be interpreted as the same as the posterior one would obtain by pooling both the prior 'sample' and the actual sample and using a very diffuse prior. If the prior information is at odds with the sample information, the use of natural conjugate priors will gloss this over because of the implicit assumption that both prior and sample information are equally valid. Many investigators would like to be aware of the situation in which the prior information is very different from the sample information. It is often difficult to check on this if the parameter space is high-dimensional. This is particularly a problem in hierarchical models in which natural conjugate results like the ones given here are used. At this point, we call attention to this as a potential weakness of natural conjugate priors. After we develop computational methods that can work with nonconjugate priors, we will return to this issue.

2.8.2 Assessing Priors for Regression Models

Expression (2.8.13) is one implementation of a particular view regarding prior information on the regression coefficients. The idea here is that our views about coefficients are dependent on the error variance or, equivalently, the scale of the regression error terms (σ). If σ is large, the prior on the regression coefficients spreads out to accommodate larger values or reflects greater uncertainty. This is very much driven by the view that prior information comes from data. In situations where the 'sample' creating the prior information comes from a regression with highly variable error terms, this sample information will be less valuable and the prior should spread out. However, it is entirely possible to imagine prior information that directly addresses the size of the regression coefficients. After all, the regression coefficients are designed to measure the effects of changes in the X variables. We may have information about the size of effects of a one-unit price change on sales, for example, that is independent of the percentage of variation in sales explained by price. For this reason, the natural conjugate prior may not be appropriate for some sorts of information.

---

[8] $\text{var}(\beta) = (\nu_1/(\nu_1 - 2))s_1^2(X'X + A)^{-1}$.

[9] In Section 2.11, we will discuss further details for simulation from the distributions required for natural conjugate Bayes models.

The more serious practical problem with the prior in (2.8.13) is the large number of prior hyperparameters that must be assessed, two for the prior distribution of σ^2, and $k + (k(k+1))/2$ for the conditional prior on β. Moreover, as the likelihood is of order n, we may find that for any standard regression data set with a modest number of observations, the hard work of prior assessment will have a low return as the prior may make little difference. Obviously, the prior will only matter with a small amount of sample information, which typically arises in situations with small data sets or lack of independent variation in the X variables. The full Bayesian machinery is rarely used directly on regression problems with one data set, but is increasingly being used with sets of related regression equations. We will develop this further in Chapter 5 when we discuss hierarchical models.

One approach to the problem of assessment of the natural conjugate prior is to use prior settings that are very diffuse. This would involve a 'large' value of s_0^2, 'small' value of ν_0, and a 'small' value of A, the prior precision. If the prior is diffuse, the mean of β, $\bar{\beta}$, is not very critical and can be set to zero. What constitutes a 'diffuse' or spread-out prior is a relative statement – relative to the diffusion of the likelihood. We can be sure that the prior has 'small' influence if we make the prior diffuse relative to the likelihood. If ν_0 is some small fraction of n such as $\max(4, 0.01n)$ and

$$A = \nu_0 S_X \qquad \text{where } S_X = \operatorname{diag}(s_1^2, \ldots, s_k^2) \text{ and } s_j^2 = 1/(n-1)\sum_i (x_{ij} - \bar{x}_j)^2, \tag{2.8.19}$$

then we have assessed a relatively diffuse prior. We have introduced the scaling of x into the prior as would seem reasonable. If we change the units of X, we might argue that our prior information should remain the same, and this sort of prior is invariant to scale changes. The g-prior of Zellner (1986) takes this idea further and sets $A = gX'X$. This means that we are using not only the scale of the X variables but their observed correlation structure as well. There is some controversy as to whether this is 'coherent' and/or desirable, but for relatively diffuse priors the difference between the prior given by (2.8.19) and the g-prior will be minimal.

Our scheme for assessing a relatively diffuse prior is not complete without a method for assessing the value of s_0^2. s_0^2 determines the location of the prior in the sense that the mean of σ^2 is

$$\begin{aligned} E[\sigma^2] &= \frac{\nu_0 s_0^2}{\nu_0 - 2}, \qquad \text{for } \nu_0 > 2, \\ \operatorname{var}(\sigma^2) &= \frac{2\nu_0^2}{(\nu_0 - 2)^2(\nu_0 - 4)}(s_0^2)^2, \qquad \text{for } \nu_0 > 4. \end{aligned} \tag{2.8.20}$$

As ν_0 increases, the prior becomes tighter and centered around s_0^2. Many investigators are rather cavalier about the value of s_0^2 and set it to some rather arbitrary number like 1. If the prior is barely proper then the argument is that there is sufficient prior mass over very large values of σ and this is all that is relevant. However, these prior settings often mean that little or no prior mass is put on small values of σ below 0.5 or so. Depending on the scale of y and the explanatory power of the X variables, this can be a very informative prior! A somewhat more reasonable, but still controversial view, is to take into account of the scale of y in assessing the value of s_0^2. For example, we could use the sample variance of y for s_0^2. At the most extreme, we could use the mean error sum of squares from the regression of y on X. Purists would argue that this is violating Bayes' theorem in that the data is being 'used' twice – once in the assessment of the prior and again in the likelihood. In the absence of true prior information on σ, it seems preferable to use some scale information on y rather than to put in an arbitrary value for s_0^2 and hope that the prior is diffuse enough so this does not affect the posterior inferences in an undesirable manner.

An alternative to assessment of an informative prior would be to use one of the many candidates for 'non-informative' priors. There are many possible priors, each derived from different principles. For the case of continuous-valued parameters, it is important to understand that there can be no such thing as an uninformative prior. For example, if we have a uniform but improper[10] prior on a unidimensional parameter θ, $p(\theta) \propto$ constant, then $\tau = e^{\theta}$ has a nonuniform density $1/\tau$. Various invariance principles have been proposed in which priors are formulated under the constraint that they are invariant to certain types of transformations. In many situations, prior information should not be invariant to transformations. Our view is that prior information is available and should be expressed through proper priors assessed in the parameterization which yields maximum interpretability. Moreover, the use of proper priors avoids mathematical pathologies and inadvertent rendering of extreme prior information. However, for the sake of completeness, we present the standard noninformative prior for the regression model:

$$p(\beta, \sigma) = p(\beta)p(\sigma) \propto \frac{1}{\sigma} \tag{2.8.21}$$

or

$$p(\beta, \sigma^2) \propto \frac{1}{\sigma^2}. \tag{2.8.22}$$

In contrast to the natural conjugate prior, β is independent of σ. The uniform distribution of β over all of $\mathbb{R}^k$ means that the prior in (2.8.21) implies that we think β is large with very high prior probability. In our view, this exposes the absurdity of this prior. We surely have prior information that the regression coefficients are not arbitrarily large! The prior on σ^2 can be motivated by appeal to the notion of scale invariance – that is, the prior should be unchanged when y is multiplied by any positive constant.

2.8.3 Bayesian Inference for Covariance Matrices

A key building block in many Bayesian models is some sort of covariance structure. The workhorse model in this area is a multivariate normal sample and the associated natural conjugate prior. Consider the likelihood for a random sample from an m-dimensional multivariate normal (for expositional purposes, we omit the mean vector; in Section 2.8.5 we will consider the general case where the mean vector is a set of regression equations):

$$p(y_1, \ldots, y_n|\Sigma) \propto \prod_{i=1}^{n} |\Sigma|^{-1/2} \exp\left\{-\frac{1}{2}y_i'\Sigma^{-1}y_i\right\} = |\Sigma|^{-n/2} \exp\left\{-\frac{1}{2}\Sigma_{i=1}^{n} y_i'\Sigma^{-1}y_i\right\}. \tag{2.8.23}$$

[10] 'Improper' means nonintegrable, that is, the integral $\int p(\theta)\, d\theta$ diverges.

We can rewrite the exponent of (2.8.23) in a more compact manner using the trace operator:

$$\Sigma_i \mathrm{tr}(y_i' \Sigma^{-1} y_i) = \Sigma_i \mathrm{tr}(y_i y_i' \Sigma^{-1}) = \mathrm{tr}(S\Sigma^{-1}),$$

with $S = \Sigma y_i y_i'$. Using the notation, $\mathrm{etr}(\bullet) \equiv \exp(\mathrm{tr}(\bullet))$,

$$p(y_1, \ldots, y_n | \Sigma) \propto |\Sigma|^{-n/2} \mathrm{etr}\left\{-\frac{1}{2} S\Sigma^{-1}\right\}. \tag{2.8.24}$$

Expression (2.8.24) suggests that a natural conjugate prior for Σ is of the form

$$p(\Sigma | \nu_0, V_0) \propto |\Sigma|^{-(\nu_0+m+1)/2} \mathrm{etr}\left(-\frac{1}{2} V_0 \Sigma^{-1}\right), \tag{2.8.25}$$

with $\nu_0 > m$ required for an integrable density.

Equation (2.8.25) is the expression for an inverted Wishart density. We use the notation

$$\Sigma \sim \mathrm{IW}(\nu_0, V_0). \tag{2.8.26}$$

If $\nu_0 \geq m+2$, then $\mathrm{E}[\Sigma] = (\nu_0 - m - 1)^{-1} V_0$. The full density function for the inverted Wishart distribution is given by

$$\begin{aligned} p(\Sigma | \nu_0, V_0) = {} & \left(2^{\nu_0 m/2} \pi^{m(m-1)/4} \prod_{i=1}^{m} \Gamma\left(\frac{\nu_0 - 1 - i}{2}\right)\right)^{-1} \\ & \times |V_0|^{\nu_0/2} |\Sigma|^{-(\nu_0+m+1)/2} \mathrm{etr}\left(-\frac{1}{2} V_0 \Sigma^{-1}\right). \end{aligned} \tag{2.8.27}$$

This implies that Σ^{-1} has a Wishart distribution,

$$\Sigma^{-1} \sim W(\nu_0, V_0^{-1}), \qquad \mathrm{E}[\Sigma^{-1}] = \nu_0 V_0^{-1}.$$

We can interpret V_0 as determining the 'location' of the prior and ν_0 as determining the spread of the distribution. However, some caution should be exercised in interpreting V_0 as a location parameter, particularly for small values of ν_0. The inverted Wishart is a highly skewed distribution which can be thought of as the matrix-valued generalization of the inverted chi-squared prior for the variance in the single-parameter case.[11] As with all highly skewed distributions, there is a close relationship between the spread and the location. As we increase V_0 for small ν_0, then we also increase the spread of the distribution dramatically.

The Wishart and inverted Wishart distributions have a number of additional drawbacks. The most important is that the Wishart has only one tightness parameter. This means that we cannot be very informative on some elements of the covariance matrix and less informative on others. We would have to use independent inverted Wisharts or some

[11] Inverted Wishart distributions have the property that the marginal distribution of a square block along the diagonal is also inverted Wishart with the same degrees of freedom parameter and the appropriate submatrix of V_0. Therefore, an inverted Wishart distribution with one degree of freedom is proportional to the inverse of a chi-squared variate.

more richly parameterized distribution to handle this situation. In some situations, we wish to condition on some elements of the covariance matrix and use a conditional prior. Unfortunately, the conditional distribution of a portion of the matrix given some other elements is not in the inverted Wishart form.

As is usual with natural conjugate priors, we can interpret the prior as the posterior stemming from some other data set and with a diffuse prior. If we use the diffuse prior suggested by Zellner (1971, p. 225) via the Jeffreys invariance principle, $p(\Sigma) \propto |\Sigma|^{-(m+1)/2}$, then we can interpret ν_0 as the effective sample size of the sample underlying the prior and V_0 as the sum of squares and cross-products matrix from this sample. This interpretation helps to assess the prior hyperparameters. Some assess V_0 by appeal to the sum of squares interpretation. Simply scale up some sort of prior notation of the variance–covariance matrix by ν_0, $V_0 = \nu_0 \hat{\Sigma}$. A form of 'cheating' would be to use the exact or a stylized version of covariance matrix of $\{y_1, \ldots, y_n\}$. In situations where a relatively diffuse prior is desired, many investigators use $\nu_0 = m + 3$, $V_0 = \nu_0 I$. This is an extremely spread-out prior which should be used with some caution.

If we combine the likelihood in (2.8.24) with the conjugate prior (2.8.25), then we have a posterior in the inverted Wishart form:

$$
\begin{aligned}
p(\Sigma|Y) &\propto p(\Sigma)p(Y|\Sigma) \\
&= |\Sigma|^{-(\nu_0+m+1)/2} \operatorname{etr}\left(-\frac{1}{2}V_0\Sigma^{-1}\right)|\Sigma|^{-n/2} \operatorname{etr}\left(-\frac{1}{2}S\Sigma^{-1}\right) \qquad (2.8.28)\\
&= |\Sigma|^{-(\nu_0+m+n+1)/2} \operatorname{etr}\left(-\frac{1}{2}(V_0+S)\Sigma^{-1}\right)
\end{aligned}
$$

or

$$
p(\Sigma|Y) \sim \mathrm{IW}(\nu_0 + n, V_0 + S) \qquad (2.8.29)
$$

with mean

$$
\mathrm{E}[\Sigma] = \frac{V_0 + S}{\nu_0 + n - m - 1}
$$

which converges to the MLE as n approaches infinity. Y is the $n \times m$ matrix whose rows are y_i, $i = 1, \ldots n$.

2.8.4 Priors and the Wishart Distribution

Although it seems more natural and interpretable to place a prior directly on Σ, there is a tradition in some parts of the Bayesian literature of choosing the prior on Σ^{-1} instead. In the Σ^{-1} parameterization, the prior (2.8.26) is a Wishart distribution:

$$
p(G = \Sigma^{-1}) \propto p(\Sigma|\nu_0, V_0)\big|_{\Sigma=G^{-1}} \times J_{\Sigma\to G}
$$

or

$$
\begin{aligned}
p(G) &\propto |G|^{(\nu_0+m+1)/2} \operatorname{etr}\left(-\frac{1}{2}V_0 G\right) \times |G|^{-(m+1)} \\
&= |G|^{(\nu_0-m-1)/2} \operatorname{etr}\left(-\frac{1}{2}A_0^{-1}G\right)
\end{aligned} \qquad (2.8.30)
$$

where $A_0 = V_0^{-1}$. This is denoted $G \sim W(\nu_0, A_0)$. If $\nu_0 \geq m + 2$, then $\mathrm{E}[G] = \nu_0 A_0$.

The real value of the Wishart parameterization comes from its use to construct simulators. This comes from a result in multivariate analysis (cf. Muirhead 1982, Theorem 3.2.1, p. 85). If we have a set of iid random vectors $\{e_1, \ldots, e_\nu\}$ and $e_i \sim N(0, I)$, then

$$\mathrm{E} = \Sigma_{i=1}^{\nu} e_i e_i' \sim W(\nu, I); \tag{2.8.31}$$

if $\Sigma = U'U$, then

$$W = U'EU \sim W(\nu, \Sigma). \tag{2.8.32}$$

We note that we do not use (2.8.32) directly to simulate Wishart random matrices. In Section 2.11, we will provide an algorithm for simulating from Wisharts.

2.8.5 Multivariate Regression

One useful way to view the multivariate regression model is as a set of regression equations related through common X variables and correlated errors:

$$\begin{aligned} y_1 &= X\beta_1 + \varepsilon_1, \\ &\vdots \\ y_c &= X\beta_c + \varepsilon_c, \\ &\vdots \\ y_m &= X\beta_m + \varepsilon_m. \end{aligned} \tag{2.8.33}$$

The subscript c denotes a vector of n observations on equation c; we use the subscript c to suggest the columns of a matrix which will be developed later. The model in (2.8.33) is not complete without a specification of the error structure. The standard multivariate regression model specifies that the errors are correlated across equations. Coupled with a normal error assumption, we can view the model as a direct generalization of the multivariate normal inference problem discussed in Section 2.8.3. To see this, we will think of a row vector of one observation on each of the m regression equations. This row vector will have a multivariate normal distribution with means given by the appropriate regression equation:

$$y_r = B'x_r + \varepsilon_r, \qquad \varepsilon_r \sim \text{iid } N(0, \Sigma). \tag{2.8.34}$$

The subscript r refers to observation r and B is a $k \times m$ matrix whose columns are the regression coefficients in (2.8.33). y_r and ε_r are m-vectors of the observations on each of the dependent variables and error term.

Expression (2.8.34) is a convenient way of expressing the model for the purpose of writing down the likelihood function for the model:

$$p(\varepsilon_1, \ldots, \varepsilon_n|\Sigma) \propto |\Sigma|^{-n/2} \exp\{-\tfrac{1}{2}\mathrm{tr} S_\varepsilon \Sigma^{-1}\}, \qquad S_\varepsilon = \sum_{r=1}^{n} \varepsilon_r \varepsilon_r'. \tag{2.8.35}$$

Since the Jacobian from ε to y is 1, we can simply substitute in to obtain the distribution of the observed data. Clearly, an inverted Wishart prior will be the conjugate prior for Σ. To obtain the form for the natural conjugate prior for the regression coefficients, it will be helpful to write down the likelihood function using the matrix form of the multivariate regression model,

$$Y = XB + E, \qquad B = [\beta_1, \ldots, \beta_c, \ldots, \beta_m]. \tag{2.8.36}$$

Both Y and E are $n \times m$ matrices of observations whose (i, j)th elements are the ith observation on equation j. X is an $n \times k$ matrix of observations on k common independent variables. The columns of Y and E are $\{y_c\}$ and $\{\varepsilon_c\}$, respectively given in (2.8.33). The rows of the E matrix are the $\{\varepsilon_r\}$ given in (2.8.35). Observing that $E'E = S_\varepsilon$, we can use (2.8.35) and (2.8.36) to write down the complete likelihood:

$$p(E|\Sigma) \propto |\Sigma|^{-n/2} \operatorname{etr}\{-\tfrac{1}{2} E'E\Sigma^{-1}\},$$
$$p(Y|X, B, \Sigma) \propto |\Sigma|^{-n/2} \operatorname{etr}\{-\tfrac{1}{2}(Y - XB)'(Y - XB)\Sigma^{-1}\}.$$

Again, we can decompose the sum of squares using the least squares projection:

$$p(Y|X, B, \Sigma) \propto |\Sigma|^{-n/2} \operatorname{etr}\{-\tfrac{1}{2}(S + (B - \hat{B})'X'X(B - \hat{B}))\Sigma^{-1}\},$$

with $S = (Y - X\hat{B})'(Y - X\hat{B})$ and $\hat{B} = (X'X)^{-1}X'Y$. To suggest the form of the natural conjugate prior, we can break up the two terms in the exponent:

$$p(Y|X, B, \Sigma) \propto |\Sigma|^{-(n-k)/2} \operatorname{etr}\{-\tfrac{1}{2}S\Sigma^{-1}\}|\Sigma|^{-k/2} \operatorname{etr}\{-\tfrac{1}{2}(B - \hat{B})'X'X(B - \hat{B})\Sigma^{-1}\}. \tag{2.8.37}$$

Expression (2.8.37) suggests that the natural conjugate prior is an inverted Wishart on Σ and a prior on B which is conditional on Σ. The term involving B is a density expressed as a function of an arbitrary $k \times m$ matrix. We can convert this density expression from a function of B to a function of $\beta = \operatorname{vec}(B)$ using standard results on vec operators[12] (see Magnus and Neudecker 1988, p. 30):

$$\operatorname{tr}((B - \hat{B})'X'X(B - \hat{B})\Sigma^{-1}) = \operatorname{vec}(B - \hat{B})'\operatorname{vec}(X'X(B - \hat{B})\Sigma^{-1})$$

and

$$\operatorname{vec}\left(X'X\left(B - \hat{B}\right)\Sigma^{-1}\right) = (\Sigma^{-1} \otimes X'X)\operatorname{vec}\left(B - \hat{B}\right).$$

Thus,

$$\begin{aligned}\operatorname{tr}((B - \hat{B})'X'X(B - \hat{B})\Sigma^{-1}) &= \operatorname{vec}(B - \hat{B})'(\Sigma^{-1} \otimes X'X)\operatorname{vec}(B - \hat{B}) \\ &= (\beta - \hat{\beta})'(\Sigma^{-1} \otimes X'X)(\beta - \hat{\beta}).\end{aligned}$$

[12] $\operatorname{tr}(A'B) = (\operatorname{vec}(A))' \operatorname{vec}(B)$ and $\operatorname{vec}(ABC) = (C' \otimes A) \operatorname{vec}(B)$.

Thus, the second term on the right-hand side of (2.8.37) is a normal kernel. This means that the natural conjugate prior for β is a normal prior conditional on this specific covariance matrix which depends on Σ.

The natural conjugate priors for the multivariate regression model are of the form

$$\begin{aligned} p(\Sigma, B) &= p(\Sigma)p(B|\Sigma), \\ \Sigma &\sim \mathrm{IW}(\nu_0, V_0), \\ \beta|\Sigma &\sim N\left(\bar{\beta}, \Sigma \otimes A^{-1}\right). \end{aligned} \tag{2.8.38}$$

Just as in the univariate regression model, the prior on the regression coefficients is dependent on the scale parameters and the same discussion applies. If we destroy natural conjugacy by using independent priors on β and Σ, we will not have analytic expressions for posterior marginals. The posterior can be obtained by combining terms from the natural conjugate prior to produce a posterior which is a product of an inverted Wishart and a 'matrix' normal kernel.

$$\begin{aligned} p(\Sigma, B|Y, X) \propto& |\Sigma|^{-(\nu_0+m+1)/2} \operatorname{etr}\left(-\tfrac{1}{2} V_0 \Sigma^{-1}\right) \\ &\times |\Sigma|^{-k/2} \operatorname{etr}(-\tfrac{1}{2}(B-\bar{B})' A (B-\bar{B})\Sigma^{-1}) \\ &\times |\Sigma|^{-n/2} \operatorname{etr}(-\tfrac{1}{2}(Y-XB)'(Y-XB)\Sigma^{-1}). \end{aligned} \tag{2.8.39}$$

We can combine the two terms involving B, using the same device we used for the univariate regression model:

$$\begin{aligned} &(B-\bar{B})' A (B-\bar{B}) + (Y-XB)'(Y-XB) \\ &\quad = (Z-WB)'(Z-WB) \\ &\quad = (Z-W\tilde{B})'(Z-W\tilde{B}) + (B-\tilde{B})' W' W (B-\tilde{B}) \end{aligned} \tag{2.8.40}$$

with

$$W = \begin{bmatrix} X \\ U \end{bmatrix}, \qquad Z = \begin{bmatrix} Y \\ U\bar{B} \end{bmatrix}, \qquad A = U'U.$$

The posterior density can now be written

$$\begin{aligned} p(\Sigma, B|Y, X) \propto& |\Sigma|^{-(\nu_0+n+m+1)/2} \operatorname{etr}(-\tfrac{1}{2}(V_0 + (V - W\tilde{B})'(\bullet)\Sigma^{-1})) \\ &\times |\Sigma|^{-k/2} \operatorname{etr}(-\tfrac{1}{2}(B-\tilde{B})' W' W (B-\tilde{B})\Sigma^{-1}), \end{aligned} \tag{2.8.41}$$

with

$$\begin{aligned} &\tilde{B} = (X'X + A)^{-1}\left(X'X\hat{B} + A\bar{B}\right), \\ &(Z-W\tilde{B})'(Z-W\tilde{B}) = (Y-X\tilde{B})'(Y-X\tilde{B}) + (\tilde{B}-\bar{B})' A (\tilde{B}-\bar{B}). \end{aligned} \tag{2.8.42}$$

Thus, the posterior is in the form the conjugate prior: inverted Wishart $\times$ conditional normal.

$$\begin{aligned}
&\Sigma|Y, X \sim IW(\nu_0 + n, V_0 + S), \\
&\beta|Y, X, \Sigma \sim N\left(\tilde{\beta}, \Sigma \otimes \left(X'X + A\right)^{-1}\right), \\
&\tilde{\beta} = \text{vec}(\tilde{B}), \qquad \tilde{B} = \left(X'X + A\right)^{-1}(X'X\hat{B} + A\overline{B}), \\
&S = (Y - X\tilde{B})'(Y - X\tilde{B}) + (\tilde{B} - \overline{B})'A(\tilde{B} - \overline{B}).
\end{aligned} \tag{2.8.43}$$

2.8.6 The Limitations of Conjugate Priors

Up to this point, we have considered some standard problems for which there exist natural conjugate priors. Although the natural conjugate priors have some features which might not always be very desirable, convenience is a powerful argument. However, the set of problems for which there exist usable expressions for conjugate priors is very small, as was pointed out in Section 2.7. In this section, we will illustrate that a seemingly minor change in the multivariate regression model destroys conjugacy. We will also examine the logistic regression model (one of the most widely used models in marketing) and see that there are no nice conjugate priors there. One should not conclude that conjugate priors are useless. Conjugate priors such as the normal and Wishart are simple useful representations of prior information which can be used even in nonconjugate contexts. Finally, many models are *conditionally conjugate* in the sense that conditional on some subset of parameters, we can use these conjugate results. This is exploited heavily in the hierarchical models literature.

The multivariate regression model in (2.8.33) with different regressors in each equation is called the *seemingly unrelated regression* (SUR) model by Zellner. This minor change to the model destroys conjugacy. We can no longer utilize the matrix form in (2.8.36) to write the model likelihood as in (2.8.37).[13] The best we can do is stack up the regression equations into one large regression:

$$\begin{aligned}
&y = X\beta + \varepsilon, \\
&y = \begin{bmatrix} y_1 \\ y_2 \\ \vdots \\ y_m \end{bmatrix}, \qquad X = \begin{bmatrix} X_1 & 0 & 0 & 0 \\ 0 & X_2 & 0 & 0 \\ 0 & 0 & \ddots & 0 \\ 0 & 0 & 0 & X_m \end{bmatrix}, \qquad \varepsilon = \begin{bmatrix} \varepsilon_1 \\ \varepsilon_2 \\ \vdots \\ \varepsilon_m \end{bmatrix},
\end{aligned} \tag{2.8.44}$$

with

$$\text{var}(\varepsilon) = \Sigma \otimes I_n. \tag{2.8.45}$$

Conditional on Σ, we can introduce a normal prior, standardize the observations to remove correlation, and produce a posterior. However, we cannot find a convenient prior to integrate out Σ from this conditional posterior. We can also condition on β,

[13] This should not be too surprising since we know from the standard econometrics treatment that the MLE for the SUR model is not the same as equation-by-equation least squares.

use an inverted Wishart prior on Σ and derive an inverted Wishart posterior for Σ. The reason we can do this is that, given β, we 'observe' the ε directly and we are back to the problem of Bayesian inference for a correlation matrix with zero mean. In Chapter 3 we will illustrate a very simple simulation method based on the so-called Gibbs sampler for exploring the marginal posteriors of the SUR model with a normal prior on β and an inverted Wishart prior on Σ.

A workhorse model in the marketing literature is the multinomial logit model. The dependent variable is a multinomial outcome whose probabilities are linked to independent variables which are alternative specific: $y_i = \{1, \ldots, J\}$ with probability p_{ij}, where

$$p_{ij} = \frac{\exp\left(x'_{ij}\beta\right)}{\sum_{j=1}^{J} \exp\left(x'_{ij}\beta\right)}. \tag{2.8.46}$$

The x_{ij} represent alternative specific attributes. Thus, the likelihood for the data (assuming independence of the observations) can be written as

$$p(y|\beta) = \prod_{i=1}^{n} p_{iy_i} = \prod_{i=1}^{n} \frac{\exp\left(x'_{iy_i}\beta\right)}{\sum_{j=1}^{J} \exp\left(x'_{ij}\beta\right)}. \tag{2.8.47}$$

Given that this model is in the exponential family, there should be a natural conjugate prior.[14] However, all this means is that the posterior will be in the same form as the likelihood. This does not assure us that we can integrate that posterior against any interesting functions. Nor does the existence of a natural conjugate prior ensure that it is interpretable and, therefore, easily assessable. Instead, we might argue that a normal prior on β would be reasonable. The posterior, of course, is not in the form of a normal or, for that matter, any other standard distribution. While we might have solved the problem of prior assessment, we are left with the integration problem. Since it is known that the log-likelihood is globally concave, we might expect asymptotic methods to work reasonably well on this problem. We explore these in the next section.

2.9 INTEGRATION AND ASYMPTOTIC METHODS

Outside the realm of natural conjugate problems, we will have to resort to numerical methods to compute the necessary integrals for Bayesian inference. The integration problem is of the form

$$I = \int h(\theta) p(\theta) p(D|\theta)\, d\theta. \tag{2.9.1}$$

This includes the computation of normalizing constants, moments, marginals, credibility intervals and expected utility.

There are three basic kinds of methods for approximating integrals of the sort given in (2.9.1). One may: (i) approximate the integrand by some other integrand which can be integrated numerically; (ii) approximate the infinite sum represented by the integral

[14] See Robert and Casella (2004, p. 146), for an example of the conjugate prior for $J = 2$.

through some finite-sum approximation such as a quadrature method; or (iii) view the integral as an expectation with respect to some density and use simulation methods to approximate the expectation by a sample average over simulations.

The first approach is usually implemented via resort to an asymptotic approximation of the likelihood. We can expand the log-likelihood in a second-order Taylor series about the MLE and use this as a basis of a normal kernel:

$$I \doteq \int h(\theta)p(\theta)\exp\{L(\theta)|_{\theta=\hat{\theta}} + \tfrac{1}{2}(\theta-\hat{\theta})'H(\theta-\hat{\theta})\}\,\mathrm{d}\theta, \tag{2.9.2}$$

where $H = [\partial^2 L/\partial\theta\partial\theta']$. This is the classic Laplace approximation. If we use the 'asymptotic' natural conjugate normal prior, we can compute approximate normalizing constants and moments:

$$I \doteq \mathrm{e}^{L(\hat{\theta})}\int h(\theta)(2\pi)^{-1/2}|A|^{1/2}\exp\{-\tfrac{1}{2}(\theta-\bar{\theta})'A(\bullet)\}\exp\{-\tfrac{1}{2}(\theta-\hat{\theta})'H^*(\bullet)\}\,\mathrm{d}\theta, \tag{2.9.3}$$

where $H^* = -H$ (negative of the Hessian of the log-likelihood). Completing the square in the exponent,[15] we obtain

$$\begin{aligned} I &\doteq \mathrm{e}^{L(\hat{\theta})-\frac{1}{2}\mathrm{SS}}|A|^{1/2}|A+H^*|^{-1/2} \\ &\quad\times\int h(\theta)(2\pi)^{-k/2}|A+H^*|^{1/2}\exp\{-\tfrac{1}{2}(\theta-\tilde{\theta})'(A+H^*)(\theta-\tilde{\theta})\}\,\mathrm{d}\theta, \end{aligned} \tag{2.9.4}$$

where $\mathrm{SS} = (\tilde{\theta}-\bar{\theta})'A(\bullet) + (\tilde{\theta}-\hat{\theta})'H^*(\bullet)$ and $\tilde{\theta} = (A+H^*)^{-1}(A\bar{\theta}+H^*\hat{\theta})$. Equation (2.9.4) implies that the normalizing constant and first two posterior moments $h(\theta)=\theta$; $h(\theta) = (\theta - E[\theta])(\theta-E[\theta])'$ are given by

$$\begin{aligned} \int p(\theta)p(\theta|D)\,\mathrm{d}\theta &\doteq \mathrm{e}^{L(\hat{\theta})-\frac{1}{2}\mathrm{SS}}|A|^{1/2}|A+H^*|^{-1/2}, \\ E[\theta|D] &\doteq \tilde{\theta}, \\ \mathrm{var}(\theta|D) &\doteq (A+H^*)^{-1}. \end{aligned} \tag{2.9.5}$$

While quadrature methods can be very accurate for integrands that closely resemble a normal kernel multiplied by a polynomial, many posterior distributions are not of this form. Moreover, quadrature methods suffer from the curse of dimensionality (the number of computations required to achieve a given level of accuracy increases to the power of the dimension of the integral) and are not useful for more than a few dimensions. For this reason, we will have to develop specialized simulation methods that will rely on approximating (2.9.1) with a sample average of simulated values. One such method is importance sampling.

---

[15] Recall the standard result for completing the square: $(x-\mu_1)'A_1(x-\mu_1) + (x-\mu_2)'A_2(x-\mu_2) = (x-\tilde{\mu})'(A_1+A_2)(x-\tilde{\mu}) + (\mu_1-\tilde{\mu})'A_1(\mu_1-\tilde{\mu}) + (\mu_2-\tilde{\mu})'A_2(\mu_2-\tilde{\mu})$, where $\tilde{\mu} = (A_1+A_2)^{-1}(A_1\mu_1 + A_2\mu_2)$.

2.10 IMPORTANCE SAMPLING

While we indicated that it is not feasible to devise efficient algorithms for producing iid samples from the posterior, it is instructive to consider how such samples could be used, if available. The problem is to compute

$$\mathrm{E}_{\theta|D}[h(\theta)] = \int h(\theta)p(\theta|D)\,\mathrm{d}\theta.$$

If $\{\theta_1, \ldots, \theta_R\}$ are a random sample[16] from the posterior distribution, then we can approximate this integral by a sample average:

$$\bar{h}_R = \frac{1}{R}\sum_r h(\theta_r). \tag{2.10.1}$$

If var$(h(\theta))$ is finite, we can rely on the standard theory of sample averages to obtain an estimate of the accuracy of $\bar{h}_R$:

$$\begin{aligned} \mathrm{STD}(\bar{h}_R) &= \frac{\sigma}{\sqrt{R}}, \qquad \sigma^2 = \mathrm{E}_{\theta|D}[(h_R - \mathrm{E}_{\theta|D}[h(\theta)])^2], \\ \mathrm{STDERR}(\bar{h}_R) &= \frac{s}{\sqrt{R}};\ s^2 = \frac{1}{R}\sum_r\left(h_r - \frac{\sum_r h_r}{R}\right)^2, \end{aligned} \tag{2.10.2}$$

where STD denotes standard deviation and STDERR denotes the estimated standard deviation or standard error. The formula in (2.10.2) has tremendous appeal since the accuracy of the integral estimate is ***independent*** of k, the dimension of θ. This is true in only a strict technical sense. However, in many applied contexts, the variation of the h function increases as the dimension of θ increases. This means that in some situations σ in (2.10.2) is a function of k. The promise of Monte Carlo integration to reduce or eliminate the curse of dimensionality is somewhat deceptive.

However, since we do not have any general-purpose methods for generating samples from the posterior distribution, the classic method of Monte Carlo integration is not of much use in practice. The method of importance sampling uses an importance sampling density to make the Monte Carlo integration problem practical. Assume that we only have access to the unnormalized posterior density (the product of the prior and the likelihood). We then must estimate

$$\mathrm{E}_{\theta|D}[h(\theta)] = \frac{\int h(\theta)p(\theta)p(D|\theta)\,\mathrm{d}\theta}{\int p(\theta)p(D|\theta)\,\mathrm{d}\theta}.$$

Suppose we have an importance distribution with density g. We can rewrite the integrals in the numerator and denominator as expectations with respect to this distribution and then use a ratio of sample averages to approximate the integral:

$$\mathrm{E}_{\theta|D}[h(\theta)] = \frac{\int h(\theta)[p(\theta)p(D|\theta)/g(\theta)]g(\theta)\,\mathrm{d}\theta}{\int[p(\theta)p(D|\theta)/g(\theta)]g(\theta)\,\mathrm{d}\theta} \tag{2.10.3}$$

[16] We use R to denote the size of the simulation sample in order to distinguish this from the sample size (n).

This quantity can be approximated by sampling $\{\theta_1, \ldots, \theta_R\}$ from the importance distribution:

$$\bar{h}_{\mathrm{IS},R} = \frac{R^{-1}\sum_r h(\theta_r)w_r}{R^{-1}\sum_r w_r} = \frac{\sum_r h(\theta_r)w_r}{\sum_r w_r}, \tag{2.10.4}$$

where $w_r = p(\theta_r)p(D|\theta_r)/g(\theta_r)$. Equation (2.10.4) is a ratio of weighted averages. Note that we do not have to use the normalized density g in computing the importance sampling estimate as the normalizing constants will cancel out from the numerator and denominator. The kernel of g is often called the *importance function.*

Assuming that $\mathrm{E}_{\theta|D}[h(\theta)]$ is finite, then the law of large numbers implies that $\bar{h}_{\mathrm{IS},R}$ converges to the value of (2.10.3). This is true for any importance density that has the same support as the posterior. However, the choice of importance sampling density is critical for proper functioning of importance sampling. In particular, the tail behavior of the posterior relative to the importance density is extremely important. If the ratio $p(\theta)p(D|\theta)/g(\theta)$ is unbounded, then the variance of $\bar{h}_{\mathrm{IS},R}$ can be infinite (see Geweke 1989, Theorem 2; or Robert and Casella 2004, p. 82). Geweke (1989) gives sufficient conditions for a finite variance. These conditions are basically satisfied if the tails of the importance density are thicker than the tails of posterior distribution.[17] The standard error of the importance sampling estimate (sometimes called the 'numerical standard error') can be computed as follows:

$$\mathrm{STDERR}(\bar{h}_{\mathrm{IS},R}) = \sqrt{\frac{\sum_r(h(\theta_r) - R^{-1}\sum_r h(\theta_r))^2 w_r^2}{\left(\sum_r w_r\right)^2}}. \tag{2.10.5}$$

One useful suggestion for an importance density is to use the asymptotic approximation to the posterior developed in (2.9.4). The thin normal tails can be fattened by scale mixing to form a multivariate Student t distribution with low degrees of freedom. This strategy was suggested by Zellner and Rossi (1984). Specifically, we develop an importance function as $\mathrm{MSt}(\nu, \hat{\theta}_{\mathrm{MLE}}, s(-H|_{\theta=\hat{\theta}_{\mathrm{MLE}}})^{-1})$. One could also use the posterior mode and Hessian evaluated at the posterior mode.

Draws from this distribution can be obtained via (2.11.6). 'Trial' runs can be conducted with different degrees of freedom values. The standard error in (2.10.5) can be used to help 'tune' the degrees of freedom and the scaling of the covariance matrix (s). The objective is to fatten the tails sufficiently to minimize variation in the weights. It should be noted that very low degree of freedom Student t distributions ($\nu < 5$) become very 'peaked' and do not make very good importance functions as the 'shoulders' of the distribution are too narrow. We recommend moderate values of the degrees of freedom parameter and more emphasis on choice of the scaling parameter to broaden the 'shoulders' of the distribution.

Geweke (1989) provides an important extension to this idea by developing a 'split-t' importance density. The 'split-t' importance function can handle posterior distributions

[17] One of the conditions is that the ratio of the posterior to the importance density is finite. Robert and Casella point out that this means that g could be used in a rejection method. This is a useful observation only in very low-dimensional parameter spaces. However, the real value of importance sampling methods is for parameter spaces of moderate to reasonably high dimension. In any more than a few dimensions, rejection methods based on the importance density would grind to a halt, rejecting a huge fraction of draws.

which are highly skewed. The standard multivariate Student importance function will be inefficient for skewed integrands as considerable fattening will be wasted in one 'tail' or principal axis of variation.

A fat-tailed importance function based on the posterior mode and Hessian can be very useful in solving integration problems in problems of moderate to large dimension (2 to 20). For example, the extremely regular shape of the multinomial logit likelihood means that importance sampling will work very well for this problem. Importance sampling has the side advantage of only requiring a maximizer and its associated output. Thus, importance sampling is almost a free good to anyone who is using the MLE.

But, there is also a sense that this exposes the limitations of importance sampling. If the posterior looks a lot like its asymptotic distribution, then importance sampling with the importance density proposed here will work well. However, it is in these situations that finite-sample inference is apt to be least valuable.[18] If we want to tackle problems with very nonnormal posteriors, then we must take greater care in the choice of importance function. The variance formulas can be deceiving. For example, suppose we situate the importance density far from where the posterior has mass. Then the weights will not vary much and we may convince ourselves using the standard error formulas that we have estimated the integral very precisely. Of course, we have completely missed the mass we want to integrate against. In high-dimensional problems, this situation can be very hard to detect. In very high dimensions, it would be useful to impose greater structure on the parameter space and break the integration problem down into more manageable parts. This is precisely what the methods and models discussed in Chapter 3 are designed to do. Finally, there are many models for which direct evaluation of the likelihood function is computationally intractable. Importance sampling will be of no use in these situations. However, importance sampling can be used as part of other methods to tackle many high-dimensional problems, even those with intractable likelihoods.

2.10.1 GHK Method for Evaluation of Certain Integrals of Multivariate Normal Distributions

In many situations, the evaluation of integrals of a multivariate normal distribution over a rectangular region may be desired. The rectangular region A is defined by $A = \{x : a < x < b\}$, where a and b are vectors of endpoints which might include infinity. Let $P = \Pr(x \in A)$ with $x \sim N(0, \Sigma)$. The GHK method (Keane 1994; Hajivassiliou *et al.* 1996) uses an importance sampling method to approximate this integral. The idea of this method is to construct the importance function and draw from it using univariate truncated normals.

We can define P either in terms of the correlated normal vector, x, or in terms of a vector of z of uncorrelated unit normals, $z \sim N(0, I)$:

$$
\begin{aligned}
x &= Lz, \qquad \Sigma = LL', \\
P &= \Pr(x \in A) = \Pr(a < x < b) = \Pr(a < Lz < b) \\
&= \Pr(L^{-1}a < z < L^{-1}b) = \Pr(z \in B), \\
B &= \{z : L^{-1}a < z < L^{-1}b\}.
\end{aligned}
$$

[18] One classical econometrician is rumored to have said that 'importance sampling just adds fuzz to our standard asymptotics'.

L is the lower triangular Cholesky root of Σ. We express the density of z as the product of a series of conditional densities which will allow us to exploit the lower triangular array which connects z and x:

$$p(z|B) = p(z_1|B)p(z_2|z_1, B) \cdots p(z_m|z_1, \ldots, z_{m-1}, B). \tag{2.10.6}$$

The region defined by A can be expressed in terms of elements of z as follows:

$$a_j < x_j < b_j,$$
$$a_j < l_{j1}z_1 + l_{j2}z_2 + \cdots + l_{j,j-1}z_{j-1} + l_{jj}z_j < b_j.$$

The l_{ij} are the elements of the lower Cholesky root, L. Given $z_{<j} = (z_1, z_2, \ldots, z_{j-1})$, this inequality implies that z is a univariate normal truncated to a particular interval given by

$$\frac{a_j - \mu_j(z_{<j})}{l_{jj}} < z_j < \frac{b_j - \mu_j(z_{<j})}{l_{jj}}, \tag{2.10.7}$$

where $\mu_j(z_{<j}) = l_{j1}z_1 + l_{j2}z_2 + \cdots + l_{j,j-1}z_{j-1}$. Inequality (2.10.7) provides an algorithm for drawing from $p(z|B)$. We simply draw $z_1|B$ and then $z_2|z_1, B$ and so on to fill out the z vector. The equation also provides us with a way to evaluate the conditional density of $z|B$.

$$p(z_j|z_{<j}, B) = \frac{\varphi(z_j)}{D_j(z_{<j})} \qquad \text{with } D_j(z_{<j}) = \Phi\left(\frac{b_j-\mu_j}{l_{jj}}\right) - \Phi\left(\frac{a_j-\mu_j}{l_{jj}}\right).$$

Using (2.10.6),

$$p(z|B) = \frac{\prod \varphi(z_j)}{\prod D_j(z_{<j})} = \frac{f(z)}{D(z)}.$$

We now write the integral defining P in terms of the density of z and use $p(z|B)$ as an importance function.

$$\begin{aligned} P = \Pr(z \in B) &= \int_B f(z)\,dz \\ &= \int_B \frac{f(z)}{p(z|B)} p(z|B)\,dz \\ &= \int_B D(z)p(z|B)\,dz \end{aligned} \tag{2.10.8}$$

Thus, the GHK algorithm can be constructed as follows:

GHK ALGORITHM

Draw z from $p(z|B)$ using (2.10.7) and truncated univariate normal draws.

Evaluate $D(z)$.

Repeat R times and form the estimate $\hat{P} = R^{-1}\Sigma D(z_r)$.

2.11 SIMULATION PRIMER FOR BAYESIAN PROBLEMS

If we could construct an iid sample directly from the posterior, the problem of summarizing the posterior could be solved to any desired degree of simulation accuracy. Unfortunately, the problem of generating random variables from an arbitrary (and possibly very high-dimensional) distribution has no general-purpose and computationally tractable solution. We will have to exploit the special structure of Bayesian models in order to develop useful methods. The basis for all of these methods are methods of simulating random variates from a set of frequently used distributions.

2.11.1 Uniform, Normal and Gamma Generation

All methods of continuous random variate generation start with a uniform pseudo-random number generator. Univariate pseudo-random number generators generate deterministic sequences of numbers (conditional on a seed) which pass various 'tests' for distributional accuracy and appearance of randomness. In particular, the sequence should have a very long 'period' before it repeats itself, exhibit minimal time dependence (sometimes measured by autocorrelation), and do a good job of 'filling' a k-dimensional hypercube constructed from subsequences of length k. R uses the Mersenne twister (Matsumoto and Nishimura 1998) as the default method. Some might argue that the KISS generator by Marsaglia (1999) is faster.

Any standard computing environment will also supply methods for generating normal as well as gamma-distributed random variates. By default, R uses the inverse cumulative distribution function (cdf) method to generate normal random variates. This certainly produces draws with excellent properties, but it may be slightly inefficient compared to method such as the Ziggurat method of Marsaglia and Tsang (2000a). However, we have not found the computations required for normal draws to be a computational bottleneck in Bayesian computations.

R also provides methods to draw gamma and chi-squared random variates using the inverse cdf method. If your computing environment does not provide high-quality Gamma random variates, the method of Marsaglia and Tsang (2000b) can be programmed in a low level language such as C.

We have seen in Section 2.8 that draws from the inverted gamma prior and posterior for σ^2 are needed. Recall that $\mathrm{IG}\left(\alpha = \nu/2, \beta = \frac{\nu s^2}{2}\right) \sim \nu s^2/\chi^2_\nu$:

$$\mathrm{IG}\left(\alpha = \nu/2, \beta = \frac{\nu s^2}{2}\right) = \frac{\nu s^2}{\mathrm{Gamma}(\nu/2, 1/2)}. \tag{2.11.1}$$

The uniform, normal and gamma methods available in R and many other computing environments can be used to construct simulators for many of the distributions needed for Bayesian inference. We will now provide a simulation primer for these distributions.

2.11.2 Truncated Distributions

The inverse cdf method can be used to draw from truncated distributions, provided that a computationally efficient method is available to evaluate the inverse cdf or quantile function. To review the inverse cdf method, let $F_X^{-1}(x)$ be the inverse of the cdf of random variable X.[19] Then $X = F_X^{-1}(U)$, $U \sim \text{Unif}(0, 1)$ has distribution with cdf F_X. An important example is the truncated normal distribution. Consider the normal distribution truncated to the interval (a, b):

$$Y = X \times I_{(a,b)}(X), \qquad I_{(a,b)}(X) = \begin{cases} 1 & \text{if } X \in (a, b) \\ 0 & \text{otherwise.} \end{cases}$$

The cdf of Y can be obtained from the cdf of X:

$$G_Y(y) = \frac{F(y) - F(a)}{F(b) - F(a)}.$$

Let $p = G_Y(y)$ and solve for G^{-1}:

$$y = F^{-1}(p(F(b) - F(a)) + F(a)). \tag{2.11.2}$$

Recall that $F(x) = \Phi((x - \mu_X)/\sigma_X)$ and $F^{-1}(p) = \mu_X + \sigma_X \Phi^{-1}(p)$. We have a simple algorithm for simulating from the truncated normal.

TRUNCATED NORMAL ALGORITHM

Draw $U \sim \text{Unif}(0, 1)$.

$$Y = \mu + \sigma\Phi^{-1}(U(\Phi((b - \mu)/\sigma) - \Phi((a - \mu)/\sigma)) + \Phi((a - \mu)/\sigma)).$$

$$Y \sim \text{trun}_{(a,b)} N(\mu, \sigma)$$

There is a legitimate argument that this algorithm is computationally inefficient due to the evaluation of the normal cdf and inverse cdf functions. A combination of rejection sampling and other methods can be used to develop more efficient algorithms (cf. McCulloch and Rossi 1994). However, for a language such as R where vectorization is essential for efficiency, the algorithm based on the cdf method is more efficient.
R In our R package `bayesm`, we provide the routine `rtrun` to simulate a vector of truncated normals.

2.11.3 *Multivariate Normal and Student t Distributions*

Given unit or standard normal draws, we can create the general normal variate by scale and location transform. Multivariate normal draws can be calculated via the following algorithm:

$$\begin{aligned} z' &= (z_1, \ldots, z_k), \qquad z_i \sim \text{iid } N(0, 1), \\ x &= U'z + \mu \sim N(\mu, \Sigma), \qquad \Sigma = U'U. \end{aligned} \tag{2.11.3}$$

[19] To avoid any technical difficulties, we must assume that there are no jumps in F or that the distribution of X is absolutely continuous with respect to Lebesgue measure.

U is the upper triangular Cholesky root of Σ. Computation of roots of positive definite real matrices has been studied closely and there are reliable and computationally efficient algorithms for doing so (cf. the implementation in LAPACK used by R).

In some cases, we must simulate from various conditional distributions of a subvector of a multivariate normal given the remainder of the normal random vector. The fact that the conditional mean and conditional covariance matrix can be computed directly from the elements of the inverse of Σ is useful: if

$$x = \begin{pmatrix} x_1 \\ x_2 \end{pmatrix} \sim N\left(\begin{pmatrix} \mu_1 \\ \mu_2 \end{pmatrix}, \Sigma = \begin{bmatrix} \Sigma_{11} & \Sigma_{12} \\ \Sigma_{21} & \Sigma_{22} \end{bmatrix}\right)$$

and

$$\Sigma^{-1} = V = \begin{bmatrix} V_{11} & V_{12} \\ V_{21} & V_{22} \end{bmatrix},$$

then

$$x_1|x_2 \sim N(\mu_1 - V_{11}^{-1}V_{12}(x_2 - \mu_2), V_{11}^{-1}). \tag{2.11.4}$$

The multivariate t distribution is an elliptically symmetric distribution closely related to the multivariate normal distribution. The major difference is that the multivariate t has algebraic tails (not exponential like the normal) and can be very peaked for very low degrees of freedom. A k-dimensional multivariate t distribution with degrees of freedom parameter ν, location parameter μ and scale parameter Σ has density

$$p(x|\nu, \mu, \Sigma) \propto |\Sigma|^{-1/2}[\nu + (x - \mu)'\Sigma^{-1}(x - \mu)]^{-(k+\nu)/2}, \tag{2.11.5}$$

with $E[x] = \mu$, $\text{var}(x) = (\nu/(\nu - 2))\Sigma$. We can simulate from this distribution by fattening the tails of a multivariate normal:

$$\begin{aligned} X &= \frac{Y}{(Z/\nu)^{1/2}} + \mu, \qquad Y \sim N(0, \Sigma), \qquad Z \sim \chi^2_\nu, \\ X &\sim \text{MSt}(\nu, \mu, \Sigma). \end{aligned} \tag{2.11.6}$$

R *bayesm* includes `rmvst` to simulate from the multivariate Student t distribution.

2.11.4 The Wishart and Inverted Wishart Distributions

In Section 2.8 we observed that prior and posteriors for the covariance matrix of the multivariate normal distribution are of the inverted Wishart form. The correspondence between the Wishart and inverted Wishart distribution can be exploited to develop a simulator. Let $G \sim W(\nu, V^{-1})$ and factor V^{-1} into the product of upper Cholesky roots, $V^{-1} = U'U$. Then $G = U'BU$, where $B \sim W(\nu, I_m)$. To simulate from the standard Wishart, we construct the following lower triangular array of random variates (note all nonzero elements are independent):

$$T = \begin{bmatrix} \sqrt{\chi^2_\nu} & 0 & \cdots & 0 \\ Z_{21} & \sqrt{\chi^2_{\nu-1}} & & \vdots \\ \vdots & & \ddots & 0 \\ Z_{m1} & \cdots & Z_{m,m-1} & \sqrt{\chi^2_{\nu-m+1}} \end{bmatrix}, \tag{2.11.7}$$

where $Z_{ij} \sim N(0, 1)$ and $\nu > m$. Then

$$TT' \sim W(\nu, I_m).$$

G can be constructed from the pieces:

$$\begin{aligned} G &= U'TT'U = C'C, \\ C &= T'U. \end{aligned} \tag{2.11.8}$$

The inverted Wishart can be computed from the draw of the corresponding Wishart using (2.11.8) by taking the inverse of C:

$$\Sigma = G^{-1} = C^{-1}(C^{-1})'. \tag{2.11.9}$$

Note that (2.11.9) is the UL decomposition of Σ.

WISHART/INVERTED WISHART DRAW ALGORITHM

To draw $\Sigma \sim \mathrm{IW}(\nu, V)$ or $\Sigma^{-1} \sim W(\nu, V^{-1})$:

1. Factor $V^{-1} = U'U$.
2. Draw random variates and compute T as in (2.11.7).
3. Compute $C^{-1} = (T'U)^{-1}$, $\Sigma = C^{-1}(C^{-1})'$, $\Sigma^{-1} = C'C$.

R *bayesm* includes the function `rwishart` to simulate from both the inverted Wishart and Wishart distribution.

2.11.5 Multinomial Distributions

The most general discrete distribution is the multinomial distribution. If θ can take one of d values, $S = \{\theta^1, \theta^2, \ldots, \theta^d\}$, each with probability p_i, then $\theta \sim \mathrm{Multinomial}(p,S)$.

MULTINOMIAL DRAW ALGORITHM

To draw from the multinomial, we only need to draw the index into S with the appropriate probability.

Draw $U \sim \mathrm{Unif}(0, 1)$.

Find k such that $\sum_{i=0}^{k-1} p_i < U \leq \sum_{i=0}^{k} p_i$ with $p_0 = 0$.
$\theta = S(k)$.

2.11.6 Dirichlet Distribution

The natural conjugate prior for the multinomial distribution is called the Dirichlet distribution with density

$$p(\theta|\alpha_1, \ldots, \alpha_k) = \frac{\Gamma(\alpha_1 + \cdots \alpha_k)}{\Gamma(\alpha_1) \cdots \Gamma(\alpha_k)} \theta_1^{\alpha_1 - 1} \cdots \theta_k^{\alpha_k - 1}, \qquad \alpha_i > 0. \tag{2.11.10}$$

θ is a k-dimensional vector that must be in the unit simplex.

DIRICHLET DRAW ALGORITHM

Draw $x_i \sim$ ind Gamma(α_i, α_i).
$\theta_i = x_i / \sum_j x_j$.

R *bayesm* includes `rdirichlet` to simulate from the Dirichlet distribution.

2.12 SIMULATION FROM THE POSTERIOR OF THE MULTIVARIATE REGRESSION MODEL

The multivariate regression model discussed in Section 2.8.4 is a very useful model not only because of direct application to situations with sets of related linear regressions but also in various hierarchical model settings in which the model is used as part of the prior structure. To analyze these hierarchical models, it will be necessary to use simulation-based methods. For this reason, it will be useful to have an efficient algorithm for sampling from the posterior of this model given in (2.8.43). Recall the basic set-up.

Model

$$Y = XB + U, \qquad U = [u_i'], \qquad u_i \sim N(0, \Sigma). \tag{2.12.1}$$

Here Y is $n \times m$, X is $n \times k$, and B is $k \times m$, with each column of B containing the regression coefficients for one of the m equations.

Prior

$$\begin{aligned} &\Sigma \sim \text{IW}(\nu, V), \\ &\beta = \text{vec}(B)|\Sigma \sim N(\text{vec}(\bar{B}), \Sigma \otimes A^{-1}). \end{aligned} \tag{2.12.2}$$

To draw from the posterior, we first draw Σ and then draw B given Σ.

Draw of Σ

$$\begin{aligned}
&\Sigma|Y, X \sim \text{IW}(\nu + n, V + S), \\
&S = E'E, \\
&E = Y - X\tilde{B} + (\tilde{B} - \overline{B})'A(\tilde{B} - \overline{B}), \\
&\tilde{B} = (X'X + A)^{-1}(X'Y + A\overline{B}).
\end{aligned} \tag{2.12.3}$$

S and $\tilde{B}$ can be computed using the QR decomposition. The QR decomposition of a matrix X is $X = QR$, where Q is an $n \times k$ matrix whose columns are orthogonal and R is a $k \times k$ upper triangular matrix that is the Cholesky root of $X'X$ up to sign differences. We can compute the relevant quantities in (2.12.3) by forming the following augmented matrices:

$$W = \begin{bmatrix} X \\ R_A \end{bmatrix}, \qquad Z = \begin{bmatrix} Y \\ R_A\overline{B} \end{bmatrix}. \tag{2.12.4}$$

If we take the QR decomposition of W, $W = Q_W R_W$, then

$$\tilde{B} = R_W^{-1} Q_W Z, \qquad S = (Z - W\tilde{B})'(Z - W\tilde{B}) = Z'(I - Q_W Q_W')Z. \tag{2.12.5}$$

However, timing experiments in R suggest that it is faster to compute these quantities by using the Cholesky root approach (between 30 % and 100 % faster than using the LAPACK QR method in R). We take the Cholesky root of $W'W$, invert this, and use this to compute $\tilde{B}$ and E.[20] The inverse of the root can be computed efficiently in R and the root can also be used to draw from the posterior as we see below:

$$\begin{aligned}
&W'W = R_{W'W}' R_{W'W} \text{ and } IR_{W'W} = R_{W'W}^{-1}, \\
&\tilde{B} = IR_{W'W}(IR_{W'W})'W'Z, \\
&E = Z - W\tilde{B}, \\
&S = E'E.
\end{aligned} \tag{2.12.6}$$

$R_{W'W}$ is the upper triangular Cholesky root of $W'W$. We note that $IR_{W'W}$ defines the UL decomposition of $(W'W)^{-1}$ whereas $R_{W'W}$ forms the LU decomposition of $W'W$.

To draw Σ, we draw from the appropriate Wishart and then invert this matrix to obtain the Σ draw:

$$\Sigma^{-1} \sim W(\nu + n, (V + S)^{-1}).$$

The Bartlett strategy outlined in Section 2.11.4 produces a draw of the Cholesky root of Σ^{-1}:

$$\begin{aligned}
&\Sigma^{-1} = C'C, \\
&\Sigma = (C^{-1})(C^{-1})' = CICI'.
\end{aligned} \tag{2.12.7}$$

---

[20] Golub and Van Loan (1989, p. 226), state that the QR decomposition will handle $W'W$ arrays that are more nearly non-singular than the Cholesky root approach. Both are very stable numerically and, in Bayesian computations with proper priors, we never approach non-singularity.

We note that we have the LU decomposition of Σ^{-1} and the UL decomposition of Σ in (2.12.7).

Draw of $\beta|\Sigma$

The direct, but naïve, approach would be to draw β from a $N(\text{vec}(\bar{B}), \Sigma \otimes (X'X + A)^{-1})$ distribution. We can exploit the special structure of the covariance matrix to develop an efficient strategy for making this draw.

$$\begin{aligned} \text{var}(\beta) &= \Omega = \Sigma \otimes (X'X + A)^{-1} \\ &= CICI' \otimes (R'_{W'W} R_{W'W})^{-1} \\ &= CICI' \otimes IR_{W'W} IR'_{W'W} \\ &= (CI \otimes IR_{W'W})(CI \otimes IR_{W'W})' \end{aligned} \tag{2.12.8}$$

$IR_{W'W} = R^{-1}_{W'W}$. Note that (2.12.8) is the UL decomposition of the covariance matrix. Thus, we can use this root directly to produce normal variates with the right covariance:

$$\nu = (CI \otimes IR_W)z, \qquad z \sim N(0, I_{m \times k}),$$
$$\text{var}(\nu) = \Omega.$$

However, we can simplify this even further by using the identity $\text{vec}(ABC) = (C' \otimes A)\text{vec}(B)$:

$$B = \tilde{B} + IR_{W'W} Z CI', \qquad \text{vec}(Z) = z. \tag{2.12.9}$$

R `rmultireg`, in *bayesm*, implements this strategy.

3
Markov Chain Monte Carlo Methods

Using this Chapter

This chapter provides an introduction to Markov chain Monte Carlo (MCMC) methods and provides detailed discussion of the MCMC methods that have proved especially useful for marketing and micro-econometric problems. In addition, several key examples are developed which help set the stage for more complicated models covered in later chapters. These include the Gibbs samplers for binary probit, mixture of normals and hierarchical linear models. In addition, Metropolis methods are introduced and illustrated with the multinomial logit model. Those readers who desire an introduction to MCMC methods without much theoretical background should skip Section 3.3 on Markov chain theory and only skim the beginning of Section 3.10 on Metropolis methods (and skip over the proof following the introduction of the continuous state space Metropolis algorithm).

Given a model (prior and likelihood) or set of models, the computational phase of Bayesian inference requires practical methods for summarizing/exploring the posterior distribution. In many cases, the posterior distribution is represented by an unnormalized density, $\pi^*(\theta)$, and the problem is to construct simulation-based estimates of various aspects of this distribution. For any problem outside the conjugate family, the posterior density will be of a form for which analytical results on marginals or moments will be unavailable. More importantly, problems in which the dimension of θ is much more than 100 can easily arise in marketing applications. For example, suppose we want to compute the response to marketing instruments for each of 100–200 customers. We might formulate a group of 100–200 regression models each with 5–10 independent variables. If there are any linkages between these models through correlated unobservables,[1] then we must explore a parameter space of dimension 500–2000. Problems of this dimension

[1] The regression errors could be correlated (as in the SUR model of Zellner) or the prior on the regression coefficients could have correlations or dependencies across equations. The later is often formulated as a hierarchical model which is introduced below in Section 3.7 and in Chapter 5.

Bayesian Statistics and Marketing P. E. Rossi, G. M. Allenby and R. McCulloch

are considerably beyond the scope of importance sampling methods. In order to tackle these problems, we will have to exploit the structure of the model and introduce new methods for simulation from arbitrary distributions.

3.1 MARKOV CHAIN MONTE CARLO METHODS

The idea behind MCMC methods is to formulate a Markov chain on the parameter space. If care is taken to ensure that this chain has $\pi(\bullet)$ as its equilibrium or 'long-run' distribution, then the chain can be used to construct simulation-based estimates of the required integrals. Starting from some point in the parameter space, we simulate the chain forward. A subsequence of these draws can be used to construct simulation-based estimates of the posterior distribution of θ or any function of θ. For example, we can simply focus on one element of θ to simulate its marginal or we can average $h(\theta)$ over sequences of draws from the chain to estimate $E_\pi[h(\theta)]$.

A Markov chain specifies a method for generating a sequence of random variables $\{\theta_1, \theta_2, \ldots, \theta_r, \ldots\}$ starting from initial point θ_0. This sequence is created by specifying a way of transitioning or moving from θ_r to θ_{r+1}. Since we are dealing with random variables, this transition process is specified by choosing the conditional distribution, $\theta_{r+1}|\theta_r$, or $\theta_{r+1}|\theta_r \sim F(\theta_r)$. This conditional distribution can be discrete, continuous, or a mixture of discrete and continuous distributions. By iterating the conditional distribution forward we construct a joint distribution on the sequence. The fact that this conditional distribution only depends on the last θ (or that θ_r completely summarizes all information up to this point) is the Markov property which greatly simplifies simulation and analysis of the chain.

Clearly, Markov chains are well designed for simulation. To simulate from the Markov chain, do the following:

Start from θ_0.

Draw $\theta_1 \sim F(\theta_0)$.

Replace θ_0 with θ_1 and Repeat a total of R times.

This will create a set of realizations of the chain given the starting point. Under some conditions on the conditional distribution F, the distribution of $\theta_r|\theta_0$ will converge to a fixed and unique distribution as r goes to infinity. This distribution is called the stationary, invariant,[2] or equilibrium distribution. If we can construct a Markov chain with stationary distribution $\pi(\bullet)$ and the conditions for convergence are met, then we can use the Markov chain method to construct a simulation method for estimating the posterior expectation of any function:[3]

$$E_\pi[h(\theta)] \doteq \frac{1}{R}\sum h(\theta_r). \tag{3.1.1}$$

[2] The meaning of the term 'invariant' will be provided in Section 3.3.

[3] Assuming, of course, that the posterior expectation of this function exists.

Use of (3.1.1) is asymptotic in the sense that we are relying on the fact that averages of the Markov chain converge (again under some conditions) to the expectation under the stationary distribution:

$$\lim_{R \to \infty} \frac{1}{R} \sum h(\theta_r) = E_\pi[h(\theta)]. \tag{3.1.2}$$

If a Markov chain satisfies (3.1.2), it is called *ergodic*.

As a practical matter, we will be using large but finite R. Although the theory does not require it, most practitioners will discard some set of initial draws out of concern for the effects of the 'initial condition' or value of θ_0. This is sometimes called the 'burn-in' period. The idea is that it will take a while for the chain to 'equilibrate' or shrug off the effects of the initial condition. We then use the draws after the burn-in period to create simulation estimates. If we 'burn in' for B draws, then the MCMC estimate is given by

$$E_\pi[h(\theta)] \doteq \frac{1}{R-B} \sum_{r=B+1}^{R} h(\theta_r). \tag{3.1.3}$$

Of course, this is an example of a Monte Carlo integration estimate. The difference now is that the draw sequence is not an iid draw sequence. Since we are allowed to condition on the last value in generating the next value in a Markov chain, the draws can be dependent. To some, this dependence is unfamiliar and troublesome. We are relying on something like the law of large numbers to ensure that the sample average converges to the 'population' average. Most standard versions of the law of large numbers assume independence. As long as the dependence is not pathologically strong, then the intuition behind the law of large numbers still goes through.[4] We are still getting more information on the stationary distribution with every draw. The practical problem is that these simulation-based estimates may have large sampling errors.

We are relying on asymptotic theory to justify our use of simulation methods with large samples. That is, we are using the fact that long-run averages of draws from the Markov chain will converge to the appropriate integral over the posterior distribution. More generally, the posterior distribution constructed from draws from the Markov chain will closely approximate the true posterior distribution for large enough samples. This does not mean we are using sampling theory as the basis for our inferences regarding the data. The data are of fixed size, giving rise to a posterior for that sample. It is only when we approximate the posterior that we appeal to asymptotics or long-run behavior. We should recognize that, up to computational limitations, the sample size used in any Monte Carlo method is under our control. In particular, we can increase the sample size as necessary and, for most problems, we can generate truly huge simulation samples at moderate computational cost. This means that the practical use of MCMC methods comes much closer to the long-run sampling experiments envisaged by the inventors of asymptotic theory than the usual application of these methods to small and fixed size samples.

[4] Those trained in time series analysis will recall that there are versions of the law of large numbers for dependent but stationary sequences.

The purpose of this introductory section was to give the reader an overview of the basic idea behind MCMC methods. We are a long way from practical application in the sense that we need to provide:

(i) methods or algorithms for specifying chains with the right stationary distribution – this amounts to specifying the conditional distribution of $\theta_{r+1}|\theta_r$ using information about $\pi(\bullet)$;

(ii) theoretical assurance that the methods in (i) will produce ergodic chains;

(iii) practical guidance on convergence, including some notion of how long we should run the chain and how long the 'burn-in' period should be.

There is good news and bad news regarding the answers to these questions. The good news is that there are families of algorithms which can deliver chains for arbitrary posterior distributions. Further, these algorithms enjoy very strong theoretical convergence properties under relatively mild and verifiable conditions. The bad news is that even a theoretically convergent chain may be very slow to converge and/or highly dependent. This means that some care and experimentation must be used in the selection and use of the MCMC algorithms. However, it is fair to say that MCMC methods have been applied to problems of dimension (exceeding 1000) and complexity (problems for which the likelihood is intractable) well beyond the original developers' wildest dreams.

3.2 A SIMPLE EXAMPLE: BIVARIATE NORMAL GIBBS SAMPLER

The ideas introduced in Section 3.1 can be better understood by considering one of the most useful and well-used MCMC methods, the Gibbs sampler. We will treat the general case of the Gibbs sampler in Section 3.4 and we will see many nontrivial examples. However, it is best to start with a very simple problem.

Consider the problem of simulating from the bivariate normal distribution

$$\begin{pmatrix} \theta_1 \\ \theta_2 \end{pmatrix} \sim N\left(\begin{pmatrix} 0 \\ 0 \end{pmatrix}, \begin{bmatrix} 1 & \rho \\ \rho & 1 \end{bmatrix}\right). \tag{3.2.1}$$

In Section 2.11.3 the standard method for drawing from the multivariate normal distribution was presented. Recall that we only have to compute the Cholesky root of the covariance matrix and use this to induce the proper level of correlation. Let Z be an $R \times 2$ matrix of iid $N(0,1)$ draws; then we simply compute

$$\Theta = ZU, \qquad \text{where } U = \begin{bmatrix} 1 & \rho \\ 0 & \sqrt{1-\rho^2} \end{bmatrix}. \tag{3.2.2}$$

The $R \times 2$ matrix Θ is a matrix whose rows are iid draws from (3.2.1).

The Gibbs sampler specifies a Markov chain whose stationary distribution is the bivariate normal. The transition mechanism in the Gibbs sampler is specified through iterative sampling from conditional distributions. At point $\theta_r = (\theta_{r1} \quad \theta_{r2})'$ (here θ_{ri} is the rth draw of component i), the next random variables $(\theta_{r+1,1} \quad \theta_{r+1,2})'$ are constructed by drawing from the two conditional distributions associated with the bivariate normal,

$$\theta_2 \Big| \theta_1 \sim N\left(\mu_2 + \rho\frac{\sigma_2}{\sigma_1}(\theta_1 - \mu_1), \sigma_2^2(1 - \rho^2)\right)$$

and

$$\theta_1 \Big| \theta_2 \sim N\left(\mu_1 + \rho\frac{\sigma_1}{\sigma_2}(\theta_2 - \mu_2), \sigma_1^2(1 - \rho^2)\right).$$

BIVARIATE NORMAL GIBBS SAMPLER (example)

Start at point θ_0.

Draw θ_1 in two steps:

$$\begin{aligned} \theta_{12} &\sim N(\rho\theta_{01}, 1 - \rho^2), \\ \theta_{11} &\sim N(\rho\theta_{12}, 1 - \rho^2). \end{aligned} \tag{3.2.3}$$

Repeat as long as desired to draw $\theta_2|\theta_1, \ldots, \theta_{r+1}|\theta_r, \ldots$.

We draw first from the conditional distribution of the second component given the previous draw of the first component. We then draw the first component given the most recently drawn value of the second component. This means that we move from one point to another in the two-dimensional parameter space in a sequence of two moves along the coordinate axes. In this particular sampler, we draw $\theta_{r+1,2}|\theta_{r,1}$ and then $\theta_{r+1,1}|\theta_{r+1,2}$. This means that we only 'use' the initial value θ_{01} and do not use θ_{02}. Of course, we could have swapped the order of these 'intermediate' steps and would still have a valid (but different) sampler.

This simple example shows that one of the problems with MCMC algorithms is that they do not lend themselves to the vectorization that is required for the most efficient use of interpreted languages such as R. There is no avoiding the 'loop' here since the arguments in the loop are recursively dependent and cannot be computed prior to vector operations. This means that there is at least one 'loop' even in the most efficient R code to implement MCMC methods. Our experience, however, has shown that this is not a serious computational constraint for many problems of interest.

Figure 3.1 illustrates this algorithm for $\rho = 0.9$ and starting value $(\theta_{01}, \theta_{02})' = (2, -2)'$. To move away from the initial condition, we first draw the second (vertical) component value. Given that we start from the first component value of 2, we draw (according to (3.2.3)) from an $N(0.9 \times 2, 0.19)$ distribution, illustrated by the vertically oriented density curve in Figure 3.1. This distribution is centered at 1.8 but with a fairly large standard deviation of 0.44. A realized value is shown on the figure by the '•' symbol. This realized value is about 1.7, near the mean of the conditional distribution. We now draw the first component given the second value of 1.7 by drawing from an $N(0.9 \times 1.7, 0.19)$ distribution. These two draws combine to move the chain from the

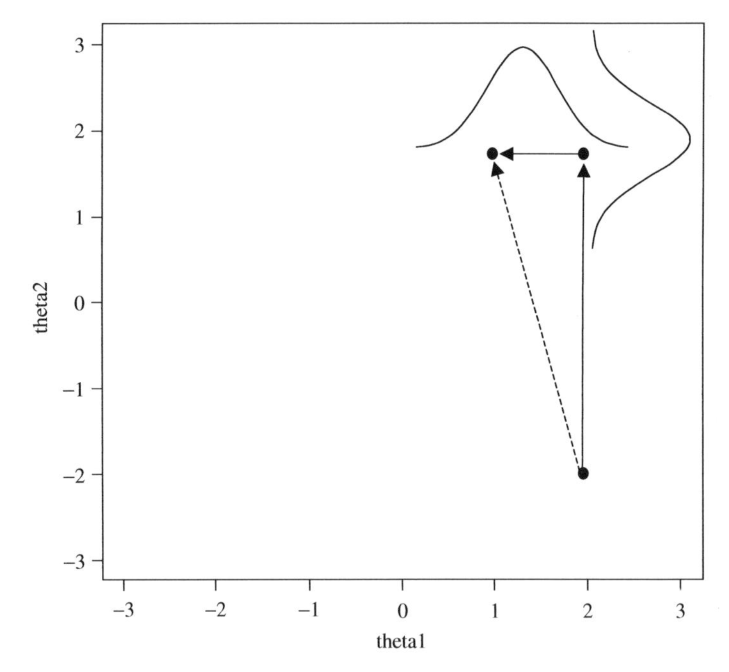

Figure 3.1 Functioning of bivariate normal Gibbs sampler

initial value of (2, −2) to a new value of approximately (1, 1.7). This move is denoted by the dotted arrow on the figure.

Figure 3.2 shows a realization of the first 20 draws from the same initial condition as in Figure 3.1. The figure shows both the 'intermediate' moves internal to the sampler (that is, the component-by-component updatings) and the final result. Each of these points are connected by lines to trace out the movements. The fact the intermediate moves are included is the reason why the trace consists of only vertical and horizontal line segments of random length. The chain moves quickly away from the initial condition to the region where the bivariate normal has substantial mass. The reason for this is that the initial value of (2, −2) is highly unusual for this bivariate normal. The conditional distributions capture this immediately by moving the second component to a value more in line with the strong positive correlation.

The average size of the move along any given coordinate axis is constant since the standard deviation of normal conditional distributions, $\sqrt{1-\rho^2}$, does not depend on the conditioning argument. Clearly, as ρ gets even closer to one in magnitude, the size of the moves of the sampler will be very limited. However, for all but extremely large values of ρ, the chain will dissipate the effects of the initial conditions rapidly although it may move slowly once it gets to the areas of high mass. After only 40 draws, the chain has started to navigate the regions where the bivariate normal distribution puts mass, but nowhere near adequately for the purpose of estimating probabilities of sets or moments. For example, we certainly would be foolish to estimate the probability of a set by simply computing the proportion of the 40 draws falling in this set. The theory of MCMC would tell us that as the number of draws tends to infinity this answer can be computed to any desired degree of accuracy. Figure 3.3 shows 1000 draws (without the

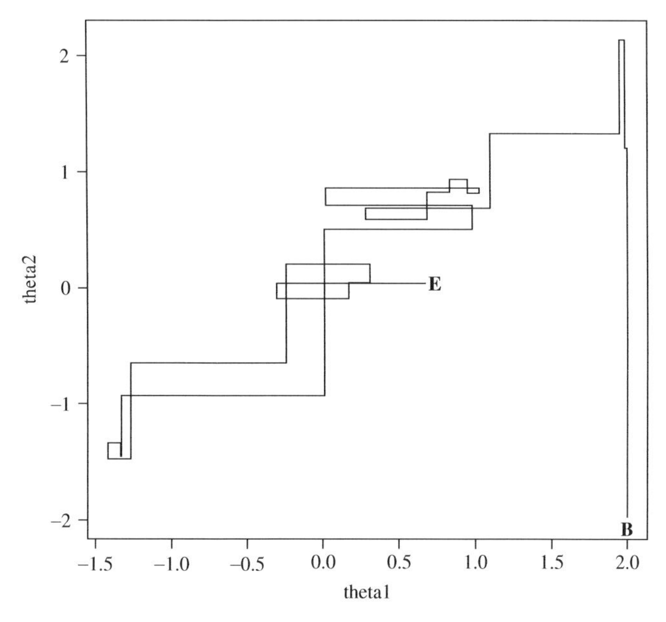

Figure 3.2 Twenty draws from bivariate Gibbs sampler showing intermediate moves

intermediate moves on the lattice) of the sampler. Clearly, the chain seems to navigate freely and the regions of high mass are dark with ink.

Figure 3.2 also shows that there is substantial dependence for replicates close together in the sequence. If the chain is in the positive orthant along the 'ridge' created by the positive correlation, then it may take several (or more) moves to navigate into the negative orthant. One way of measuring this serial or 'time' dependence is to compute the sample autocorrelation function (ACF),

$$s_{\theta_1}(k) = \frac{\sum_{r=k+1}^{R}\left(\theta_{r1} - \bar{\theta}_1\right)\left(\theta_{r-k,1} - \bar{\theta}_1\right)}{\sum_{r=1}^{R}\left(\theta_{r1} - \bar{\theta}_1\right)^2}. \tag{3.2.4}$$

The top panel of Figure 3.4 shows the ACF for the first component of θ. The lag or 'order' of the autocorrelation is the number of 'periods' or draws separating random variables in the chain (k in (3.2.4)). As we might expect, there is a fair amount of autocorrelation in the sequence of draws. However, by lag 10 or 12, there is no appreciable dependence.

If this sampler is ergodic, we can use sample averages of functions evaluated on the chain draws to estimate the expectations of those functions with respect to the stationary distribution of the chain. Figure 3.3 suggests that the stationary distribution for this chain is actually the bivariate normal distribution. We can use the sample correlation of the draws of θ_1 and the draws of θ_2 to estimate the 'population' correlation or the expectation of the function $h(\theta) = \theta_1\theta_2$ with respect to the bivariate normal. In this example, we know that $\mathrm{E}[h(\theta)] = \mathrm{E}[\theta_1\theta_2] = \rho$. The bottom panel of Figure 3.4 shows the sample correlation based on the Gibbs sampler draw sequence (dotted line) for samples of successively larger size, starting from the beginning of the chain. For example,

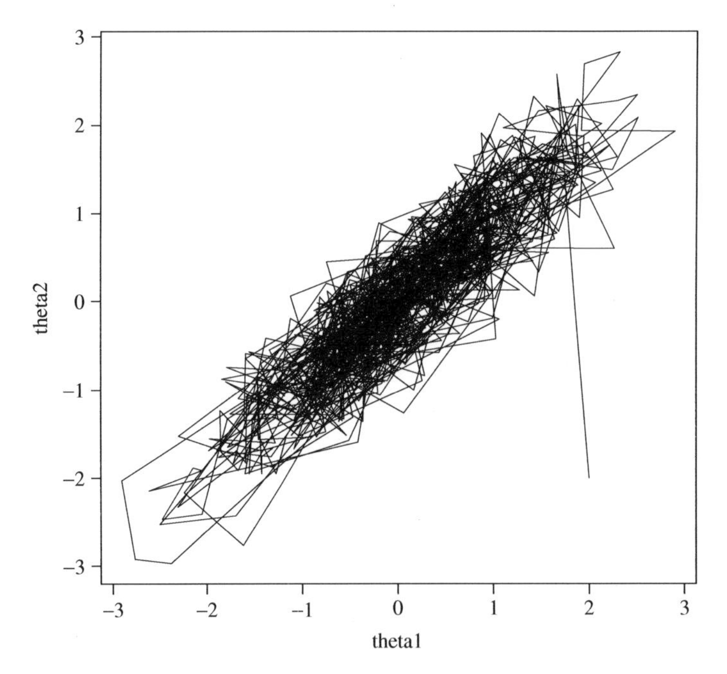

Figure 3.3 One thousand draws from bivariate Gibbs sampler

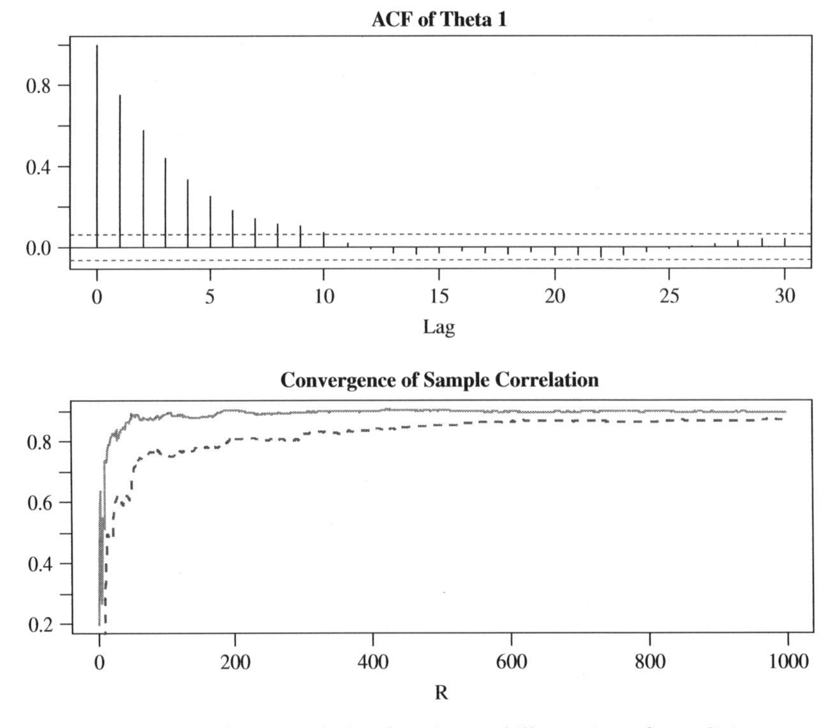

Figure 3.4 Autocorrelation function and illustration of ergodicity

the value of the line corresponding to the horizontal axis value of 100 is the sample correlation based on replicates 0, 1, ..., 100. The figure illustrates the ergodicity of this chain. These sample averages rapidly converge to the true value of 0.9. The figure also illustrates the importance of a 'burn-in' period. It takes a good 100 or so iterations to 'work off' the effects of our extreme initial condition. Of course, with modern algorithms for normal random number generators, we can generate in excess of 50 million univariate normals per second on garden-variety PCs. This means we can run out this sampler to a million or more draws in less time than it takes us to graph the results. However, the problems of dissipation of initial conditions and serial dependence demonstrated for this example can be found in higher-dimensional situations with tens or even hundreds of thousands of draws.

The serial dependence in the Gibbs sampler draws is the price of this method. The estimates of integrals using these dependent draws can be substantially less efficient (in the sense of sampling error) than estimates based on iid draw sequences. For the bivariate normal, we have an iid sampler and the solid line in the bottom panel of Figure 3.4 shows the convergence of estimates of ρ based on a sequence of iid draws. These estimates converge much more rapidly than those based on the Gibbs sampler draws. This, of course, is not really a fair comparison since the Gibbs sampler is used on problems for which iid draw algorithms are not available or are computationally infeasible.

The Gibbs sampler for the bivariate normal problem is a nice illustration of the general idea of MCMC as well as an introduction to one important method. However, this method exploits the very special structure of the normal distribution and the ease with which the conditional distributions can be drawn from. It remains to be seen how generalizable this approach is and what alternatives are available if a strict Gibbs sampler cannot easily be constructed.

3.3 SOME MARKOV CHAIN THEORY

Before outlining some of the more useful algorithms for constructing Markov chains, we will discuss some of the basic theory. While Markov chains are easy to invent, analysis of their convergence and distributional properties can be involved. This literature has a complex notation and requires at least some familiarity with measure-theoretic probability for full access. Tierney (1994) and Robert and Casella (2004, Chapters 7 and 10) have distilled much of the relevant theory from this literature. However, both Tierney and Robert and Casella still require measure theory and a considerable time investment to digest. One view is that theory is largely irrelevant for the practitioner, so that, one should only provide a menu of algorithms along with assurances that they will work. Our view is that a practitioner should understand the basic intuition as to why his methods work. This basic intuition can help diagnose algorithmic and programming errors. In addition, some notation and vocabulary may make it easier to follow the practical implications of the burgeoning MCMC literature, some of which is highly technical. For this reason, we provide a minimal set of concepts and illustrate these with discrete and continuous state space chains.

In most instances, the parameters in models (remember this includes all unobservables) will be continuous random variables and $\pi(\bullet)$ will be a standard density. Thus, most of the Markov chains we will consider generate random variables with a continuous component in their transition distribution and the chain navigates in some subset of $\mathbb{R}^k$. This sort of Markov chain is called a continuous state space Markov chain. We will start, however, by considering discrete state space[5] chains. Much of the intuition we develop for discrete state space chains carries over to the continuous case with some technical difficulties which we will note.

Interest in discrete state space chains can be motivated by considering a discrete approximation to the posterior distribution. We could lay a grid down along each of the coordinate axes and, therefore, construct a discrete approximation to the posterior distribution using the heights of the posterior density on this grid of points. Let g_i be a grid of values $(g_{i1}, \dots, g_{im})$ for the ith component of the parameter vector, θ, where m is the number of grid points. If we lay a grid of points on each of the k coordinate axes in the parameter space, Θ, then we have constructed the product set, $G = g_1 \times g_2 \times \cdots \times g_k$. G has m^k elements in it. An element $\theta \in G$ takes on one of the values of the grid for each of the k axes, $\theta'_{i_1, i_2, \dots, i_k} = (g_1(i_1), g_2(i_2), \dots, g_k(i_k))$. i_j is the index of the grid for θ component j and $g_d(i_d)$ is the i_d value of the g_d grid at this index. For example, consider the two-dimensional case: θ_{23} is the value corresponding to the 2nd element of the grid on the first coordinate and the 3rd element of the grid on the second component, $\theta'_{23} = (g_1(2), g_2(3))$; as both indices range over the m possible values, all possible discrete values for θ are enumerated.

Figure 3.5 illustrates discretization for the two-dimensional case with a continuous bivariate $\pi(\bullet)$. In the left-hand panel, the bivariate double exponential density is plotted, $\pi(\theta) \propto \exp(-|\theta_1| - |\theta_2|)$. This density has a mode at (0, 0) and a scale of 1. If we lay down an equal-spaced grid of 10 values between -2 and 2 on each coordinate axis, we have the basis for a $10 \times 10 = 100$-point discretization of $\pi(\bullet)$. The right-hand panel of Figure 3.5 shows the discrete approximation to the density based on this grid. A simple discrete approximation would be to normalize the 100 values of π on the grid.

We have seen that we can motivate an interest in discrete state space Markov chains by discretizing the parameter space. If grids of m points are used for each component of θ, then any Markov chain would be defined on a state space with m^k elements. This grid

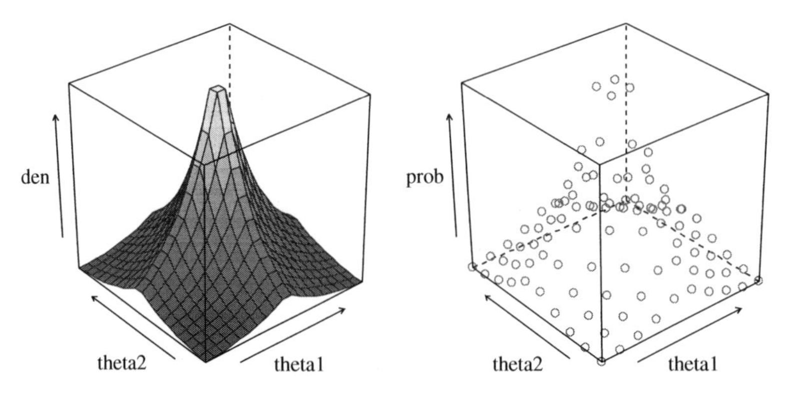

Figure 3.5 Double exponential density and discrete approximation

[5] The term 'state space' itself conjures up a discrete world in which the chain can only take on a finite, or at least countable, number of values or 'states'. The current 'state' of the chain is nothing more than the current realized value which corresponds to one of the finite number of possibilities that define the state space.

is also the basis for a discrete approximation to π and an iid sampler from this discrete approximation:

DISCRETE APPROXIMATION ALGORITHM

Lay down grids on each axis, $(g_1, g_2, \ldots, g_k)$.

Develop a mapping[6] from the integers $j = 1, 2, \ldots, m^k$ to each of the m^k points in $G = g_1 \times g_2 \times \cdots g_k$, $i_1(j), \ldots, i_k(j)$.

Evaluate the unnormalized posterior,[7] π^*, at each of the m^k grid points and normalize the vector

$$p_j = \left(\pi^* \left(\theta_{i_1(j),\ldots,i_k(j)}\right)\right) / \sum_{l=1}^{m^k} \pi^* \left(\theta_{i_1(l),\ldots,i_k(l)}\right).$$

θ is multinomial with probabilities given above.

This seems like a very appealing general idea to construct iid samplers. In particular, it would work extremely well for the two-dimensional example in Figure 3.5 and a grid of 100 points on each axis. With $k = 10$ and a grid of 100 points, we would have to make 10 billion evaluations of the posterior density. For simple densities and small data sets, this is very feasible on modern one-gigaflop computers. However, for complicated densities or larger dimensions this quickly becomes a computational nightmare since the computations required rise to the power of k. Moreover, there is an assumption that we would know where $\pi(\bullet)$ concentrates its mass. Since the point of doing MCMC in the first place is to explore the posterior and see where its mass is, it may be difficult to lay down grids so that there is enough detail where the posterior is concentrated. Even if the posterior is unimodal, we do not know much about its spread and shape except by resort to asymptotic approximations.

Even on discrete parameter spaces, we need methods of exploring that are capable of handling high-dimensional problems. Markov chains are one class of these methods. We now introduce general notation for a discrete space Markov chain. Let $S = \{\theta^1, \theta^2, \ldots, \theta^d\}$ be the state space; we define a Markov chain as the sequence of random variables, $\{\theta_1, \theta_2, \ldots, \theta_r, \ldots\}$ given θ_0 generated by the following transition:

$$\Pr\left[\theta_{r+1} = \theta^j | \theta_r = \theta^i\right] = p_{ij}. \tag{3.3.1}$$

Given the current realization of the chain, the rows of the matrix P, formed from p_{ij}, specify the conditional distribution of the chain at the next iteration. The Markov property states that the conditional distribution of θ_{r+1} depends only on θ_r and is independent of the 'earlier' history of the chain. Up to this point, we have conditioned

[6] Usually this is done by establishing an order in which the subscripts denoting the components of θ are allowed to vary and then considering a base m representation of the integer j which gives the grid elements for each of the k components as digits.

[7] Not only is it unnecessary to evaluate the normalized posterior but to the extent that any normalizing constants are expensive to compute, this could be very inefficient. Recall that the normalizing constants can be functions of the data, making this a real possibility.

on a specific initial value θ_0. This is certainly how we simulate chains. One way of thinking about the chain is how this initial value is transformed. However, in analyzing the behavior of Markov chains, it is more useful to consider the more general case in which we specify a distribution for the initial value, $\theta_0 \sim \pi_0$. By convention in the Markov chain literature, distributions over states are denoted by the *row* vector of probabilities. The chain transforms this distribution into a new distribution in each iteration of the chain. Consider the distribution of θ_1 given that $\theta_0 \sim \pi_0$:

$$\Pr\left[\theta_1 = \theta^j\right] = \sum_{i=1}^{d} \Pr\left[\theta_0 = \theta^i\right] p_{ij} = \sum_{i=1}^{d} \pi_{0i} p_{ij}$$

In matrix form, the above equation states that $\pi_1 = \pi_0 P$. After r iterations, we have $\theta_r \sim \pi_0 P^r$.

As the number of iterations increases, we might expect that the effects of the initial distribution π_0 will 'wear off'. In addition, we might expect some chains to 'settle down' to some sort of equilibrium distribution. Here we are ruling out chains that have 'absorbing' states or sets of states that they get trapped in and never get out of, or, conversely, that they never visit some states (these are called 'reducible' chains). If $p_{ij} > 0$ for all i and j, then all states will communicate with one and other and there can be no subset of states to get trapped in. If you get into state i, there is some positive probability that you will get out of it. However, we can immediately see the distinction between the theory and practice. If p_{ij} is small for all i and a specific j, then the chain might only visit state j very infrequently in a finite sequence of draws[8] even though, in theory, this state will be visited infinitely often!

If $p_{ij} > 0$ for all i, j, then the chain is called *irreducible* and there exists a stationary distribution, π, such that

$$\lim_{r\to\infty} \pi_0 P^r = \pi. \tag{3.3.2}$$

Equation (3.3.2) states that, if we start from any distribution, we will get to π eventually. If we start in π, then we must stay in π, otherwise π would not be the stationary distribution:

$$\pi P = \pi. \tag{3.3.3}$$

Equation (3.3.3) is the reason why the stationary distribution is also called the *invariant* distribution. For discrete state space chains, irreducibility also implies ergodicity (chain averages of functions converge to their expectation under π).

If presented with a discrete Markov chain that claims to have π as the stationary distribution, it is straightforward to check that (3.3.3) holds. However, for more general state space chains, it will turn out to be useful to have an equivalent property called *time reversibility*. Time reversibility states that, if we reverse the sequence order of a Markov chain, the resulting chain will have the same transition behavior. First, we will reverse the order of the chain and check to see that it is still Markov. Then we will compute the transition probabilities for the reversed chain in terms of the standard forward chain. We want to compute the probability of being in state j at 'time' r given the future history.

[8] Resulting in a very poor estimate of the marginal probability of state j.

Using the standard definition of conditional probability, we can write the 'backward' transition probability as follows:

$$\Pr[\theta_r = \theta^j | \theta_{r+1} = \theta^{i_1}, \theta_{r+2} = \theta^{i_2}, \ldots, \theta_{r+s} = \theta^{i_s}]$$
$$= \frac{\Pr[\theta_r = \theta^j, \theta_{r+1} = \theta^{i_1}, \theta_{r+2} = \theta^{i_2}, \ldots, \theta_{r+s} = \theta^{i_s}]}{\Pr[\theta_{r+1} = \theta^{i_1}, \theta_{r+2} = \theta^{i_2}, \ldots, \theta_{r+s} = \theta^{i_s}]}.$$

To see that the 'reversed' or 'backward' chain is Markov, we can write this ratio using terms which involve the future $r + 2$ periods and beyond and other terms which involve only the rth and $(r + 1)$th period:

$$\Pr[\theta_r = \theta^j | \theta_{r+1} = \theta^{i_1}, \theta_{r+2} = \theta^{i_2}, \ldots, \theta_{r+s} = \theta^{i_s}]$$
$$= \frac{\Pr[\theta_r = \theta^j]\Pr[\theta_{r+1} = \theta^{i_1} | \theta_r = \theta^j]\Pr[\theta_{r+2} = \theta^{i_2}, \ldots, \theta_{r+s} = \theta^{i_s} | \theta_r = \theta^j, \theta_{r+1} = \theta^{i_1}]}{\Pr[\theta_{r+1} = \theta^{i_1}]\Pr[\theta_{r+2} = \theta^{i_2}, \ldots, \theta_{r+s} = \theta^{i_s} | \theta_{r+1} = \theta^{i_1}]}.$$

The Markov property of the forward chain implies that

$$\Pr[\theta_{r+2} = \theta^{i_2}, \ldots, \theta_{r+s} = \theta^{i_s} | \theta_r = \theta^j, \theta_{r+1} = \theta^{i_1}]$$
$$= \Pr[\theta_{r+2} = \theta^{i_2}, \ldots, \theta_{r+s} = \theta^{i_s} | \theta_{r+1} = \theta^{i_1}].$$

The conditional probabilities for periods $r + 2, \ldots, r + s$ cancel from the numerator and denominator. This means that the reversed chain is also Markov and, further, that

$$\Pr[\theta_r = \theta^j | \theta_{r+1} = \theta^{i_1}, \theta_{r+2} = \theta^{i_2}, \ldots, \theta_{r+s} = \theta^{i_s}] = \Pr[\theta_r = \theta^j | \theta_{r+1} = \theta^{i_1}]$$
$$= \frac{\Pr[\theta_r = \theta^j]\Pr[\theta_{r+1} = \theta^{i_1} | \theta_r = \theta^j]}{\Pr[\theta_{r+1} = \theta^{i_1}]}.$$

If P^* represents the transition matrix of the reverse chain, then the above equation is the relationship

$$p^*_{ij} = \frac{\pi_j p_{ji}}{\pi_i}. \tag{3.3.4}$$

Time reversibility requires that $p^*_{ij} = p_{ij}$. This means that time reversibility is equivalent to

$$p_{ij} = \frac{\pi_j p_{ji}}{\pi_i} \text{ or } \pi_i p_{ij} = \pi_j p_{ji}. \tag{3.3.5}$$

Roughly speaking, the property of time reversibility implies that the chance of seeing a transition from state i to state j is the same as the chance of seeing a transition from state j to state i. Some say that (3.3.5) means that the chain described by P is 'reversible with respect to π'.

There is a complete equivalence between reversibility and the stationarity of π in the sense that if a chain is reversible with respect to some distribution ω then ω is also the stationary distribution of the chain. Reversibility with respect to ω means $\omega_i p_{ij} = \omega_j p_{ji}$. Summing both sides over i, we obtain

$$\sum_i \omega_i p_{ij} = \sum_i \omega_j p_{ji} = \omega_j \sum_i p_{ji} = \omega_j \times 1$$

or

$$\omega P = \omega,$$

and ω is the stationary distribution of the chain.

The bivariate normal Gibbs sampler discussed in Section 3.2 is an example of a continuous state space chain. It will, therefore, be important to extend the ideas we have developed for discrete state space chains to the continuous case. Fortunately, the basic ideas of reversibility and invariant distributions extend without much difficulty. There are some technical difficulties in establishing convergence and ergodic results, but central intuition that the chain must freely navigate is at the core of these results just as in the discrete case.

In the continuous state space case, the transitional conditional distribution of $\theta_{r+1}|\theta_r$ must have a continuous component. Rather than specifying this distribution via the probabilities of each of the singletons $\{\theta^i\}$ that comprise the state space, we must specify the conditional distribution by associating probabilities with sets in the state space. For example, consider a set $A \in \Theta$[9] then the chain is specified by the probabilities of the set A given the value of the chain on the previous iteration. This is sometimes called the *kernel* of the chain. $K(\theta, A)$ is the probability of set A given the chain is at value θ. Some kernels (such as the one corresponding to the Gibbs sampler, but not all kernels) can be represented using a standard density,

$$K(\theta, A) = \int_A p(\theta, \vartheta)\, d\vartheta.$$

$p(\theta, \vartheta)$ is a density for fixed θ. To distinguish p from K, we will call $p(\bullet, \bullet)$ the transition function of the kernel.

Analogous to the discrete case (see (3.3.3)), we can define the concept of an invariant distribution. A distribution with density $\pi(\bullet)$ is an invariant distribution if the probability of $A \in \Theta$ computed under $\pi(\bullet)$ is the same as the one-step-ahead probability of A given that $\theta \sim \pi$:

$$\int_A \pi(\theta)\, d\theta = \int_\Theta K(\theta, A)\pi(\theta)\, d\theta = \int_\Theta \left[\int_A p(\theta, \vartheta)\, d\vartheta\right] \pi(\theta)\, d\theta.$$

The principle of time reversibility can also be defined for the continuous state space chain. The chain is said to be reversible with respect to a distribution with density $\omega(\theta)$ if the transition function satisfies

$$\omega(\theta)p(\theta, \vartheta) = \omega(\vartheta)p(\vartheta, \theta). \tag{3.3.6}$$

If (3.3.6) is satisfied, then the stationary distribution of the chain with transition function $p(\bullet, \bullet)$ has density $\omega(\theta)$:

$$\begin{aligned}\int_\Theta \omega(\theta)K(\theta, A)\, d\theta &= \int_\Theta \omega(\theta) \int_A p(\theta, \vartheta)\, d\vartheta\, d\theta \\ &= \int_\Theta \int_A \omega(\theta)p(\theta, \vartheta)\, d\vartheta\, d\theta.\end{aligned}$$

[9] Technically, we can only assign probabilities to certain subsets, but we will gloss this over.

Reversing the order of integration and using the reversibility condition,

$$\int_{\Theta} \omega(\theta) K(\theta, A)\,\mathrm{d}\theta = \int_A \int_{\Theta} \omega(\vartheta) p(\vartheta, \theta)\,\mathrm{d}\theta\,\mathrm{d}\vartheta = \int_A \omega(\vartheta) \left[\int_{\Theta} p(\vartheta, \theta)\,\mathrm{d}\theta \right] \mathrm{d}\vartheta$$

$$= \int_A (\omega(\vartheta) \times 1)\,\mathrm{d}\vartheta.$$

ω, therefore, is the invariant distribution of the chain. Thus, reversibility and invariance are equivalent.

Finally, the concept of *irreducibility* can be extended to the continuous state space setting. Irreducibility requires that the chain navigate the state space freely so that it cannot get trapped in a subset of the entire state space. In the discrete case, all that is required is strict positivity of the transition probabilities, $P > 0$. In the continuous case, the definition of irreducibility is straightforward but verification that a given kernel produces an irreducible chain is not always a simple matter. A chain with kernel K is irreducible with respect to $\pi(\bullet)$ if every set A with positive π probability can be reached with positive probability after a finite number of steps. If $\int_A \pi(\theta)\,\mathrm{d}\theta > 0$ then there exists $n \geq 1$ such that $K^n(\theta, A) > 0$.

3.4 GIBBS SAMPLER

The Gibbs sampler is a Markov chain obtained by cycling through a set of conditional distributions of π. If we break θ into p separate 'groups' or 'blocks'[10] of parameters,

$$\theta' = (\theta_1, \theta_2, \ldots, \theta_p) \tag{3.4.1}$$

then the Gibbs sampler is defined by iterative sampling from each of these p conditional distributions:

GIBBS SAMPLER

Set θ_0.

Sample from

$$\begin{aligned} \theta_{11} &\sim \mathrm{f}_1(\theta_1 | \theta_{02}, \ldots, \theta_{0p}) \\ \theta_{12} &\sim \mathrm{f}_2(\theta_2 | \theta_{11}, \theta_{03}, \ldots, \theta_{0p}) \\ &\vdots \\ \theta_{1p} &\sim f_p(\theta_p | \theta_{11}, \ldots, \theta_{1,p-1}) \end{aligned} \tag{3.4.2}$$

to obtain the first iterate.

Repeat, as necessary.

---

[10] We will see examples where each block is only one-dimensional and others in which each block corresponds to subsets of the complete set of unobservables.

$f_1, \ldots, f_p$ are the appropriate conditional densities derived from π:

$$f_i = \frac{\pi(\theta)}{\int \pi(\theta)\, d\theta_i}.$$

Implementation of the Gibbs sampler requires the ability to sample from the set of conditional posterior distributions. In many situations, it is possible to define the groups or blocks of parameters so that the conditional distributions are of known form and can be sampled from very efficiently using the algorithms defined in Chapter 2. As a 'fall-back' or default alternative, one could always define a Gibbs sampler based on the k univariate conditionals implied by π. This would require a generic method for making draws from univariate distributions whose unnormalized density can be evaluated.

The Gibbs sampler defined by (3.4.2) is clearly a Markov chain. It is also easy to verify that the invariant distribution of this chain is π. To see this, consider the bivariate case, $\theta' = (\theta_1, \theta_2)$. The rth iteration of the bivariate Gibbs sampler draws successively from two conditional distributions,

$$\begin{aligned} \theta_{r+1,2} &\sim \pi_{2|1}(\theta_{r,1}), \\ \theta_{r+1,1} &\sim \pi_{1|2}(\theta_{r+1,2}). \end{aligned} \tag{3.4.3}$$

To check that π is the invariant distribution, we must verify that if $\theta_r \sim \pi(\bullet)$, then $\theta_{r+1} \sim \pi(\bullet)$. The notation $\pi_{i|j}(\theta)$ means the distribution of component i given that component j takes on the value θ.

Suppose $\theta_r \sim \pi(\bullet)$. This means that $\theta_{r,1} \sim \pi_1(\theta_1) = \int \pi(\theta_1, \theta_2)\, d\theta_2$. $\theta_{r+1,2}$ is a draw from the conditional distribution $\pi_{2|1}$. Therefore, the distribution of $\theta_{r+1,2}$ from one iteration of the sampler is a draw from the marginal distribution of the second component, $\theta_{r+1,2} \sim \pi_2 = \int \pi_{2|1}(\theta_2|\theta_1)\pi_1(\theta_1)\, d\theta_1$. The same argument can be used to show that $\theta_{r+1,1} \sim \pi_1$ using the fact that $\theta_{r+1,2} \sim \pi_2$. Thus, the $(r+1)$th iteration of the chain reproduces a draw from the invariant distribution.

Convergence of the Gibbs sampler is assured under very mild conditions.[11] If the Gibbs sampler is irreducible,[12] then Theorem 1 of Tierney (1994, p. 1712) ensures convergence of the n-step-ahead distribution to the invariant distribution for almost all starting points. Most examples of reducible Gibbs samplers involve some sort of constraint on the state space Θ.[13] If the state space is the Cartesian product of intervals on each coordinate axis, then a Gibbs sampler whose conditional densities are strictly positive everywhere and

[11] These conditions are so mild that some, among them Liu (2001), assume convergence and focus attention on the rate of convergence, extent of autocorrelation and various ideas for improving performance.

[12] Also, it is required that the sampler be aperiodic. Periodic chains require deterministic constraints on movement of the chain, something not found in MCMC algorithms.

[13] The classic example involves two disjoint disks located on the 45-degree line. The sampler would get 'trapped' in one disk, depending on the initial condition. A closely related is the example in Hobert *et al.* (1997) in which the state space consists of two boxes that are oriented along the 45-degree line and are tangent at one vertex. In this example, the state space is a subset of a product set which defines a larger box enclosing the two smaller boxes. Finally, an example in Geweke (2005) has a state space consisting of a solid polygon with an acute angle for one vertex and, again, oriented on the 45-degree line. In this example, this vertex is an absorbing state. Again, the product set rules this example out while keeping what is needed for virtually all applications.

whose marginal densities exist will be irreducible.[14] Furthermore, by Tierney's Theorem 2, Corollary 1, this sampler will converge to the stationary distribution from all initial points and will be ergodic. This justifies the practical use of the Gibbs sampler – to start from an arbitrary initial condition and use sample averages to approximate integrals of the posterior. Marginal densities will not exist for improper posteriors. If proper priors are used and if the likelihood is bounded, we can avoid the problem of improper posteriors and Gibbs samplers which attempt to approximate quantities that do not exist! We regard this as yet another (and not even the most important) reason to use proper priors.

An appreciation for the power of the Gibbs sampler as well as better feel for implementation can be obtained by considering some important and nontrivial examples. We will consider first the generalization of the multivariate regression model introduced in Chapter 2.

3.5 GIBBS SAMPLER FOR THE SEEMINGLY UNRELATED REGRESSION MODEL

In the SUR model, a system of m regression equations are related through correlated error terms:

$$\begin{aligned} y_i &= X_i\beta_i + \varepsilon_i, \\ (\varepsilon_{k,1}, \varepsilon_{k,2}, \ldots, \varepsilon_{k,m})' &\sim N(0, \Sigma), \\ i = 1, \ldots, m, &\quad k = 1, \ldots, n. \end{aligned} \tag{3.5.1}$$

It will be convenient to stack up the m regressions in (3.5.1) into one large regression:

$$y = X\beta + \varepsilon, \qquad \varepsilon \sim N(0, \Sigma \otimes I_n) \tag{3.5.2}$$

with

$$y' = (y_1', \ldots, y_m'), \quad X = \begin{bmatrix} X_1 & 0 & \cdots & 0 \\ 0 & X_2 & \ddots & \vdots \\ \vdots & \ddots & \ddots & 0 \\ 0 & \cdots & 0 & X_m \end{bmatrix}, \quad \beta' = (\beta_1', \ldots, \beta_m'), \quad \varepsilon' = (\varepsilon_1', \ldots, \varepsilon_m').$$

As discussed in Section 2.8, there is no convenient natural conjugate joint prior on $\{\beta_i\}, \Sigma$. Recall that the natural conjugate prior for the multivariate regression model (MRM) has the prior on β depending on Σ. This prior embodies the notion that information on β can never be *scale-independent*. In all three situations (SUR, MRM and univariate regression), we may have prior information which is non-data-based and,

[14] This is equivalent to the positivity condition (cf. Robert and Casella 2004, p. 345).

hence, would not always be scale-dependent. A simple prior specification would be to make β and Σ a priori independent:

$$\begin{aligned} p(\beta, \Sigma) &= p(\beta)p(\Sigma) \\ \beta &\sim N(\bar{\beta}, A^{-1}) \\ \Sigma &\sim \mathrm{IW}(\nu_0, V_0) \end{aligned} \tag{3.5.3}$$

These priors are not conjugate but they are *conditionally* conjugate. Given Σ, the SUR likelihood in (3.5.2) can be written in the standard normal regression form and is conjugate with a normal prior on the stacked vector of regression coefficients. Given β, the SUR likelihood is in a form that has an inverted wishart conjugate prior. The Gibbs sampler simply alternates between draws from these two sets of conjugate distributions.

Given Σ, we can transform (3.5.2) into a system with uncorrelated errors using the root of the cross-equation covariance matrix, $\Sigma = U'U$ and $(U^{-1})'\Sigma U^{-1} = I_m$. This means that, if we premultiply both sides of (3.5.2) by $(U^{-1})' \otimes I_n$, the transformed system has uncorrelated errors

$$\tilde{y} = \tilde{X}\beta + \tilde{\varepsilon}, \qquad \mathrm{var}(\tilde{\varepsilon}) = \mathrm{E}\left[\left((U^{-1})' \otimes I_n\right)\varepsilon\varepsilon'\left(U^{-1} \otimes I_n\right)\right] = I_m \otimes I_n,$$
$$\tilde{y} = \left((U^{-1})' \otimes I_n\right)y, \qquad \tilde{X} = \left((U^{-1})' \otimes I_n\right)X.$$

A normal prior for β, $\beta \sim N(\bar{\beta}, A^{-1})$, is conjugate with the conditional likelihood for the transformed system. This means that we can apply the results from Section 2.8 and the posterior of β given Σ is normal:

$$\beta|\Sigma, y, X \sim N\left(\tilde{\beta}, \left(\tilde{X}'\tilde{X} + A\right)^{-1}\right), \qquad \tilde{\beta} = \left(\tilde{X}'\tilde{X} + A\right)^{-1}(\tilde{X}'\tilde{y} + A\bar{\beta}). \tag{3.5.4}$$

As A gets small, the prior becomes flat and we recognize that the mean of this distribution is the generalized least squares estimator.

The posterior of $\Sigma|\beta$ is in the inverted Wishart form. To see this, first recognize that, given β, we 'observe' or can compute the errors ε. This means that, given β, the problem is the standard one of inference regarding a covariance matrix using a multivariate normal sample. The inverted Wishart prior is, therefore, conditional conjugate. If $\Sigma \sim \mathrm{IW}(\nu_0, V_0)$, the posterior is in the form

$$\Sigma|\beta, y, X \sim \mathrm{IW}(\nu_0 + n, S + V_0), \qquad S = E'E, \qquad E = [\varepsilon_1, \ldots, \varepsilon_m]. \tag{3.5.5}$$

Again, if we let the prior precision go to zero, the posterior on Σ is centered over the sum of squared residuals matrix.

GIBBS SAMPLER FOR SUR MODEL[15]

Pick starting values, β_0, Σ_0 (note: Σ_0 must be a positive definite matrix).

[15] As in Chib and Greenberg (1995b).

Draw $\beta_1|\Sigma_0$ from (3.5.4).[16]
Draw $\Sigma_1|\beta_1$ from (3.5.5).
Repeat.

This sampler can be related to the non-Bayesian approach to estimating this model. Zellner originally proposed a feasible GLS procedure in which an estimate of Σ is formed by using residuals from equation-by-equation least squares estimates, $\hat{\Sigma} = n^{-1}\hat{E}'\hat{E}$, where $\hat{E} = [e_1, \ldots, e_m]$ and $e_i = y_i - X_i\hat{\beta}_{\text{LS},i}$. If we start the Gibbs sampler at this point and if we have a very diffuse prior on β, then the first iteration on β will be a draw from a distribution centered on the Zellner feasible GLS estimator. The Gibbs sampler takes this a step further and uses simulation to capture the uncertainty in both β and Σ. The finite-sample distribution of the feasible GLS estimator is a nightmare due to the nonlinearities introduced by matrix inversion and multiplication. For this reason, econometricians have had to resort to asymptotic approximations. The sampling error in $\hat{\Sigma}$ does not figure in the asymptotic distribution of the 'plug-in' or two-stage feasible GLS estimator. This shows the weakness of asymptotics. However, we no longer have to utilize these approximations. The Gibbs sampler for the SUR model performs extremely well with relatively trivial computation costs. Finally, note that if $m = 1$, we have a sampler for the univariate regression model with a nonconjugate prior.

3.6 CONDITIONAL DISTRIBUTIONS AND DIRECTED GRAPHS

One of the most common applications of the Gibbs sampler is to hierarchical models. Hierarchical models are models constructed from a sequence of conditional distributions. More generally, we can construct a model by 'connecting' or piecing together a set of conditional distributions in some sort of network or 'graph'. In this section we provide a brief introduction to the basics of directed graphs. We will explain how to write a model as a directed graph and how to 'read off' the Gibbs sampler from a graph.

The Bayesian paradigm starts with a prior and a likelihood. We can think of the prior as the first step and then we consider the distribution of the data given the model parameters. One way of remembering this 'ordering' is to think about how we would simulate from the model (here 'model' refers to the joint distribution of the unknowns and the data). First we would draw from the prior and then we would draw the data given the prior. This can be represented by a 'directed acyclic' graph (DAG). A graph is a set of connected nodes. A directed graph has a notion of direction from node to node. An acyclic graph must have a direction from top to bottom with no 'recirculation':

$$\begin{array}{ccc} p(\theta) & & p(y|\theta) \\ \theta & \to & y \end{array} \tag{3.6.1}$$

[16] Some care should be taken in the computations to draw β. The transformation of y and X involves very sparse matrices and can be optimized dramatically by taking advantage of the structure of these matrices.

A hierarchical model is specified through a sequence of two or more conditional distributions which specify the prior. This case of two conditional distributions can be represented as a directed graph as follows:

$$\begin{array}{ccccc} p(\theta_2) & & p(\theta_1|\theta_2) & & p(y|\theta_1) \\ \text{1st stage} & & \text{2nd stage} & & \\ \theta_2 & \rightarrow & \theta_1 & \rightarrow & y \end{array} \tag{3.6.2}$$

Typically, θ_2 is of much lower dimension than θ_1. The sequence of two prior distributions can be thought of as a device to induce a marginal prior over θ_1:

$$p(\theta_1) = \int p(\theta_1, \theta_2)\, d\theta_2 = \int p(\theta_2)p(\theta_1|\theta_2)\, d\theta_2. \tag{3.6.3}$$

The hierarchical model in (3.6.2) specifies that θ_2 and y are independent conditional on θ_1 or that all dependence comes through θ_1. We can easily verify this by writing down the joint distribution:

$$p(\theta_1, \theta_2, y) = p(\theta_2)p(\theta_1|\theta_2)p(y|\theta_1) = f(\theta_1, \theta_2)g(y, \theta_1).$$

There are two ways to see that this implies conditional independence. First, given θ_1, the joint distribution factors into two terms (represented by the functions f and g). Therefore, we have conditional independence. The other way to see this is to observe that there is no term involving all three variables, only a term (f) involving θ_1 and θ_2. This means that

$$p(\theta_2|\theta_1, y) \propto f(\theta_1, \theta_2) \Rightarrow p(\theta_2|\theta_1, y) = p(\theta_2|\theta_1).$$

The hierarchical structure in (3.6.2) immediately suggests a 'two-stage' Gibbs sampler to simulate from the distribution of (θ_1, θ_2) given y:

$$\begin{array}{c} \theta_2|\theta_1 \\ \theta_1|\theta_2, y \end{array} \tag{3.6.4}$$

It is possible to write down more complicated directed graphs. However, some simple rules can help understand the structure of dependence implied by the graph. There are three sorts of local node arrangements.

The first is a linear set of three nodes

$$\theta_1 \rightarrow \theta_2 \rightarrow \theta_3. \tag{3.6.5}$$

We have already seen an example of this in (3.6.2) (except that we do not 'draw' y but condition on it). This structure has the basic conditional independence in it. A Gibbs sampler for (3.6.5) is given by

$$\begin{array}{c} \theta_1|\theta_2 \\ \theta_2|\theta_1, \theta_3 \\ \theta_3|\theta_2 \end{array} \tag{3.6.6}$$

Here θ_1, θ_3 are independent conditional on θ_2.

The next structure looks different but has the same feature of conditional independence.

$$\begin{array}{ccc} & \nearrow & \theta_2 \\ \theta_1 & & \\ & \searrow & \theta_3 \end{array} \tag{3.6.7}$$

The joint distribution implied by the graph in (3.6.7) is

$$p(\theta_1, \theta_2, \theta_3) = p(\theta_1)p(\theta_2|\theta_1)p(\theta_3|\theta_1).$$

θ_2, θ_3 are independent conditional on θ_1.

However, the structure formed from two nodes pointing into one node does not display conditional independence:

$$\begin{array}{ccc} \theta_1 & \searrow & \\ & & \theta_3 \\ \theta_2 & \nearrow & \end{array} \tag{3.6.8}$$

Here we have full dependence among all three random variables. The joint for (3.6.8) would be written $p(\theta_1, \theta_2, \theta_3) = p(\theta_1)p(\theta_2)p(\theta_3|\theta_1, \theta_2)$. The Gibbs sampler requires the full set of complete conditionals:

$$\begin{array}{l} \theta_1|\theta_3, \theta_2 \\ \theta_3|\theta_1, \theta_2 \\ \theta_2|\theta_3, \theta_1 \end{array} \tag{3.6.9}$$

It is obvious that the 'middle' conditional in (3.6.9) belongs in the sampler. What is less obvious is that the 'top' and 'bottom' conditionals depend on a node that is more than one node away. But inspection of the joint distribution that the graph represents indicates that there is no conditional independence at all as there is one term in the joint involving all three parameters.

All directed graphs are made up of some combination of three examples above. This suggests three rules for 'reading' the dependence structure from a graph. A node depends on:

(i) any node it points to;

(ii) any node that points to it;

(iii) any node that points to the node directly 'downstream'.

For example, consider the graph

$$\begin{array}{ccccc} & \theta_2 & \searrow & & \\ & & & \theta_4 \longrightarrow \theta_5 & \\ \theta_1 \longrightarrow & \theta_3 & \nearrow & & \end{array} \tag{3.6.10}$$

The Gibbs sampler for the graph in (3.6.10) is

$$\begin{aligned} &\theta_1 | \theta_3 \\ &\theta_2 | \theta_4, \theta_3 \\ &\theta_3 | \theta_1, \theta_4, \theta_2 \\ &\theta_4 | \theta_2, \theta_3, \theta_5 \\ &\theta_5 | \theta_4 \end{aligned} \tag{3.6.11}$$

3.7 HIERARCHICAL LINEAR MODELS

In Section 3.5, we considered systems of regressions that are related through correlated errors. An alternative approach would be to relate regression equations through correlations between the regression coefficient vectors. This amounts to specifying a prior structure. Consider the set of regression equations

$$y_i = X_i\beta_i + \varepsilon_i, \quad \varepsilon_i \sim \text{iid } N\left(0, \sigma_i^2 I_{n_i}\right), \qquad i = 1, \ldots, m. \tag{3.7.1}$$

We specify a different error variance for each equation but consider each regression to be independent of others. We tie together the equations by assuming that the $\{\beta_i\}$ have a common prior distribution

$$\beta_i = \Delta' z_i + v_i, \qquad v_i \sim \text{iid } N\left(0, V_\beta\right). \tag{3.7.2}$$

Equation (3.7.2) specifies a normal prior with mean $\Delta' z_i$ for each β. The n_z variables in the z vector represent characteristics of each of the m 'cross-sectional' units or regression equations. A special case has $z_i = 1$ and $\Delta = \mu'$, which would have a common mean vector for all betas. This prior can be written as a multivariate regression model:

$$B = Z\Delta + V, \quad B = \begin{bmatrix} \beta_1' \\ \vdots \\ \beta_m' \end{bmatrix}, \quad Z = \begin{bmatrix} z_1' \\ \vdots \\ z_m' \end{bmatrix}, \quad \Delta = [\, \delta_1 \; \cdots \; \delta_k \,], \quad v_i' \sim N(0, V_\beta); \tag{3.7.3}$$

B is $m \times k$, Z is $m \times n_z$, where n_z is the number of z variables, and Δ is $n_z \times k$. Each column of Δ has coefficients which describe how the mean of the k regression coefficients varies as a function of the variables in z. We also need a prior on the regression error variances. It is convenient to take a prior which specifies that each of the error variances is independent:

$$\sigma_i^2 \sim \frac{v_i s_{0i}^2}{\chi_{v_i}^2} \tag{3.7.4}$$

The prior in (3.7.3) specifies a fixed Δ matrix which determines the mean of the β distribution and a fixed V_β matrix which specifies the variance. Assessment of these priors could be difficult. Early 'empirical Bayes' approaches simply estimate these parameters and then, conditional on these estimates, perform an approximate Bayesian analysis of each regression. A full Bayesian solution can be obtained by specifying a further 'stage' of priors on Δ and V_β:

$$\begin{aligned} V_\beta &\sim \text{IW}(\nu, V), \\ \text{vec}(\Delta)|V_\beta &\sim N(\text{vec}(\bar{\Delta}), V_\beta \otimes A^{-1}). \end{aligned} \tag{3.7.5}$$

The priors in (3.7.5) are the natural conjugate priors for the multivariate regression model, $B = Z\Delta + V$.

The prior on the collection of βs is specified through a two-stage process. First, we specify a normal prior on β and then a second-stage prior on the parameters of this distribution. We can write out this model as a sequence of conditional distributions:

$$\begin{aligned} &y_i|X_i, \beta_i, \sigma_i^2 \\ &\beta_i|z_i, \Delta, V_\beta \\ &\sigma_i^2|\nu_i, s_{0i}^2 \\ &V_\beta|\nu, V \\ &\Delta|V_\beta, \bar{\Delta}, A \end{aligned} \tag{3.7.6}$$

The model above can also be written as a directed graph. The rules of directed graphs given in Section 3.6 can be used to write down the Gibbs sampler for this model. In particular, a key observation is that all dependence between V_β, Δ and the data comes through the regression coefficients, $\{\beta_i\}$. The directed graph is shown below:

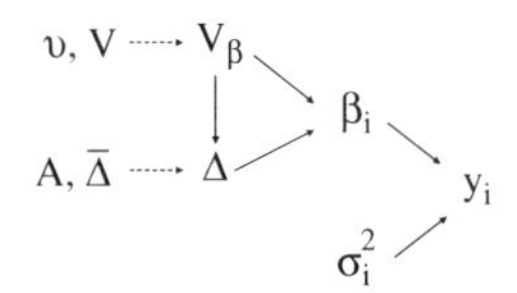

Model (3.7.6) converts the problem of assessing a prior on the $(m \times k)$-dimensional joint distribution of the βs into the problem of assessing hyperparameters, $\bar{\Delta}, A, \nu, V$. A combination of the data and these hyperparameters (plus the functional forms of the distributions) will influence the posterior on Δ and V_β. The Bayes estimators of the $\{\beta_i\}$ are of a shrinkage variety and will exhibit less variation than least squares estimates computed equation by equation. The amount of shrinkage will be dictated both by the prior hyperparameters and the data. If we assess a tight prior on a 'small' covariance matrix by setting ν large and V to a small location value, then there will be a great deal of shrinkage. In addition, if the data suggests little variation in the β vectors from equation to equation, then the Bayes estimator will 'adapt' to a posterior centered around a small value of V_β and the Bayes estimates of each beta vector will be shrunk even if our prior is relatively uninformative. As discussed in Chapter 5, there are many subtle aspects of this normal hierarchical model. We will defer a full discussion of the prior used in this model until Chapter 5.

Carl Morris was the first to observe that the Gibbs samplers were ideal for analysis of hierarchical models. The key observation is that given Δ and V_β, the $\{\beta_i, \sigma_i^2\}$ are independent with a prior which is the product of a normal prior on β_i and the inverse of a χ^2 prior. This can be analyzed via a Gibbs sampler by drawing $\beta_i|\sigma_i^2$ and then $\sigma_i^2|\beta_i$. Once the $\{\beta_i\}$ are drawn, they are sufficient for V_β and Δ. Given $\{\beta_i\}$, V_β and Δ can be drawn using the algorithm for the multivariate regression model given in Section 2.12. Thus, a Gibbs sampler for this model can be constructed by first drawing the regression parameters, $\{\beta_i, \sigma_i^2\}$, given the parameters of the first-stage prior, Δ, V_β and then drawing the prior parameters conditional on $\{\beta_i, \sigma_i^2\}$.

GIBBS SAMPLER FOR HIERARCHICAL LINEAR MODEL

We use τ_i to denote σ_i^2 to reduce notational clutter.

Start with $\{\tau_i^0\}$, Δ^0, V_β^0.

Draw

$$\beta_i^1|y_i, X_i, (\Delta^0)'z_i, V_\beta^0, \tau_i^0$$

and

$$\tau_i^1|y_i, X_i, \beta_i^1, \nu_i, s_{0i}^2, \qquad i = 1, \ldots, m.$$

Draw

$$V_\beta^1|\{\beta_i^1\}, \nu, V, Z, \bar{\Delta}, A$$

and

$$\Delta^1|\{\beta_i^1\}, V_\beta^1, Z, A, \bar{\Delta}.$$

Repeat, as necessary.

The draw of $\{\beta_i, \sigma_i^2\}$ is conducted with a prior that specifies that β_i and σ_i^2 are a priori independent rather than dependent as in the natural conjugate prior, $\beta_i \sim N(\Delta' z_i, V_\beta)$ and $\sigma_i^2 = \nu_i s_{0i}^2/\chi_{\nu_i}^2$. We have

$$\begin{aligned} \beta_i^1|y_i, X_i, (\Delta_0)'z_i, V_\beta^0, \tau_i^0 &\sim N(\tilde{\beta}, (\tilde{X}_i'\tilde{X}_i + (V_\beta^0)^{-1})^{-1}), \\ \tilde{\beta} &= (\tilde{X}_i'\tilde{X}_i + (V_\beta^0)^{-1})^{-1}(\tilde{X}_i'\tilde{y}_i + (V_\beta^0)^{-1}(\Delta_0)'z_i), \end{aligned} \tag{3.7.7}$$

with $\tilde{y}_i = y_i/\sigma_i$ and $\tilde{X}_i = X_i/\sigma_i$.

R The R code for this sampler is in `rhierLinearModel`, which is available in *bayesm*. This code makes use of the function to draw from the posterior of a multivariate regression model given at the end of Chapter 2.

To illustrate the functioning of this sampler, consider a very typical problem in promotional response modeling. Most consumer packaged goods (CPG) manufacturers have data on the pricing and promotional activities of many of their 'key' accounts. A 'key' account is a combination of a retailer and market area. For example, Safeway–Denver is a key account that receives attention from the manufacturer's salesforce. The data is usually weekly data for 1–2 years. The manufacturer may define several hundred key accounts. In order to allocate funds and salesforce effort over these accounts, it would be useful to understand how the customers in these accounts respond to various types of promotional activity. We can use data on sales and measure of price and promotional activity to estimate a simple sales response model for each of these accounts.

As an example, consider the sliced cheese product manufactured by Borden. This data set can be loaded from *bayesm* using the command `data(cheese)`. Data is available on some 88 key accounts for an average of 65 weeks. Weekly observations are recorded of unit volume (number of units sold in all stores in this account during each week), price in dollars, and a measure of display activity. Displays are a form of in-store advertising that usually consists of special signage or display of the merchandise in a prominent location. The measure of display activity is 'percentage of ACV on display'. Borden would like to see retailers use the optimal combination of display and pricing to promote and sell this product. One straightforward approach would be to run 88 separate regressions each with about 60 observations. However, the independent variables do not always have much variation at the account level. For example, two of the accounts have no display activity in this period. Even ignoring the fact that the display coefficient is not estimable for two of the 88 regressions, the least squares coefficients are not very usable, as shown in Figure 3.6. These are coefficients from a regression of ln(Volume) on Display and ln(Price). Some of the display coefficients are absurdly large. To interpret the coefficient, recall that the Display variable reaches a maximum of 1, which means 100 % display coverage. Thus, a display coefficient of 5 implies a multiple of sales volume of e^5 or almost a 150-fold increase. Clearly, these coefficients have been influenced by sampling error and, perhaps, outlying observations.

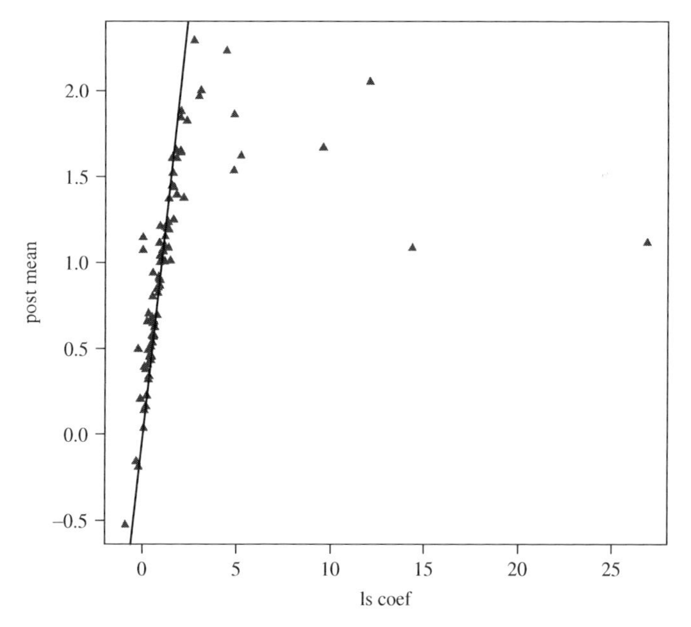

Figure 3.6 Linear hierarchical models: failure of least squares estimates

Figure 3.6 plots the least squares coefficients against the posterior means obtained from the linear hierarchical model Gibbs sampler.[17] The prior settings are

$$\nu_i = 3, \quad s_{0i}^2 = \text{var}(y_i), \qquad \nu = k + 3, \qquad V = \nu \times 0.1 I_k, \qquad \bar{\Delta} = 0, \quad A = 0.01.$$

These prior settings represent a proper but very diffuse prior. Even so, the posterior means display a strong shrinkage effect. The absurdly large values of the least squares coefficients are shrunk in towards more reasonable values. This shrinkage stems from two

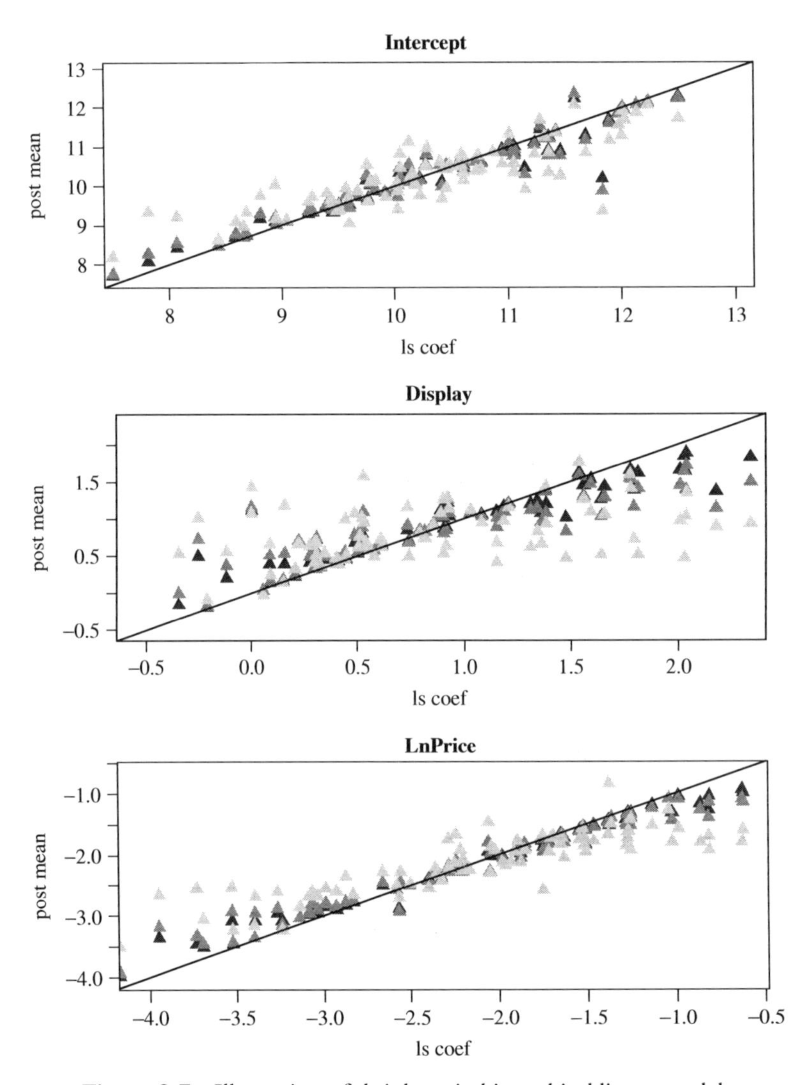

Figure 3.7 Illustration of shrinkage in hierarchical linear models

[17] Here there are no 'z' variables, so Z is simply a vector of ones and Δ is a common mean vector for the coefficients.

forces: the adaptive nature of the hierarchical model, which adapts V_β to the observed variation in the data; and the normal first-stage prior which has very thin tails. It should be noted that, for most accounts, the least squares and posterior means are similar. This means that the sample information dominated the prior for these accounts (note that the prior is centered over 0 for the mean of the regression coefficients). The prior is set to a very diffuse setting so this is not very surprising. There are several coefficients for which both the posterior mean and the least squares estimates are negative. A negative coefficient would imply the displays depressed sales. This might be something that one would want to rule out a priori. Unfortunately, the conditional conjugate priors employed in this model are not capable of imposing a sign restriction on the regression coefficients. We will consider this in Chapter 5.

To illustrate how the prior settings affect the degree of shrinkage, we compute posterior means for three different values of ν: $k + 3$, $k + 0.5\bar{n}$, $k + 2\bar{n}$. $\bar{n}$ is the average number of observations in each data set. The three values of ν tighten down the inverted Wishart prior on V_β. The V location matrix has already been set to a small value. This means that if ν is large, our prior on V_β is highly informative and located over small values, inducing a great deal of shrinkage. Figure 3.7 plots the least squares coefficients against the posterior means for three levels of ν and for each of the three coefficients in the model. The small value of ν is black, the medium dark gray and the high value light gray. The shrinkage is rather dramatic for ν representing prior information obtained from a sample roughly twice the size of each regression sample. Although the inverted Wishart and Wishart distributions only have one tightness parameter, this does not imply that the degree of shrinkage is the same for each coefficient. The degree of shrinkage depends on the amount of sample information available for each coefficient. The display coefficient is most difficult to estimate as displays are relatively rare compared to price changes and, therefore, displays the greatest shrinkage. The intercept is the easiest to estimate and displays the least shrinkage.

3.8 DATA AUGMENTATION AND A PROBIT EXAMPLE

The examples that we have seen so far show that the Gibbs sampler is extremely well suited to sets of linear models for which the conditional distributions are known and for which standard methods can be used to make direct draws from these conditional distributions. We enlarge the set of models that can be analyzed by requiring only conditional and not full conjugacy. This allows for analysis of systems of linear models which, heretofore, required approximate methods. However, the Gibbs sampler can be applied to a much wider class of models once the principle of data augmentation is introduced. The idea of data augmentation has its origin in the literature on missing values and the EM algorithm. The idea that missing values are unobserved and, therefore, should properly be considered as part of the 'parameter' vector comes naturally to a Bayesian. In the EM algorithm for missing-data models, the missing data is replaced with its expectation conditional on the observed data and the 'complete' data likelihood is maximized over the 'parameters'. To a Bayesian, it is much more natural to compute the joint posterior of the missing values and the 'parameters' and simply margin down to the parameters if this is all that is of interest. As first pointed out by Tanner and Wong (1987), the idea of data augmentation extends to any situation in which there are unobservable constructs. For example, many distributions can be written as mixtures of other distributions so that data augmentation can be used to form a Gibbs sampler for these problems. Of particular relevance for marketing problems is the use of latent variables to formulate models with discrete lumps of probability.

To illustrate the usefulness of the data augmentation concept for discrete dependent variables models, consider the latent variable formulation of the binary probit model:

$$\begin{aligned} z_i &= x_i'\beta + \varepsilon_i \qquad \varepsilon_i \sim N(0, 1), \\ y_i &= \begin{cases} 0, & \text{if } z_i < 0, \\ 1, & \text{otherwise.} \end{cases} \end{aligned} \tag{3.8.1}$$

We observe (X, y). If this model is used to represent the choice between two alternatives, then z has the interpretation as the difference in utility between the two alternatives. We choose alternative A if it is more attractive than B. We only partially observe latent utility in the sense that only x is observed. Other influences on utility are represented by the 'error' term. Given that the latent structure in (3.8.1) is a standard regression model, we can use a normal prior for β, $\beta \sim N(\bar{\beta}, A^{-1})$.

Data augmentation proceeds by considering the entire n-dimensional vector of z values as part of the parameter vector, $\theta' = (z', \beta)$. Given the normal prior for β, the model is complete in the sense that (3.8.1) specifies the joint distribution[18] of z and β:

$$p(z, \beta|X) = p(z|\beta, X)p(\beta).$$

We note that this is a highly correlated distribution, particularly if the prior on β is diffuse. We can write the directed graph for this model as follows:

$$\beta \to z \to y \tag{3.8.2}$$

This directed graph immediately reveals that β, y are independent conditional on z.

The posterior distribution of θ can easily be computed by using a Gibbs sampler:[19]

$$z|\beta, X, y, \tag{3.8.3}$$

$$\beta|z, X. \tag{3.8.4}$$

This Gibbs sampler recognizes that θ separates into two natural groups or 'blocks'. Given β and the data, the z_i s are independent truncated univariate normal distributions. Given z, inference on β is just a Bayes linear regression analysis with a normal prior and no scale parameter (note that z is sufficient for β and we do not need to add y to the conditioning arguments (3.8.4)).

GIBBS SAMPLER FOR BINARY PROBIT

Start with β_0.

---

[18] If you regard the z values as meaningful objects for inference, then this is a prior distribution. If not, then the augmented parameters are simply devices by which one gets at the posterior distribution of β.

[19] Albert and Chib (1993) were the first to propose this sampler (for a closely related model, see also Chib 1992).

Draw z from (3.8.3) by making n independent draws from $\text{trun}_{(a_i,b_i)}(-x_i'\beta_0, 1)$, with $a_i = 0$ if $y_i = 1$, $-\infty$ otherwise, and $b_i = 0$ if $y_i = 0$, ∞ otherwise.

Draw β_1 from (3.8.4) using standard normal theory:

$$\beta|z, X \sim N(\tilde{\beta}, (X'X + A)^{-1}); \tilde{\beta} = (X'X + A)^{-1}(X'z + A\bar{\beta}).$$

Repeat, as necessary.

The dependence between draws of β comes entirely through the z vector. This will be useful Gibbs sampler to the extent to which the latent variables and β are not too highly correlated. R code for this binary probit sampler is available as the function

R `rbprobitGibbs` in *bayesm*.

Figure 3.8 shows the results of running this Gibbs sampler with a simulated data set with $n = 100$, and two regressors which are uniform on (0,1) and independent.

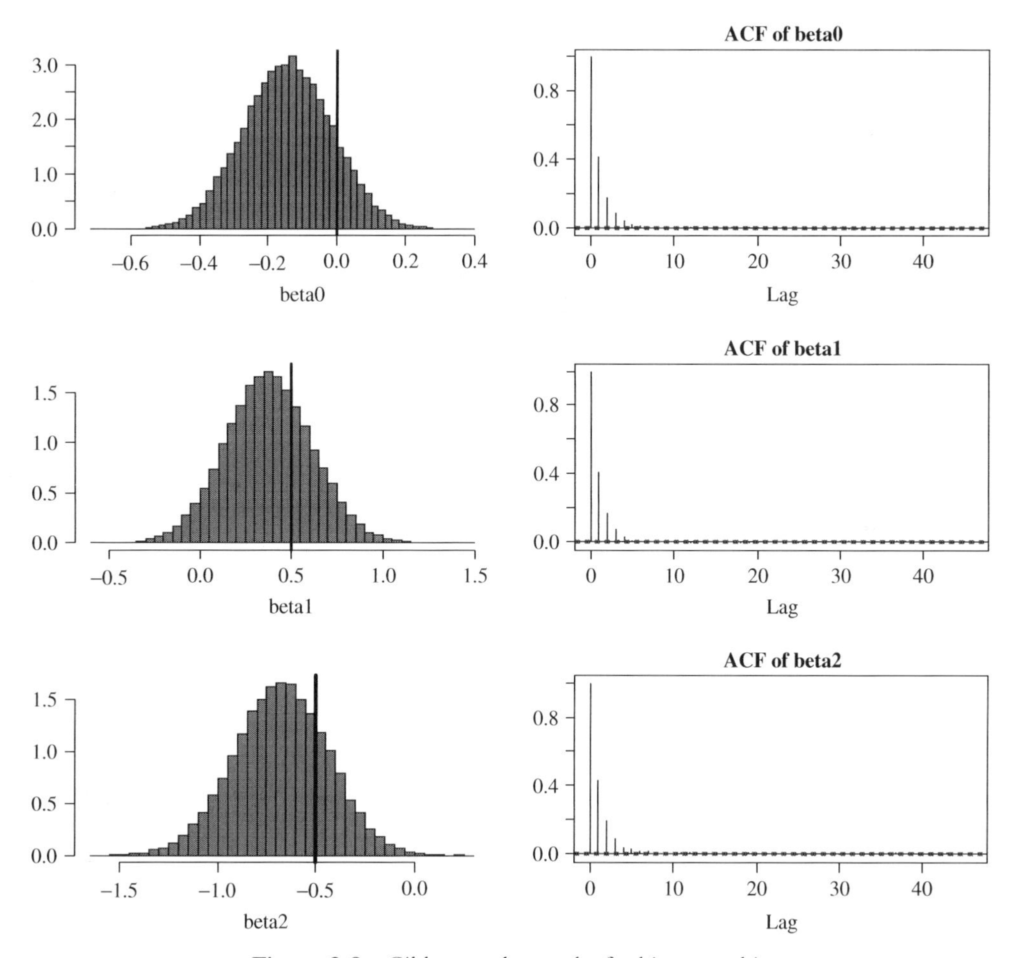

Figure 3.8 Gibbs sampler results for binary probit

The left-hand side shows histograms of the marginal posterior distribution of each parameter. The vertical line marks the 'true' value of the parameter underlying the simulated data. The right-hand panels show the autocorrelation functions. The marginal posterior distributions of the model parameters are very normal and the ACFs for this data are very reasonable even though there is not much information in this data on the betas.

One might be tempted to conclude that all of this machinery is an elaborate way of producing results from asymptotic theory (although one would have to verify that the posterior covariance is close to the asymptotic covariance). However, the parameters of this model are not as directly interpretable and relevant as in the linear regression context. In marketing applications, we are often interested in predicted probabilities for given values of x. Figure 3.9 shows the posterior distribution of the probability $y = 1$ for various x vectors, that is, the posterior distribution of $\Phi(x'\beta)$. Since probabilities are bounded and there is posterior uncertainty, these distributions are very nonnormal. Superimposed on the histogram are normal densities evaluated at the posterior mean and variance.[20] Even this small problem and highly regular model provides a powerful motivation to eliminate asymptotic approximations.

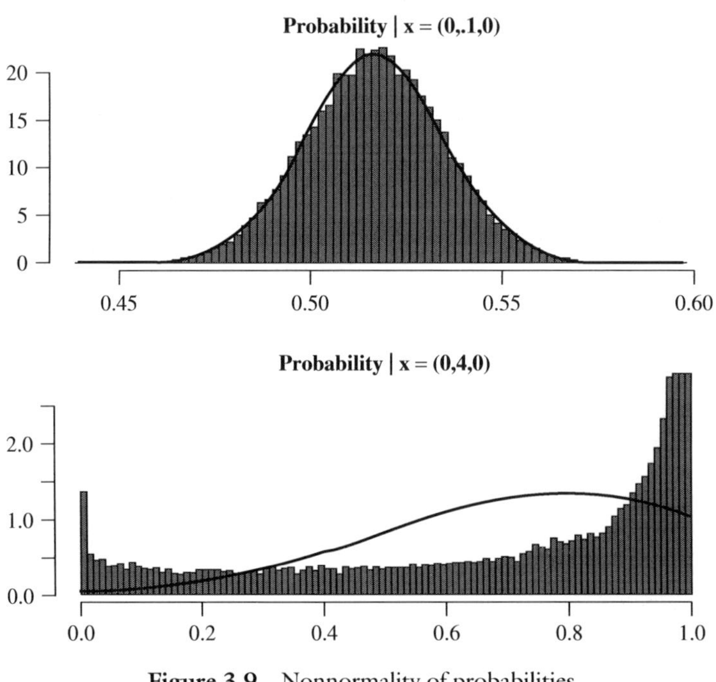

Figure 3.9 Nonnormality of probabilities

[20] Asymptotic theory would be even worse since it would keep the normal distribution assumption but insert asymptotic estimates of the mean and variance.

Finally, we should note that this model can easily be investigated by importance sampling performed on the posterior obtained by integrating out z.

$$p(\beta|X, y) \propto p(\beta) \prod_i \Phi(x_i'\beta)^{y_i}(1 - \Phi(x_i'\beta))^{1-y_i}. \tag{3.8.5}$$

In fact, per unit of computing time, a direct approach via importance sampling will surely yield more information than the binary Gibbs sampler due to the autocorrelation in the draw sequence. The binary probit Gibbs sampler is of interest primarily because it suggests that data augmentation strategies can be useful for the multinomial probit and other problems for which the likelihood over the model parameters is difficult to evaluate.

3.9 MIXTURES OF NORMALS

Finite mixtures of multivariate normal distributions can provide a very flexible model for multivariate data. Mixtures of normals can accommodate heavy-tailed and skewed distributions. However, in the multivariate case, the possibilities are even broader. For example, we can create a joint distribution with 'banana'-shaped contours by arranging closely spaced normal distributions along a curve. There is a sense that with enough mixture components, one can approximate any multivariate distribution in the same sense that one can build any shaped hill by piling up small mounds of gravel.

The basic mixture of normals model can be written

$$\begin{aligned} y_i &\sim N(\mu_{\text{ind}_i}, \Sigma_{\text{ind}_i}), \\ \text{ind}_i &\sim \text{Multinomial(pvec)}. \end{aligned} \tag{3.9.1}$$

Here y_i is a p-dimensional vector and pvec is a vector of K mixture probabilities. This model is referred to as a mixture of normals with K components. Expression (3.9.1) is

R a direct model for simulation from a mixture of normals (see `rmixture` in *bayesm*). First, we draw a multinomial distributed indicator of which component is 'active' and then we draw a multivariate normal vector from this component. This representation of the model also suggests the basis of a Gibbs sampler by augmenting the parameters with the vector of n indicators.

Priors for the mixture of normals model can be taken in convenient conditionally conjugate forms:

$$\begin{aligned} \text{pvec} &\sim \text{Dirichlet}(\alpha), \\ \mu_k &\sim N(\bar{\mu}, \Sigma_k \otimes a_\mu^{-1}), \qquad k = 1, \ldots, K, \\ \Sigma_k &\sim \text{IW}(\nu, V). \end{aligned} \tag{3.9.2}$$

In (3.9.2), the joint prior on the normal component parameters is independent conditional on pvec and in the form of the natural conjugate prior for multivariate regression (see Section 2.8.5). The DAG for the mixture of normals model can be written as

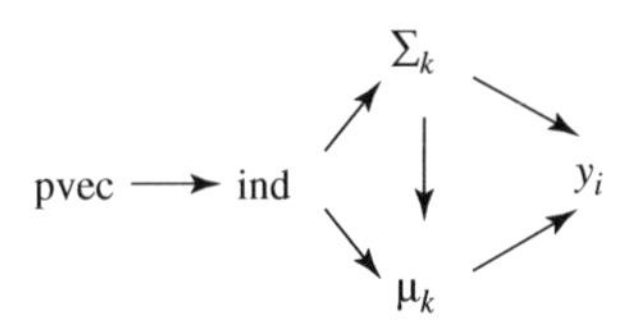

Given this DAG, we can easily write down the Gibbs sampler as consisting of the following sets of conditionals:

$$\begin{aligned} &\text{ind}|\text{pvec}, \{\mu_k, \Sigma_k\}, Y \\ &\text{pvec}|\text{ind} \\ &\{\mu_k, \Sigma_k\}|\text{ind}, Y, k = 1, \ldots, K \end{aligned} \tag{3.9.3}$$

Y is the $n \times p$ matrix of multivariate observations. This sampler was introduced by Diebolt and Robert (1994). The key idea is that once the indicators are drawn, the observations are classified by normal component and then one can proceed with K independent conjugate draws of the normal component parameters.

The draw of the indicators is a multinomial draw based on the likelihood ratios with pvec_k as the prior probability of membership in each component:

$$\begin{aligned} &\text{ind}_i \sim \text{multinomial}(\pi_i); \qquad \pi' = (\pi_{i,1}, \ldots, \pi_{i,K}), \\ &\pi_{i,k} = \frac{\text{pvec}_k \varphi(y_i|\mu_k, \Sigma_k)}{\sum_m \text{pvec}_m \varphi(y_i|\mu_m, \Sigma_m)}. \end{aligned} \tag{3.9.4}$$

Here $\varphi(\bullet)$ is the multivariate normal density.

The draw of pvec given the indicators is a Dirichlet draw

$$\begin{aligned} &\text{pvec} \sim \text{Dirichlet}(\tilde{\alpha}), \\ &\tilde{\alpha}_k = n_k + \alpha_k, \\ &n_k = \sum_{i=1}^{n} I(\text{ind}_i = k). \end{aligned} \tag{3.9.5}$$

The draw of each (μ_k, Σ_k) can be made using the algorithm to draw from the multivariate regression model as detailed in Section 2.8.5. For each subgroup of observations, we have a multivariate regression model of the form

$$Y_k = \iota\mu_k' + U, \qquad U = \begin{bmatrix} u_1' \\ \vdots \\ u_{n_k}' \end{bmatrix}, \qquad u_i \sim N(0, \Sigma_k). \tag{3.9.6}$$

Here Y_k is the submatrix of Y that consists of the n_k rows where $\text{ind}_i = k$; ι is a vector of ones. The results of Chapter 2 simplify to the following draws:

$$\begin{aligned} &\Sigma_k|Y_k, \nu, V \sim \text{IW}(\nu + n_k, V + S), \\ &\mu_k|Y_k, \Sigma_k, \bar{\mu}, a_\mu \sim N\left(\tilde{\mu}_k, 1/(n_k + a_\mu)\Sigma_k\right), \end{aligned} \tag{3.9.7}$$

where

$$S = (Y_k - \iota\tilde{\mu}_k')'(Y_k - \iota\tilde{\mu}_k'),$$
$$\tilde{\mu}_k = (n_k + a_\mu)^{-1}(n_k\bar{y}_k + a_\mu\bar{\mu}), \tag{3.9.8}$$
$$\bar{y}_k = (Y_k'\iota/n_k).$$

This Gibbs sampler for mixtures of normals is available as function, `rnmixGibbs`, in
R *`bayesm`*.

3.9.1 Identification in Normal Mixtures

It is well known that the likelihood for the normal mixture model can have up to $K!$ symmetric modes. This is due to what is referred to as the label-switching 'problem'. We can simply interchange or permute the labels for each of the components and have the same value of the likelihood. For example, consider the mixture of two univariate normal distributions. There are two equal-height posterior modes (assuming the priors are identical). We can simply interchange the labels, calling mode 1 '2' and mode 2 '1', and leave the likelihood unchanged. This means that the marginal posteriors of the mean parameters will often have two modes. This will create navigation problems for many MCMC algorithms, particularly those based on the Gibbs sampler. It is possible that the algorithm may only investigate one of the modes or some subset of modes, leaving others completely untouched. This will occur when there is strong separation or good classification information in the 'data'. However, when differences between components are small, there can be a good deal of switching from mode to mode. This switching from mode to mode by MCMC methods is often what researchers in this area mean by 'label switching'.

Figure 3.10 illustrates the label-switching problem by considering the problem of inference about a mixture of two normals. Component '1' is an $N(1,1)$ random variable and component '2' is $N(2,1)$. The mixture probability is 0.5. This is an example where there is little distance between the modes for each component. The mixture is a symmetric and unimodal density centered at 1.5. If we have a modest number of observations, the Gibbs sampler will 'flip' the component labels so that what had been labeled a draw of the mean for the first component is now the mean of the second component ($\mu_1 \leftrightarrow \mu_2$). Figure 3.10 plots the Gibbs sampler output for μ_1 (labeled '1') and μ_2 (labeled '2'). We can see several label switches, including one around draw number 50.

This means that we cannot simply look at the marginal distribution of μ_1 from our Gibbs sampler output. This parameter is not identified. However, the joint density is identified:

$$p(y) = p\varphi(y|\mu_1, \sigma_1) + (1 - p)\varphi(y|\mu_2, \sigma_2). \tag{3.9.9}$$

For any given value of y, we can compute the posterior distribution of the density at this value. If we use a grid of y values, we can compute the posterior distribution of the entire density of the data. In particular, we can average the posterior to obtain the posterior

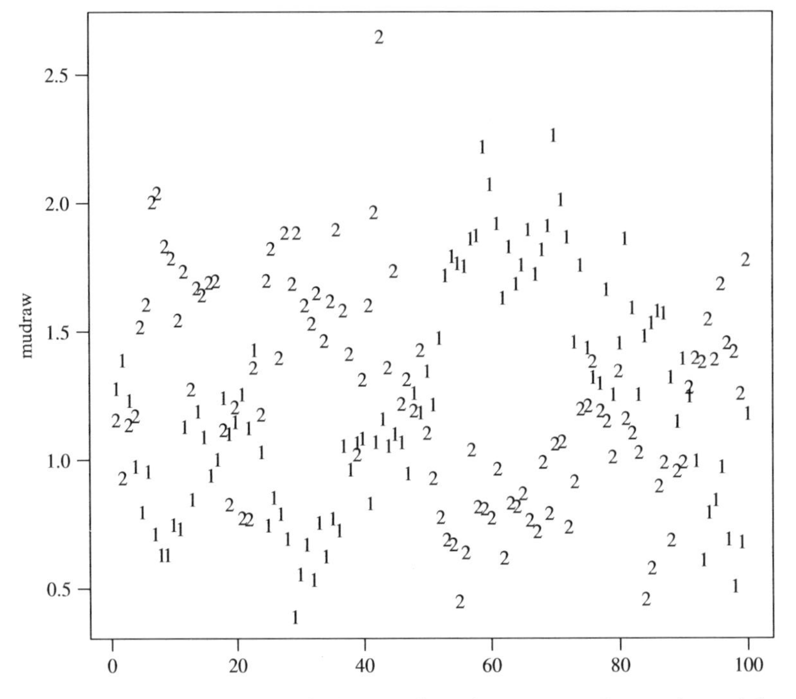

Figure 3.10 Illustration of label switching for mixture of normals model

mean as an estimator of the density. Figure 3.11 shows ten posterior draws of the fitted density in (3.9.9). The solid density is the true density.

Much of the early work with mixture models attempted to solve the identification problem for individual component parameters by using various a priori ordering restrictions. For example, some advocate ordering the components by prior probability:

$$\text{pvec}_1 > \text{pvec}_2 > \cdots > \text{pvec}_K. \tag{3.9.10}$$

Unfortunately, imposing this restriction does not necessarily remove the identification problem (cf. Stephens 2000). Other proposals include ordering by the normal mixture component parameters. Obviously, it may be difficult to define an ordering in the case of mixtures of ***multivariate*** normal distributions. Moreover, even in the cases of univariate normals, Stephens (2000) has noted that identification cannot always be achieved. Choosing the right way to divide the parameter space so as to ensure that only one mode remains can be somewhat of an art form and may be close to impossible in high-dimensional problems (see Frühwirth-Schnatter (2001) for a suggested method that may work in low dimensional situations).

Instead of imposing what can been termed 'artificial' identification constraints, Stephens (2000) advocates post-processing via relabeling of MCMC draws so as to minimize some sort of statistical criterion such as divergence of the estimated marginal posteriors from some unimodal distribution. This means that post-simulation optimization methods must be used to achieve a relabeling in the spirit of clustering algorithms.

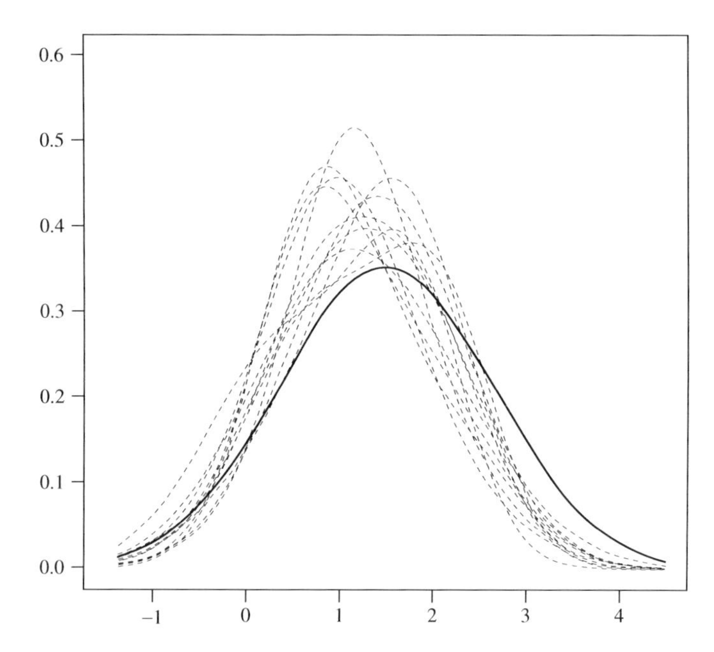

Figure 3.11 Posterior draws of density: example of univariate mixture of two normals

There is no guarantee that such post-processing will necessarily uncover the true structure of the data and there is still 'art' in the choice of objective function and tuning of the relabeling.

The label-switching phenomenon is only a problem to the extent to which the investigator wishes to attach meaning to specific normal components. If one interprets each normal component as a subpopulation, then we might want to make inferences about the means and covariances for that component. For example, we might think of the population of respondents to a questionnaire as comprised of distinct groups each with some heterogeneity within group but where this may be small relative to across-group heterogeneity. In this situation, label switching becomes an issue of identification for the parameters of interest. The only way to 'solve' this identification problem is by use of prior information. However, if we regard the mixture of normals as a flexible approximation to some unknown joint distribution, then the label-switching 'problem' is not relevant. Ultimately, the data identify the joint distribution and we do not attach substantive importance to the mixture component parameters. We should recognize that this runs somewhat counter to the deeply ingrained tradition of identifying consumer segments. We do not really think of segments of homogeneous consumers as a reality but merely as a convenient abstraction for the purpose of marketing strategy discussions. The empirical evidence, to date, overwhelmingly favors the view that there is a continuum of consumer tastes.

Thus, we view the object of interest as the joint density of the parameters. This density and any possible functions defined on it, such as moments or probabilities of regions, are identified without a priori restrictions or ad hoc post-processing of MCMC draws. Once you adopt this point of view, the identification and label-switching literature becomes irrelevant and, instead, you are faced with the problem of summarizing a fitted multivariate density function. Lower–dimensional summaries of the mixture of normals density are required – as Frühwirth-Schnatter *et al.* (2004) point out, any function of this density will also be immune to the label-switching identification issues. One possible summary would be the univariate marginal densities for each component of θ. However, this does not capture co-movement of different elements. It is certainly possible to compute posterior distributions of covariance but, as pointed out above, these lose interpretability for multimodal and nonnormal distributions.

3.9.2 Performance of the Unconstrained Gibbs Sampler

The Gibbs sampler outlined in (3.9.3) is referred to as the 'unconstrained' Gibbs sampler in the sense that no prior constraints have been imposed to achieve identification of the mixture component parameters. As such, this sampler may exhibit label switching. As pointed out above, any function which is invariant to label switching such as the estimated mixture density will not be affected by this problem. There also may be algorithmic advantages to not imposing identification constraints. As pointed out by Frühwirth-Schnatter (2001), identification constraints hamper mixing in single-move constrained Gibbs samplers. For example, if we imposed the constraint that the mixture probabilities must be ordered, then we must draw from a prior distribution restricted to a portion of the parameter space in which the ordering is imposed. A standard way to do this is to draw each of the K probabilities, one by one, given the others. The ordering constrains mean that as we draw the kth probability, it must lie between the $(k-1)$th and $(k+1)$th probabilities. This may leave little room for navigation. Thus, the unconstrained Gibbs sampler will often mix better than constrained samplers. Frühwirth-Schnatter introduces a 'random permutation' sample to promote better navigation of all modes in the unidentified parameter space of the mixture component parameters. This may improve mixing in the unidentified space but will not improve mixing for the identified quantities such as the estimated density and associated functions. The fact that the unconstrained Gibbs sampler may not navigate all or even more than one of the $K!$ symmetric modes does not mean that it does not mix well in the identified space. As Gilks (1997, p. 771) comments: 'I am not convinced by the ... desire to produce a unique labeling of the groups. It is unnecessary for valid Bayesian inference concerning identifiable quantities; it worsens mixing in the MCMC algorithm; it is difficult to achieve in any meaningful way, especially in high dimension, and it is therefore of dubious explanatory or predictive value'.

Our experience is that the unconstrained Gibbs sampler works very well, even for multivariate data and with a large number of components. To illustrate this, consider five-dimensional data simulated from a three-component normal mixture, $n = 500$:

$$\mu_1 = \begin{pmatrix} 1 \\ 2 \\ 3 \\ 4 \\ 5 \end{pmatrix}, \qquad \mu_2 = 2\mu_1, \qquad \mu_3 = 3\mu_1, \qquad \Sigma_k = \begin{bmatrix} 1 & 0.5 & \cdots & 0.5 \\ 0.5 & 1 & \ddots & \vdots \\ \vdots & \ddots & \ddots & 0.5 \\ 0.5 & \cdots & 0.5 & 1 \end{bmatrix},$$

$$\text{pvec} = \begin{pmatrix} 1/2 \\ 1/3 \\ 1/6 \end{pmatrix}. \tag{3.9.11}$$

We started the unconstrained Gibbs sampler with nine normal components. Figure 3.12 represents the distribution of the indicator variable across the nine components for each of the 400 first draws. The width of each horizontal line is proportional to the frequency with which that component number occurs in the draw of the indicator variable. The sampler starts out with an initial value of the indicator vector which is split evenly among the nine components. The sampler quickly shuts down a number of the components. By 320 or so draws, the sampler is visiting only three components with frequency corresponding to the mixture probabilities that were used to simulate the data.

Figure 3.13 shows each of the five marginal distributions for particular draws in the MCMC run shown in Figure 3.12. The solid lines are the 'true' marginal distributions implied by the normal mixture given in (3.9.11). The top panel shows the draw of the five marginal distributions for the fourth draw of the sampler. The marginals vary in the extent to which they show 'separation' in modes. The marginal of the fifth component has the most pronounced separation. After only four draws, the sampler has not yet 'burned in' to capture the multimodality in the marginal distribution of the third through fifth components. However, as shown in the bottom panel, by 100 draws the sampler has found the rough shape of the marginal distributions. Variation from draw to draw after the 100th draw simply reflects posterior uncertainty.

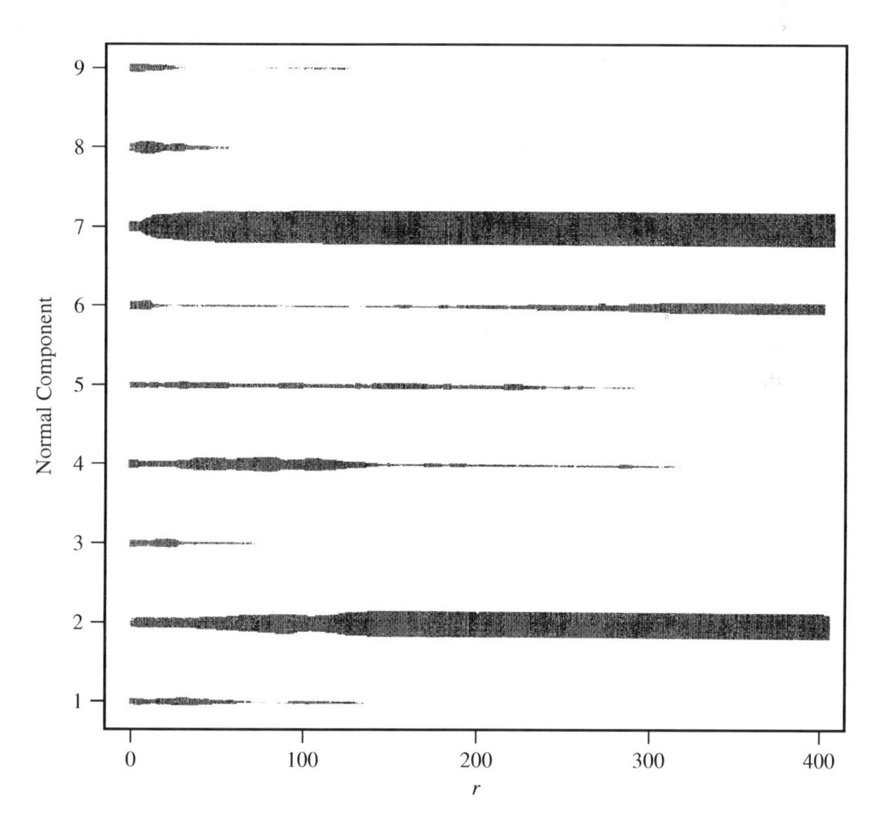

Figure 3.12 Plot of frequency of normal components: multivariate normal mixture example

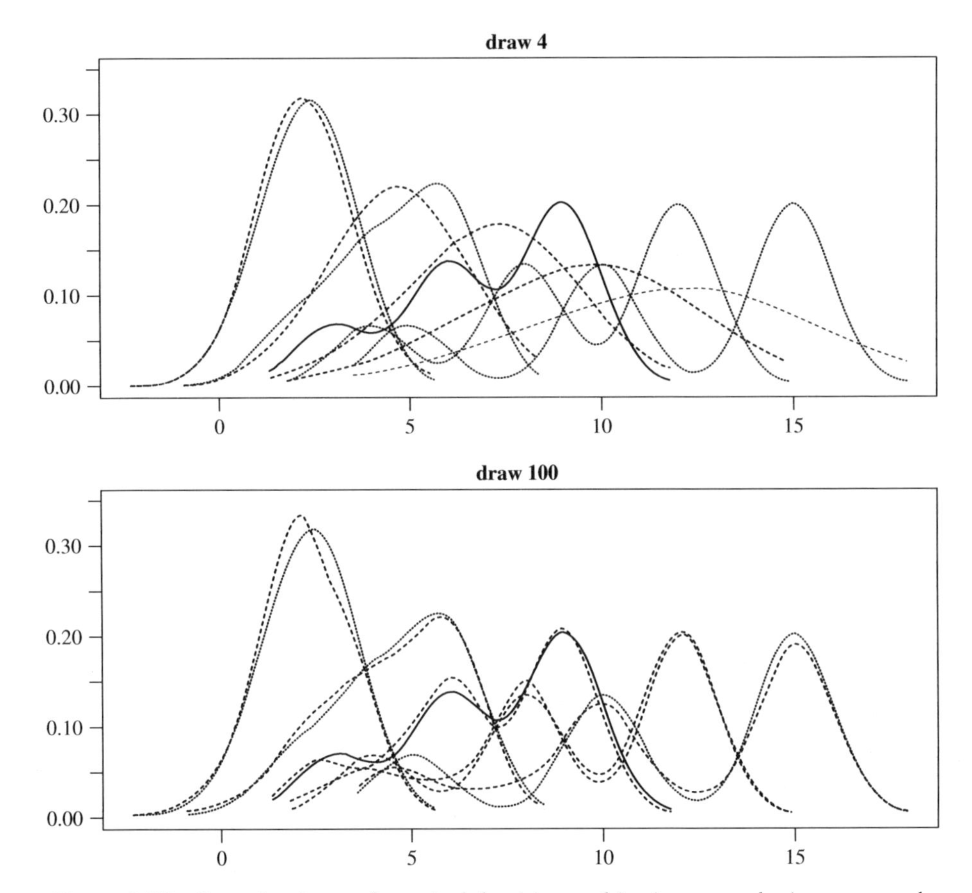

Figure 3.13 Posterior draws of marginal densities: multivariate normal mixture example

3.10 METROPOLIS ALGORITHMS

The Gibbs sampler is enormously useful, particularly for models built up from hierarchies of relatively standard distributions. However, there are many problems for which the conditional distributions are not of a known form that is easy to simulate from. For this reason, it is useful to have a more general-purpose tool. By this we mean a tool that can be applied to, at least in principle, any posterior distribution. Of course, in practice, there will be problems for which the general-purpose algorithm will produce a poorly performing MCMC sampler. This means that the general-purpose algorithm means an approach to generating a candidate MCMC method for virtually any problem. The performance of this candidate will not be assured and must still be investigated.

The Metropolis class of algorithms is a general-purpose approach to producing Markov chain samplers.[21] The idea of the Metropolis approach is to generate a Markov chain

[21] Although the Metropolis method was developed for discrete distributions and discussed in the statistics literature as early as 1970 (Hastings 1970), the popularity of this method is due in large part to Tierney (1994) and Chib and Greenberg (1995a) who provided a tutorial in the method as well as many useful suggestions for implementation.

with the posterior, $\pi(\bullet)$, as its invariant distribution by appropriate modifications to a related Markov chain that is relatively easy to simulate from. Viewed in this way, Metropolis algorithms are similar in spirit to the accept/reject method of iid sampling. Accept/reject methods sample from a proposal distribution and then reject draws to modify the proposal distribution to achieve the desired target distribution. The proper invariant distribution is achieved by constructing a new chain that is time-reversible with respect to π.

We will start with a discrete state space version of the Metropolis algorithm which will illustrate the essential workings of the algorithm. We have a transition matrix, Q, which we want to modify to ensure that the resultant chain has a stationary distribution given by the vector π. The dimension of the state space is d.

DISCRETE METROPOLIS ALGORITHM

Start in state i, $\theta_0 = \theta^i$.

Draw state j with probability given by $q_{ij}, j = 1, \ldots, d$ (multinomial draw).

Compute $\alpha = \min\{1, \pi_j q_{ji}/\pi_i q_{ij}\}$.

With probability α, $\theta_1 = \theta^j$ (move), else $\theta_1 = \theta^i$ (stay).

Repeat, as necessary.

Note that only the ratios π_j/π_i are required and we only need to know the posterior distribution up to a constant.

The unique aspect of this algorithm is the possibility that it will not move on a given iteration of the chain. With probability $1 - \alpha$, the chain will repeat the value from the rth to $(r+1)$th iteration. These repeats are to ensure that new chain is reversible. If $\pi_i q_{ij} > \pi_j q_{ji}$, then there will be 'too many' transitions from state i to state j and not enough reverse transitions from state j to i. For this reason, the Metropolis chain only accepts the α fraction of the transitions from i to j and all of the moves from j to i.

This algorithm is constructed to be time-reversible with respect to π. To see this, recall that time reversibility requires $\pi_i p_{ij} = \pi_j p_{ji}$. The transition probability matrix for the chain defined by the Metropolis algorithm is defined by

$$p_{ij} = q_{ij}\alpha(i, j).$$

Therefore,

$$\pi_i p_{ij} = \pi_i q_{ij} \min\left\{1, \frac{\pi_j q_{ji}}{\pi_i q_{ij}}\right\} = \min\{\pi_i q_{ij}, \pi_j q_{ji}\},$$

$$\pi_j p_{ji} = \pi_j q_{ji} \min\left\{1, \frac{\pi_i q_{ij}}{\pi_j q_{ji}}\right\} = \min\{\pi_j q_{ji}, \pi_i q_{ij}\},$$

and the condition for time reversibility is satisfied.

The continuous version of the Metropolis algorithm has exactly the same formulation as the discrete case except that the analysis of reversibility and convergence is slightly more complex. In particular, the transition kernel for the Metropolis does not have a standard density but rather has a mixture of discrete and continuous components. The continuous state space Metropolis algorithm starts with a proposal transition kernel defined by the transition function $q(\theta, \vartheta)$. Given θ, $q(\theta, \bullet)$ is a density. The continuous state space version of the Metropolis algorithm is as follows:

CONTINUOUS STATE SPACE METROPOLIS

Start at θ_0.
Draw $\vartheta \sim q(\theta_0, \bullet)$.
Compute $\alpha(\theta, \vartheta) = \min\{1, \pi(\vartheta)q(\vartheta, \theta)/\pi(\theta)q(\theta, \vartheta)\}$.
With probability α, $\theta_1 = \vartheta$, else $\theta_1 = \theta_0$.
Repeat, as necessary.

To see that the continuous version of the Metropolis algorithm has π as its invariant distribution,[22] we first define the kernel. Recall that the kernel provides the probability that the chain will advance to a set A given that it is currently at point θ:

$$K(\theta, A) = \int_A p(\theta, \vartheta)\, d\vartheta + r(\theta)\delta_A(\theta) \tag{3.10.1}$$

where

$$p(\theta, \vartheta) = \alpha(\theta, \vartheta)q(\theta, \vartheta) \tag{3.10.2}$$

and

$$\delta_A(\theta) = \begin{cases} 1, & \theta \in A, \\ 0, & \text{otherwise.} \end{cases}$$

p defined in (3.10.2) is the transition function for the Metropolis chain. The probability that the chain will move away from θ is given by

$$\int_\Theta p(\theta, \vartheta)\, d\vartheta = \int_\Theta \alpha(\theta, \vartheta)q(\theta, \vartheta)\, d\vartheta.$$

Since $K(\theta, \Theta) = 1$, this implies that $r(\theta)$ is the probability that the chain will stay at θ. Since there is a possibility for the chain to repeat the value of θ, we cannot, given θ, provide a standard density representation for the distribution of ϑ since there is a mass point at θ. The conditional distribution defined by the Metropolis kernel can be interpreted as a mixture of a mass point at θ and a continuous density $p(\theta, \vartheta)/(1 - r(\theta))$.

The Metropolis transition function p satisfies the 'detailed balance' condition

$$\pi(\theta)p(\theta, \vartheta) = \pi(\vartheta)p(\vartheta, \theta) \tag{3.10.3}$$

or $\pi(\theta)\alpha(\theta, \vartheta)q(\theta, \vartheta) = \pi(\vartheta)\alpha(\vartheta, \theta)q(\vartheta, \theta)$ which is true by the construction of the α function. Equation (3.10.3) ensures that the Metropolis chain will be time-reversible. We can now show that (3.10.3) implies that π is the invariant distribution of the

[22] Here we follow the exposition of Chib and Greenberg (1995a).

chain. Recall that, for a continuous state chain, π is the invariant distribution if $\int_A \pi(\theta)\,d\theta = \int_\Theta \pi(\theta)K(\theta, A)\,d\theta$. We have

$$\begin{aligned}\int_\Theta \pi(\theta)K(\theta, A)\,d\theta &= \int_\Theta \pi(\theta)\left[\int_A p(\theta, \vartheta)\,d\vartheta + r(\theta)\delta_A(\theta)\right]d\theta \\ &= \int_\Theta \int_A \pi(\theta)p(\theta, \vartheta)\,d\vartheta\,d\theta + \int_\Theta \pi(\theta)r(\theta)\delta_A(\theta)\,d\theta.\end{aligned}$$

Interchanging the order of integration and applying the detailed balanced equation,

$$\begin{aligned}\int_\Theta \pi(\theta)K(\theta, A)\,d\theta &= \int_A \int_\Theta \pi(\vartheta)p(\vartheta, \theta)\,d\theta\,d\vartheta + \int_\Theta \pi(\theta)r(\theta)\delta_A(\theta)\,d\theta \\ &= \int_A \pi(\vartheta)\left[\int_\Theta p(\vartheta, \theta)\,d\theta\right]d\vartheta + \int_A \pi(\theta)r(\theta)\,d\theta.\end{aligned}$$

The integral of the Metropolis function in the first term above is simply the probability of moving:

$$\begin{aligned}\int_\Theta \pi(\theta)K(\theta, A)\,d\theta &= \int_A \pi(\vartheta)[1 - r(\vartheta)]\,d\vartheta + \int_A \pi(\theta)r(\theta)\,d\theta \\ &= \int_A \pi(\vartheta)\,d\vartheta.\end{aligned}$$

Convergence of the Metropolis algorithm is assured by positivity of the proposal transition function $q(\theta, \vartheta)$,[23] which assures that the chain is irreducible (see Robert and Casella 2004, Section 7.3.2). The challenge is to choose a 'proposal' or candidate distribution that is relatively easy to evaluate and simulate from and yet produces a Metropolis chain with acceptable convergence properties. There are a wide variety of different styles of proposal densities. We will review some of the most useful.

3.10.1 Independence Metropolis Chains

Importance sampling relies on having a reasonable approximation to π. Usually, the importance function is based on an asymptotic approximation to the posterior with fattened tails. This idea can be embedded in a Metropolis chain by taking q to be independent of the current value of the chain, $q(\theta, \vartheta) = q_{\text{imp}}(\vartheta)$, and based on the same sort of importance function ideas. We denote the independent Metropolis transition density by q_{imp} to draw the close analogy with importance sampling and with the criterion for a useful importance function.

INDEPENDENCE METROPOLIS

Start with θ_0.

[23] An additional condition is required to ensure that the Metropolis chain is aperiodic. This condition requires that there be a nonzero probability of repeating.

Draw $\vartheta \sim q_{\text{imp}}$.
Compute $\alpha = \min\{1, \pi(\vartheta)q_{\text{imp}}(\theta)/\pi(\theta)q_{\text{imp}}(\vartheta)\}$.
With probability α, $\theta_1 = \vartheta$, else $\theta_1 = \theta_0$.
Repeat, as necessary.

If q is an excellent approximation to π, then most draws will be accepted since the ratio in the α computation will be close to one. This means that we will have a chain with almost no autocorrelation. To understand how the chain handles discrepancies between q and π, rewrite the ratio in the α computation as

$$\frac{\pi(\vartheta)q_{\text{imp}}(\theta)}{\pi(\theta)q_{\text{imp}}(\vartheta)} = \frac{\pi(\vartheta)/q_{\text{imp}}(\vartheta)}{\pi(\theta)/q_{\text{imp}}(\theta)}.$$

If π has more relative mass at ϑ than at θ, the chain moves to ϑ with probability one to build up mass at ϑ. On the other hand, if π has less relative mass at ϑ than at θ, then there is a positive probability that the chain will repeat θ, which builds up mass at that point and introduces dependence in the sequence of chain draws. This is really the opposite of accept/reject sampling in which the proposal distribution is 'whittled down' to obtain π by rejecting draws.

It is important that the q proposal distribution has fatter tails than the target distribution as in importance sampling. If the target distribution has fatter tails than the proposal, the chain will wander off into the tails and then start repeating values to build up mass. If the proposal distribution dominates the target distribution in the sense that we can find a finite number M such that $\pi \leq Mq_{\text{imp}}$ for all $\theta \in \Theta$, then the independence Metropolis chain is ergodic in an especially strong sense in that the distance between the repeated kernel and π can be bounded for any starting point θ_0 (see Robert and Casella 2004, Theorem 7.8). It should be noted that if a normal prior is used to form the posterior target density, then the conditions for uniform ergodicity are met unless the likelihood is unbounded.

For problems of low dimension with very regular likelihoods, the independence Metropolis chain can be very useful. However, for higher-dimensional problems, the same problems which plague importance sampling apply to the independence Metropolis chain. Even if the dominance condition holds, the proposal distribution can miss where the target distribution has substantial mass and can give rise to misleading results. For this reason, independence Metropolis chains are primarily useful in hybrid samplers (see Section 3.12), in which the independence chain is embedded inside a Gibbs sampler.

3.10.2 Random-Walk Metropolis Chains

A particularly appealing Metropolis algorithm can be devised by using a random walk (RW) to generate proposal values:

$$\vartheta = \theta + \varepsilon. \tag{3.10.4}$$

This proposal corresponds to the proposal transition function, $q(\theta, \vartheta) = q_\varepsilon(\vartheta - \theta)$, which depends only on the choice of the density of the increments. This proposal function is symmetric, $q(\theta, \varphi) = q(\varphi, \theta)$. A 'natural' choice might be to make the increments normally distributed.[24]

GAUSSIAN RANDOM-WALK METROPOLIS

Start with θ_0.

Draw $\vartheta = \theta + \varepsilon$, $\varepsilon \sim N(0, s^2\Sigma)$.

Compute $\alpha = \min\{1, \pi(\vartheta)/\pi(\theta)\}$.

With probability α, $\theta_1 = \vartheta$, else $\theta_1 = \theta_0$.

Repeat, as necessary.

The simplicity of this algorithm gives it great appeal. It can be implemented for virtually any model and it does not appear to require the same in-depth a priori knowledge of π as the independence Metropolis does. In addition, there is an intuitive argument that the RW algorithm should work better than the independence Metropolis in the sense that it might roam more freely in the parameter space (due to the drifting behavior of a random walk) and 'automatically' seek out areas where π has high mass and then navigate those areas. From a theoretical point of view, however, strong properties such as uniform ergodicity cannot be proven for the RW Metropolis without further restrictions on the tail behavior of π. Jarner and Tweedie (2001) show that exponential tails of π are a necessary condition for geometric ergodicity[25] irrespective of the tail behavior of the proposal density q. While there is research into the relative tail behavior of the target and proposal density, there are few definitive answers. For example, it is debatable whether it is desirable to have the proposal (increment) density have thicker tails than the target. There are some results suggestive that this can be helpful in obtaining faster convergence, but only for special cases (see Robert and Casella 2004, Theorem 6.3.6, which applies only to log-concave targets, and Jarner and Tweedie 2001, which only applies to univariate problems). In almost all practical applications, a Gaussian RW is used. Again, these theoretical discussions are not terribly relevant to practice since we can always use a normal prior which will provide exponential behavior for the target density.

The problem with the RW Metropolis is that it must be tuned by choosing the increment covariance matrix. This is often simplified by choosing Σ to be either I or the asymptotic covariance matrix. Computation of the asymptotic covariance matrix can be problematic for models with nonregular likelihoods. We must also choose the scaling factor, s. The scaling factor should be chosen to ensure that the chain navigates the area where π has high mass while also producing a chain that is as informative as possible regarding the posterior summaries of interest. These goals cannot be met

[24] Interestingly enough, the original paper by Hastings includes examples in which the increments are uniformly distributed.

[25] Roughly speaking, geometric ergodicity means that the distance between the n-step kernel and the invariant distribution decreases to the power of n.

without some prior knowledge of π which removes some of the superficial appeal of the RW Metropolis. Even so, the RW Metropolis has found widespread application and appears to work well for many problems.

3.10.3 Scaling of the Random-Walk Metropolis

The scaling of the RW Metropolis is critical to its successful use. If we scale back on the variance of the increments by taking small values of the scaling factor, s, then we will almost always 'move' or accept the proposed draw. This will mean that the algorithm will produce a chain that behaves (locally, at least) like an RW. The chain will exhibit extremely high autocorrelation and, therefore, can provide extremely noisy estimates of the relevant posterior quantities. Given that small values of the scale factor will result in a slowly moving chain which can fail to properly explore regions of high target mass, we might take the opposite approach and take scale factors considerably larger than one. This can result in a chain of proposed moves which are far away from the region of posterior mass (overshoot the target) and a high rejection rate. In the extreme case, overscaled RW Metropolis chains can get stuck.

'Optimal' scaling of the RW Metropolis is a balancing act in which the scaling factor is chosen to optimize some performance criterion. A reasonable criterion is a measure of information obtained about a posterior quantity per unit of computing time. Since we are using averages of a function of the draws to approximate the integral of that function with respect to the target posterior, the reciprocal of the variance of the sample average can be used as a measure of information. This clearly depends on the particular function used. Given a choice of function (such as the identity function for posterior means), we must be able to calculate the variance of the sample average of this function where the sample comes from a stationary but autocorrelated process. Let $\mu = E_{\pi}[g(\theta)]$. Then we estimate μ by

$$\hat{\mu} = \frac{1}{R}\sum_r g^r,$$

$$g^r \equiv g(\theta^r)$$

with

$$\begin{aligned}\text{var}(\hat{\mu}) = \frac{1}{R^2}[&\text{var}(g^1) + \text{cov}(g^1, g^2) + \cdots + \text{cov}(g^1, g^R) \\ &+ \text{cov}(g^2, g^1) + \text{var}(g^2) + \cdots + \text{var}(g^R)].\end{aligned} \tag{3.10.5}$$

Since $\{g^r\}$ is a stationary process, we write (3.10.5) using the autocorrelations of the process:

$$\text{var}(\hat{\mu}) = \frac{\text{var}(g)}{R}\left[1 + 2\sum_{j=1}^{R-1}\left(\frac{R-j}{R}\right)\rho_j\right] = \frac{\text{var}(g)}{R}f_R. \tag{3.10.6}$$

The 'price' of autocorrelation in the MCMC draws is represented by the factor f_R which is the multiple by which the variance is increased over the estimate based on an iid sample. Some use the reciprocal of f as a measure of the *relative numerical efficiency* of the sampler.

In practice, we can use the sample moments of the g process to estimate (3.10.6). Some guidance is required for the choice of the number of autocorrelations to include in the computation of f. This is the subject of a considerable literature in time series. We should not take this formula literally and use all $R-1$ computable autocorrelations. This would create a noisy and inconsistent estimate of f. There are two strategies in the time series literature. Some put a 'taper' or declining weights on the sample autocorrelations used in estimating f. Others simply advocate truncating the sum at some point m:

$$\hat{f}_R = 1 + \sum_{j=1}^{m} \left(\tfrac{m+1-j}{m+1}\right) \hat{\rho}_j. \tag{3.10.7}$$

Some guidance is required in the choice of m. In order for the estimator to be consistent, we must promise to increase m as the sample size (R) increases (albeit at a slower rate than R). This still does not provide guidance in the choice of m for a fixed sample size. Many MCMC applications will typically involve 10,000 or more draws. The autocorrelation structure of many chains used in practice is complicated and can have significant autocorrelation out to lags of order 100 or more. The exponential decline in autocorrelation associated with linear time series models is often not present in the MCMC chains. For these reasons, m should be chosen to be at least 100.

Given a measure of the information content, we can consider the problem of optimal scaling of the RW Metropolis chain. That is, we can optimize the choice of s to maximize relative numerical efficiency or to minimize f. Clearly, the optimal choice of s depends on the target distribution. Gelman *et al.* (1996) and Roberts and Rosenthal (2001) provide some results on optimal scaling but only for target distributions that are products of identical normal densities. They also consider the interesting asymptotic experiment of increasing the dimension of the parameter vector. This analysis presents both the optimal scaling factor as well as the acceptance rate of the optimally scaled RW chain. For the case of a target distribution consisting of only one normal univariate density, the optimal scaling factor is $s = 2.38$. As the dimension increases, an asymptotic result has the scaling reduced at the rate of the square root of the dimension, $s = 2.3/\sqrt{d}$. Corresponding to these scaling results are implied optimal acceptance rates which are around 0.23.[26] In practice, it is impossible to determine how differences between the target for our problem and an iid normal distribution will translate into differences in optimal scaling. It is true that we are scaling the asymptotic covariance matrix so that if the target density is close to a normal with this covariance matrix then we can expect the Roberts and Rosenthal results to apply. However, if our target density differs markedly from the asymptotic normal approximation, it is possible that the optimal scaling is quite different from this rule of thumb. We recommend that, where possible, shorter runs of the RW chain be used to help tune or choose the scaling constant to maximize numerical relative efficiency for parameters of interest. Of course, this assumes that the researcher has already determined the 'burn-in' period necessary to dissipate initial conditions. Contrary to current practice, choice of scaling should not be made on the basis of the

[26] Gelman *et al.* (1996) refer to optimal acceptance rates closer to 0.5 for $d = 1$. This is because they only considered the first-order autocorrelation. This has led to the incorrect interpretation that the optimal acceptance rate declines from around 0.5 to 0.23 or so as d increases.

acceptance rate of the chain but rather on the measure of numerical efficiency which is the more directly relevant quantity.

3.11 METROPOLIS ALGORITHMS ILLUSTRATED WITH THE MULTINOMIAL LOGIT MODEL

As discussed in Chapter 2, the multinomial logit (MNL) model is arguably the most frequently applied model in marketing applications. Individual data on product purchases often has the property that individuals are seldom observed to purchase more than one product of a specific type on one purchase occasion. The logit model has also been applied in a variety of forms to aggregate market share data. The MNL model has a very regular log-concave likelihood but it is not in a form that is easily summarized. Moments of functions of the parameter vector are not computable using analytic methods. In addition, the natural conjugate prior is not easily interpretable, so that it is desirable to have methods which would work with standard priors such as the normal prior. If we assess a standard normal prior, we can write the posterior as

$$\begin{aligned} \pi(\beta|X, y) &\propto \ell(\beta|X, y)\pi(\beta), \\ \pi(\beta) &\propto |A|^{1/2} \exp\{-\tfrac{1}{2}(\beta - \bar{\beta})' A(\beta - \bar{\beta})\}. \end{aligned} \tag{3.11.1}$$

The likelihood in (3.11.1) is just the product of the probabilities of the observed choices or discrete outcomes over the n observations:

$$\begin{aligned} \ell(\beta|X, y) &= \prod_{i=1}^{n} \Pr(y_i = j|X_i, \beta), \\ \Pr(y_i = j|X_i, \beta) &= \frac{\exp(x_{ij}'\beta)}{\sum_{j=1}^{J} \exp(x_{ij}'\beta)}. \end{aligned} \tag{3.11.2}$$

y is a vector with the choices $(1, \ldots, J)$ and X is an $nJ \times k$ matrix of the values of the X variables for each alternative on each observation.

Experience with the MNL likelihood is that the asymptotic normal approximation is excellent. This suggests that Metropolis algorithms based on the asymptotic approximation will perform extremely well. We also note that the MNL likelihood has exponential tails (even without the normal prior) and this should provide very favorable theoretical convergence properties. We implement both an independence and an RW Metropolis algorithm for the MNL model. Both Metropolis variants use the asymptotic normal approximation

$$\pi(\beta|X, y) \dot{\propto} |H|^{1/2} \exp\left\{\frac{1}{2}(\beta - \hat{\beta})' H(\beta - \hat{\beta})\right\}. \tag{3.11.3}$$

We have a number of choices for $\hat{\beta}$, H. We can simply use the MLE for $\hat{\beta}$ or we could find the posterior mode (preferable for truly informative priors) at about the same computational cost. H could be minus the actual Hessian of the likelihood (alternatively, the posterior) evaluated at $\hat{\beta}$ or the sample information matrix. Alternatively, we can use expected sample information which can be computed for the MNL:

$$H = -\mathrm{E}\left[\frac{\partial^2 \log \ell}{\partial\beta\partial\beta'}\right] = \sum_i X_i A_i X_i',$$

$$X = \begin{bmatrix} X_1 \\ \vdots \\ X_n \end{bmatrix}, \qquad A_i = \mathrm{diag}(p_i) - p_i p_i'. \tag{3.11.4}$$

p_i is a J-vector of the probabilities for each alternative for observation i.

We illustrate the functioning of Metropolis algorithms using a small sample of simulated data, chosen to put the asymptotic approximations to a severe test, with $N = 100, J = 3$, and $k = 4$ (two intercepts and two independent variables that are produced by iid Unif(0,1) draws). We set the beta vector to $(-2.5, 1.0, 0.7, -0.7)$. The first two elements of β are intercepts for alternatives 2 and 3 that are expressed (in the usual manner) relative to alternative 1 which is set to 0. These parameter settings imply that the probability of alternative 2 will be very small. In our simulated sample, we observed alternative one 26 times, alternative two only 2 times, and alternative three 72 times.

To implement the independence Metropolis, we use a multivariate student candidate sampling distribution. That is, we draw candidate parameter vectors using $\beta \sim$ MSt $(\nu, \hat{\beta}, H^{-1})$. The only 'tuning' required is the choice of ν. Too small values of ν will be inefficient in the sense of producing such fat tails that we will reject draws more often than for smaller values of ν. In addition, very small values of ν, such as 4 or less, produce a distribution which has fat tails but is also very 'peaked' without the 'shoulders' of the normal distribution. This also implies that the Metropolis algorithm would suffer inefficiencies from repeating draws to build up mass on the shoulders of the peaked t distribution. Clearly, one could tune the independence Metropolis by picking ν so as to maximize a numerical efficiency estimate or minimize f in (3.10.7). Using the mean of the parameters as the function whose numerical efficiency should be assessed, we find that numerical efficiency is relatively flat in the range of 5 to 15. Very small values of ν result in only slightly reduced numerical efficiency. All results reported here are for

R $\nu = 6$. `rmnlIndepMetrop` provides the R implementation of this algorithm (available in `bayesm`).

The RW Metropolis must be scaled in order to function efficiently. In particular, we propose β values using the equation

$$\beta_{\text{cand}} = \beta_{\text{old}} + \varepsilon, \qquad \varepsilon \sim N(0, s^2 H^{-1}). \tag{3.11.5}$$

According to the Roberts and Rosenthal guidelines, s values close to $2.93/\sqrt{d} = 2.93/2$ should work well. Given the accuracy of the normal approximation, we might expect the Roberts and Rosenthal guidelines to work very well since we can always transform the asymptotic normal approximation into a product of identical normal densities. However, the result is 'asymptotic' in d. Figure 3.14 shows numerical efficiencies as measured by the square root of f as a function of the scaling constant s. We use the square root of f as we are interested in minimizing the numerical standard error and the square root of f is the multiple of the iid standard error. Each curve in Figure 3.14 shows numerical efficiency for each of the four parameters and 'tuning' runs of 20,000 iterations. We consider only the estimation of the posterior mean in Figure 3.14. The small inset figure shows the acceptance rate as a function of s. As s increases, we expect the acceptance rate to decline as the chain navigates more freely. However, the numerical efficiency is ultimately the more relevant criterion. Numerical efficiency is maximized at around 1.25 which is close to the Roberts and Rosenthal value of 1.47. It is worth noting that minimum numerical efficiency of the RW Metropolis is around 4.

In comparison to the RW, the independence Metropolis functions much more efficiently with an acceptance rate of 0.70 and a numerical efficiency of 1.44 or barely less than iid sampling. Figure 3.15 shows the estimated posterior distribution of β_1 constructed from 50,000 draws of the independence and RW Metropolis chains. To the right of the distribution are the corresponding ACFs. Differences between the ACFs of the RW and independence chains result in a numerical efficiency ratio of 3:1. The gray line on the distributions is the diffuse normal prior. The solid density curve is the asymptotic normal approximation to the posterior. Even in this extreme case, the actual posterior is only slightly skewed to the left from the asymptotic approximation.

For highly correlated chain output, there is a practice of 'thinning' the output by selecting only every mth draw. The hope is that the "thinned" output will be approximately iid so that standard formulas can be used to compute numerical standard errors. Not only is this practice unnecessary in the sense that the investigator is literally

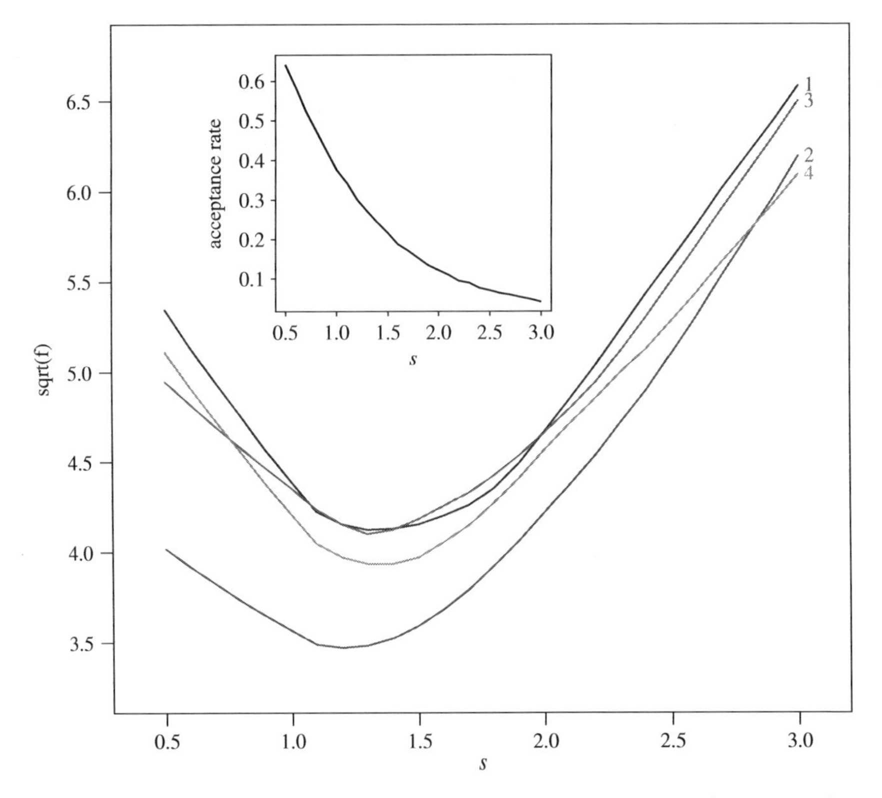

Figure 3.14 Optimal scaling of the RW Metropolis for the MNL logit example

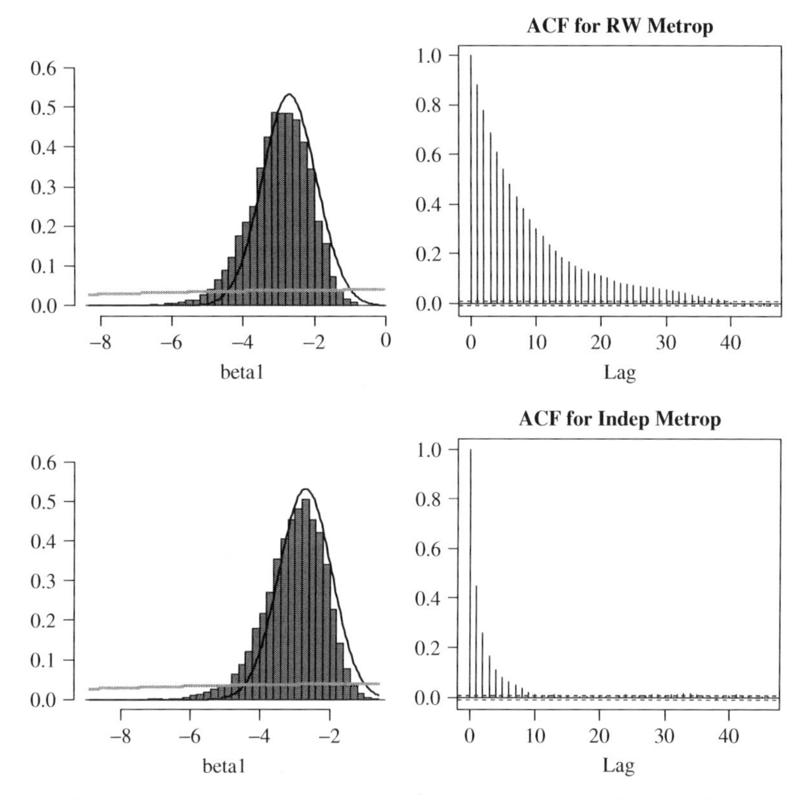

Figure 3.15 Comparison of RW and independence Metropolis for MNL example

throwing away information, but it can be misleading. In the case of the RW chain here, keeping every 10th observation (a very standard practice) will still produce a chain with nontrivial autocorrelation. The only possible reason to thin output is for convenience in storage. Given that correct numerical standard errors are trivial to compute, it seems odd that this practice continues.

3.12 HYBRID MARKOV CHAIN MONTE CARLO METHODS

In practice, many problems are solved with a 'Gibbs-style' strategy for constructing the Markov chain. In particular, hierarchical models have a structure that allows for an efficient strategy which involves various conditional draws. This is because of conditional independence assumptions which are employed in constructing hierarchical models. However, in many cases, direct draws from the requisite conditional distribution are not available and some combination of a Gibbs-style chain and Metropolis steps must be used. For example, consider the hierarchical structure given below:

$$\theta_2 \rightarrow \theta_1 \rightarrow \text{Data} \tag{3.12.1}$$

The Gibbs sampler for (3.12.1) would consist of

$$\theta_2|\theta_1 \tag{3.12.2}$$

and

$$\theta_1|\theta_2, \text{Data}. \tag{3.12.3}$$

If conjugate priors are used, the draw defined by (3.12.2) is often a direct one-for-one draw from the conditional distribution. However, for many models the draw in (3.12.3) can be difficult to achieve. One possible strategy would be to break down the draw of the θ_1 vector into one-by-one draws of each of its elements using griddy Gibbs, sliced sampling or adaptive rejection Metropolis sampling (ARMS; see Gilks *et al.* 1995). However, in many real problems, the dimension of θ_1 will be large so that this is not practical.

A useful idea is to replace the 'Gibbs' draw in (3.12.3) with a Metropolis step. This algorithm is called by some a 'hybrid' chain or 'Metropolis within Gibbs'. To construct the Metropolis chain all that is required is to evaluate the conditional density in (3.12.3) up to a normalizing constant. Of course, the joint density evaluated at the draw of θ_2 is proportional to the conditional density and can be easily evaluated:

$$p(\theta_1|\theta_2, \text{Data}) \propto p(\text{Data}|\theta_1, \theta_2)\text{p}(\theta_1, \theta_2).$$

If we implement a Metropolis step with the conditional distribution, $\theta_1|\theta_2$, Data, as its invariant distribution, then it can be shown that the hybrid chain has the full posterior as its equilibrium distribution. To see this, we first return to a discussion of the Gibbs sampler and develop the notion of the composition of two Markov chains.

The standard 'two-stage' Gibbs sampler can be written:

$$\theta_1|\theta_2, \text{Data} \tag{3.12.4}$$

and

$$\theta_2|\theta_1, \text{Data}. \tag{3.12.5}$$

The Markov chain that (3.12.4) and (3.12.5) represent can be thought of as the combination of two chains, each one of which updates one of the components of θ. If we denote K_1^{G} as the kernel of a chain formed by the conditional distribution in (3.12.4) and K_2^{G} as the kernel of the chain formed by (3.12.5), then the two-stage Gibbs sampler is the composition of these two chains:

$$K^{\text{G}} = K_1^{\text{G}} \circ K_2^{\text{G}}. \tag{3.12.6}$$

The kernel of the composed chain is

$$\begin{aligned} K(\theta, A) &= \int_A p(\theta, \varphi)\,\mathrm{d}\varphi, \\ p(\theta, \varphi) &= p((\theta_1, \theta_2), (\varphi_1, \varphi_2)) = \pi_{1|2}(\varphi_1|\theta_2)\pi_{2|1}(\varphi_2|\varphi_1). \end{aligned} \tag{3.12.7}$$

While each of the subchains is reducible (that is, they only navigate on the subspace formed by holding one component of θ fixed), the full chain is irreducible as we have seen before. What is less obvious is that the full posterior is the invariant distribution of both K_1^G and K_2^G, viewed separately. To see this, consider K_1^G. If we start with a draw from the joint posterior, we must then show that the one iteration of K_1^G reproduces the joint posterior. The joint posterior factors into the conditional and marginal densities:

$$\pi(\theta_1, \theta_2|\text{Data}) = \pi_2(\theta_2|\text{Data})\pi_{1|2}(\theta_1|\theta_2, \text{Data}).$$

This means that the 'initial' value is a draw that can always be viewed as obtained by first drawing from the appropriate marginal and then drawing from the conditional distribution:

$$\theta_2^0 \sim \pi_2 \quad \text{and} \quad \theta_1^0 \sim \pi_{1|2}.$$

To show that K_1^G produces a draw from the joint posterior, we observe that

$$\theta_1^1 \sim \pi_{1|2}(\theta_2^0).$$

The notation $\pi_{1|2}(\theta_2^0)$ means the conditional distribution of $\theta_1|\theta_2 = \theta_2^0$. Thus, the draw from K_1^G is also a draw from the joint distribution

$$(\theta_1^1, \theta_2^0) \sim \pi_{1|2}\pi_2 = \pi_{1,2}.$$

The same argument applies to K_2^G and we can see that the full two-stage Gibbs sampler is irreducible and has the joint posterior as its variant distribution.

The hybrid chain formed by substituting a Metropolis step for first step in the two-stage Gibbs sampler is also the composition of two reducible chains, each having the joint posterior as its invariant distribution, $K^H = K_1^M \circ K_2^G$. This follows directly by exactly the same argument that we applied to the Gibbs sampler. There is one additional important point. The Metropolis step will update the value of θ_1 given the previous iteration value of θ_2. This means that the values of θ_2 from the previous iteration can be used in the computation of candidate or proposal values of θ_1 in the Metropolis step. As long as we obey the proper conditioning and use only values of the *last* iteration, we will still have a Markov chain with the correct invariant distribution. This means that we can 'automatically' adjust the Metropolis candidate sampling density depending on the last value of θ_2. We should also note that this is not the same as 'adaptive' schemes that use information from a past subsequence to adjust the Metropolis step. These schemes are not Markovian and require additional analysis to establish that they have the proper invariant distribution.

3.13 DIAGNOSTICS

The theoretical properties of MCMC methods are quite appealing. For many algorithms and problems, especially with normal priors, it is easy to establish ergodicity and even stronger results such as geometric rates of convergence and uniform ergodicity. The problem with these rates of convergence results is that they do not specify the constants that govern the actual rate of convergence. In practice, we can produce an MCMC sampler which has desirable theoretical properties but poor performance given our finite computer resources. In addition, errors in formulating the model or the MCMC algorithm can be very important and difficult to diagnose.

The performance of MCMC algorithms is related to the speed at which the chain navigates the state space. Highly autocorrelated samplers require long periods of time to navigate the parameter space fully enough to properly represent the level of uncertainty in the posterior. A related problem is the dissipation of initial conditions. In practice, we start our samplers from reasonably arbitrary points and hope that they dissipate the effect of these initials after a burn-in period. Draws after the burn-in period are viewed as draws from the stationary distribution. Clearly the speed of navigation and dissipation of initial conditions are related. Near iid MCMC samplers will both dissipate initial conditions and rapidly navigate the parameter space. However, in practice we see examples of samplers that dissipate the initial conditions rapidly but are slow to navigate regions of high posterior mass. For example, if we consider the bivariate normal Gibbs sampler, high (much higher than 0.95) correlation will create a narrow ridge which the Gibbs sampler may require many iterations to navigate. However, given the thin tails of the normal distribution, if we start the sampler at a point far from this ridge, the sampler will quickly move to the ridge and dissipate the initial condition. The classic 'witch's hat' examples are examples where a Gibbs sampler can get stuck in one mode of a very peaked distribution and fail to break out to the other mode even though this is theoretically possible. In our experience with models relevant to micro data and marketing applications, these situations are rarely encountered. More common are situations in which the MCMC sampler navigates slowly with a high degree of autocorrelation. In these situations, it may take days of computing to properly navigate the posterior.

Unfortunately, there is a sense in which there can be no powerful diagnostic or 'test' for failure of convergence without substantial prior knowledge of the posterior. If the asymptotic approximation to the posterior is poor and/or if the parameter space is of high dimensions, then we are engaging in MCMC sampling in order to learn about the posterior. This means that we can never be very sure that we have gotten the right result. Many of the proposals for convergence diagnostics utilize a subsequence of an MCMC run in order to assess convergence. This requires confidence that the sampler has already navigated the relevant regions of the parameter space and the issue is one of the information content of the sampler sequence rather than whether or not navigation is complete. The information content of the sequence can easily be gauged by computing numerical standard errors for quantities of interest using the formula in (3.10.7). This does not address the convergence question. Other proposals in the convergence diagnostics area involve starting the sampler from a number of initial conditions (Gelman and Rubin 1992) to check for dissipation of initial conditions. Again, choice of these initial conditions requires information about the posterior. In problems where the parameter space is more than a few dimensions, it may be impractical to choose a comprehensive set of initial conditions. Others have suggested using multiple 'parallel' runs of the MCMC sampler to compare the distribution obtained within one run with the cross-section simulation of many runs. This is not practical in the large-scale problems considered here.

In practice, we rely on sequence or time series plots of MCMC output as well as autocorrelations and associated standard errors to monitor convergence. The assumption is that a slowly navigating chain (one with near RW behavior) must be monitored very closely for possible convergence failure. These highly autocorrelated chains should be run as long as possible to investigate the sensitivity of the estimation of the posterior variability to run length. In many situations, the MCMC sampler may exhibit dependence of an extremely long-lasting or persistent variety, often with autocorrelations that do not damp off quickly. In these situations, more attention should be focused on monitoring convergence. We have found that one of the most effective ways to assess convergence properties is to conduct sampling experiments that have been calibrated to match the characteristics of the actual data under analysis. While large-scale sampling experiments may not be possible due to the size of the parameter space or due to computing limitations, we have found that a small-scale experiment can be very useful.

In most situations, slowly navigating chains are produced by high correlation between groups of parameters. The classic example of this is the normal Gibbs sampler. Other examples include latent variables added for data augmentation. Here high correlation between the latent variables and the model 'parameters' can produce slowly navigating chains. There are two possible solutions to high correlation between two subsets of parameters. If the two subsets can be 'blocked' or combined into one iid draw, then the autocorrelation can be reduced substantially. However, care must be exercised in blocking as there are examples where the blocked sampler gets stuck. If one set of parameters is not of direct interest and can be integrated out (sometimes called 'collapsing'), then the autocorrelation can also be reduced – see Liu (2001) for a theoretical comparison of collapsed and blocked samplers. We will see an example of this technique in Case Study 3. Of course, whether these are practical strategies depends on the model and priors and our ability to perform the integration. Finally, we have seen that, with a good proposal density, the independence Metropolis can outperform the RW.

One neglected, but important, source of concern are errors made in the formulation and coding of an MCMC method. One can easily formulate an MCMC sampler with the wrong posterior for the model under consideration. In addition, errors can be made in the coding of the densities used in Metropolis methods or in the draws from the conditional distributions. While simulation experiments will often detect the presence of these errors, Geweke (2004) has proposed an additional useful check on the formulation of MCMC samplers. The idea is to simulate from the joint distribution of the observables and the parameters in two independent ways. One way will use the MCMC method to be tested and the other will simply require drawing from the prior. Draws made in these two ways can then be compared to see if they are similar.

To draw from the joint distribution of the observations and parameters, we simply need the ability to simulate from the prior and the model:

$$p(y, \theta) = p(\theta)p(y|\theta). \tag{3.13.1}$$

Independent and identically distributed draws from this joint density can be achieved by drawing from the prior and then drawing $y|\theta$ from a model simulator. Code to simulate from the prior and the model is not usually needed to implement the MCMC method.[27] This is the sense in which coding errors in the MCMC method could be independent of coding errors in the direct draws from (3.13.1). To indirectly draw from the joint distribution of the observables and θ, we can use the MCMC method to draw from the posterior in conjunction with the model simulator. These two draws can be used to construct a hybrid chain with the joint distribution of (y, θ) as the invariant distribution.

EXTENDED SAMPLER

Start from some value θ^0.

Draw $y^1|\theta^0$ using the model simulator.

Draw $\theta^1|y^1$ using the MCMC method to be tested.

Repeat, as necessary.

Denote the sequence of iid draws as $\{y^r, \theta^r\}$ and the sequence from the hybrid sampler as $\{\hat{y}^r, \hat{\theta}^r\}$. Geweke suggests that we compare $\overline{g} = R^{-1}\sum_r g(y^r, \theta^r)$ with $\overline{\overline{g}} = R^{-1}\sum_r g(\hat{y}^r, \hat{\theta}^r)$ as a diagnostic check on the MCMC algorithm. Differences between $\overline{g}$ and $\overline{\overline{g}}$ could represent conceptual errors in the MCMC algorithm which result in a different invariant distribution than the posterior of this model and prior or coding errors in implementing various aspects of the MCMC algorithm. Clearly, the power of this diagnostic to detect errors depends on the scope of the g function. There will be a trade-off between power for certain types of errors and ability to perform omnibus duty against a wide spectrum of errors. In particular, g does not have to be a function of the observables at all. In many contexts, investigators use very diffuse, but proper priors, and these prior hyperparameter settings will result in low power for the Geweke method. In order to obtain better power, it is useful to tighten up the priors by using hyperparameter settings very different from those used in analysis of data. Finally, we prefer to compare the entire distribution rather than the particular moments implied by the g function (at the risk of focusing on univariate marginals). For example, Q–Q plots using elements of (y, θ) drawn by the two methods can be very useful.

[27] For conditionally conjugate models, draws from the prior are usually achieved by using the same functions required to draw from the conditional posteriors (for example, normal and Wishart draws). Thus, the Geweke method will not have 'power' to detect errors in these simulation routines. However, simulating from the model is truly an independent coding task.

4

Unit-Level Models and Discrete Demand

Using this Chapter

This chapter reviews models for discrete data. Much of the disaggregate data collected in marketing has discrete aspects to the quantities of goods purchased. Sections 4.1–4.3 review the latent variable approach to formulating models with discrete dependent variables, while Section 4.4 derives models based on a formal theory of utility maximization. Those interested in multinomial probit or multivariate probit models should focus on Sections 4.2 and 4.3. Section 4.2.1 provides material on understanding the difference between various Gibbs samplers proposed for these models and can be omitted by those seeking a more general appreciation. Section 4.4 forges a link between statistical and economic models and introduces demand models which can be used for more formal economic questions such as welfare analysis and policy simulation.

We define the 'unit-level' as the lowest level of aggregation available in a data set. For example, retail scanner data is available at many levels of aggregation. The researcher might only have regional or market-level aggregate data. Standard regression models can suffice for this sort of highly aggregated data. However, as the level of aggregation declines to the consumer level, sales response becomes more and more discrete. There are a larger number of zeros in this data and often only a few integer-valued points of support. If, for example, we examine the prescribing behavior of a physician over a short period of time, this will be a count variable. Consumers often choose to purchase only a small number of items from a large set of alternatives. The goal of this chapter is to investigate models appropriate for disaggregate data. The common characteristic of these models will be the ability to attach lumps of probability to specific outcomes. It should also be emphasized that even if the goal is to analyze only highly aggregate data, the researcher could properly view this data as arising from individual-level decisions aggregated up to form the data observed. Thus, individual-level demand models and models for the distribution of consumer preferences (the focus of Chapter 5) are important even if the researcher only has access to aggregate data.

Bayesian Statistics and Marketing P. E. Rossi, G. M. Allenby and R. McCulloch

4.1 LATENT VARIABLE MODELS

We will take the point of view that discrete dependent variables arise via partial observation of an underlying continuous variable. For example, the binary probit model discussed in Chapter 3 was formulated as a binary outcome which is simply the sign (0 if negative, 1 if positive) of the dependent variable in a normal regression model. There are three advantages of a latent variable formulation. First, the latent variable formulation is very flexible and is capable of generating virtually any sort of outcome that has a discrete component. Second, the latent variable formulation allows for easy formulation of MCMC algorithms using data augmentation. Third, in some situations, the latent variable can be given a random utility interpretation which relates latent variable models to the formal econometric specification of demand models based on utility maximization.

We can start with a normal regression model as the model for the latent variables:

$$z_i = X_i\delta + \nu_i. \tag{4.1.1}$$

The outcome variable, y_i, is modeled as a function of the latent variable z_i:

$$y_i = f(z_i). \tag{4.1.2}$$

In order to create a discrete y, the function f must be constant over some regions of the space in which z resides. We distinguish between the cases in which z is univariate and multivariate. For the case of univariate z, standard models such as the binary logit/probit and ordered probit models can be obtained. The binary probit model assumes $f(z) = I(z > 0)$ and ν has a normal distribution. $I(\cdot)$ denotes an indicator function; $I(x) = 1$ if condition x is met, 0 otherwise. The binary logit model assumes the same f function, but uses an extreme value type I distribution for ν.

Ordered models can be obtained by taking f to be an indicator function of which of several intervals z resides in:

$$f(z) = \sum_{c=1}^{C+1} c \times I(\gamma_{c-1} < z \leq \gamma_c). \tag{4.1.3}$$

y is multinomial with values $1, \ldots, C$, and $\gamma_0 = -\infty$ and $\gamma_{C+1} = \infty$. The ordinal nature of y is enforced by the assumption of a unidimensional latent variable. As $x_i'\delta$ increases, the probability of obtaining higher values of y increases. By allowing for arbitrary cut-off points, the ordered model has flexibility to allow for different probabilities of each integer value of y conditional on x. For convenience in implementation of MCMC algorithms, we typically use an ordered probit model in which the latent variable is conditionally normal. As with the binary logit/probit models there is little difference between ordered probits and models obtained by other distributional assumptions on ν.

In other situations, y is multinomial but we do not wish to restrict the multinomial probabilities to be monotone in $x_i'\delta$. In these cases, the underlying latent structure must be multivariate in order to provide sufficient flexibility to accommodate general patterns of the influence of the X variables on the probability that y takes on various values. That is, if we increase X, this may increase the probability of some value of y and decrease the probability of others in no particular pattern. Consider the multinomial case in which y can assume one of p values associated with the integers $1, \ldots, p$. In this case, we assume that z has a continuous, p-dimensional distribution. A convenient choice of f is the indicator of the component with the maximum value of z:

$$f(z) = \sum_{j=1}^{p} j \times I(\max(z) = z_j). \tag{4.1.4}$$

The multinomial distribution is appropriate for choice situations in which one of p alternatives is chosen. However, in some situations, multiple choices are a possible outcome. For example, we may observe consumers purchasing more than one variety of a product class on the same purchase occasion. In a survey context, a popular style of question is the 'm choose k' in which m alternatives are given and the questionnaire requests that the respondent 'check all that apply'. For these situations, a multivariate discrete choice model is appropriate. We represent this situation by allowing y to be p-dimensional, with each component having only two values: $y = (y_1, \ldots, y_p)$, $y_i = 0, 1$. For this model, we must also use a multivariate latent variable and this becomes the direct generalization of the 'univariate' binary model:

$$f : \mathbb{R}^p \to \mathbb{R}^p, \qquad f(z) = \begin{pmatrix} I(z_1 > 0) \\ \vdots \\ I(z_p > 0) \end{pmatrix}. \tag{4.1.5}$$

In some contexts, the response variable might reasonably be viewed as consisting of both discrete and continuous components. For example, we may observe both product choice and quantity. On many occasions, the consumer does not purchase the product but on some occasions a variable quantity is purchased. The quantity purchased is also discrete in the sense that most products are available in discrete units. However, we might assume that a continuous variable is a reasonable approximation to the quantity demanded conditional on positive demand. These situations can also be accommodated in the latent variable framework. For example, we can specify a simple tobit model as

$$f(z) = z \times I(z > 0). \tag{4.1.6}$$

However, we will defer a formal discussion of the mixed discrete–continuous models until the introduction of a utility maximization model and a random utility framework.

Each of the models introduced in (4.1.1)–(4.1.5) has associated identification problems. These problems stem from the invariance of the f functions to location or scale transformations of z, that is, $f(cz) = f(z)$ or $f(z + k) = f(z)$. In the binary models, there is a scaling problem for the error, ν. Changing the variance of ν will not change the likelihood as multiplication of z by a positive constant in (4.1.2) will not alter y. Typically, this is solved by imposing a scale restriction such as $\text{var}(\nu) = 1$. In the ordered model in (4.1.3), there is both a scaling and location problem. Even if we fix the scale of ν, we can add a constant to all of the cut-offs and subtract the same constant from the intercept in δ and leave the likelihood unchanged. For this reason, investigators typically set the first cut-off to zero, $\gamma_1 = 0$. In the multinomial model (4.1.4), there is both a location and scale problem. If we add a constant to every component of z or if we scale the entire z vector by a positive constant, then the likelihood is unchanged.

We will discuss identification of multinomial models in Section 4.2. For the multivariate model in (4.1.5), we recognize that we can scale each of the components of z by a *different* positive amount and leave the likelihood unchanged. This means that only the correlation matrix of z is identified.

4.2 MULTINOMIAL PROBIT MODEL

If latent utility is conditionally normal and we observe outcome value j (of p) if $\max(z) = z_j$, then we have specified the multinomial probit (MNP) model:

$$\begin{aligned} y_i &= f(z_i), \\ f(z_i) &= \sum_{j=1}^{p} j \times I(\max(z_i) = z_{ij}), \qquad (4.2.1) \\ z_i &= X_i\delta + \nu_i, \qquad \nu_i \sim \text{iid } N(0, \Omega). \end{aligned}$$

If the multinomial outcome variable y represents consumer choice among p mutually exclusive alternatives, then the X matrix consists of information about the attributes of each of the choice alternatives as well as covariates which represent the characteristics of the consumer making choices.

The general X would have the structure $X_i = [(1, d_i') \otimes I_p, A_i]$, where d is a vector of 'consumer' characteristics and A is a matrix of choice attributes. `createX`, in *bayesm*, can be used to create the X array with this structure. For example, we might have information on the price of each alternative as well as the income of each consumer making choices. Obviously, we can only identify the effects of consumer covariates with a sample taken across consumers. In most marketing contexts, we obtain a sample of consumers in panel form. We will consider modeling differences between consumers in Chapter 5. In a modern hierarchical approach, covariates are included in the distribution of parameters across consumers and not directly in the X matrix above. For expositional purposes, we will consider the case in which X contains only alternative specific information. In typical applications, δ contains alternative-specific intercepts as well as marketing mix variables for each alternative. It is common to assume that the coefficient on the marketing mix variables is the same across choice alternatives. The random utility derivation of multinomial choice models provides a rationale for this which we will develop in Section 4.4. The linear model for the latent variable in (4.2.1) is a SUR model, written in a somewhat nonstandard form, and with restrictions that some of the regression coefficients are the same across regressions. For example, with a set of intercepts and one marketing mix variable, $X_i = [\, I_p \quad m_i \,]$, where I_p is a $p \times p$ identity matrix and m is a vector with p values of one marketing mix variable for each of p alternatives.

As indicated in Section 4.1, the model in (4.2.1) is not identified. In particular, if we add a scalar random variable to each of the p latent regressions, the likelihood remains unchanged. That is, $\text{var}(z_i|X_i, \delta) = \text{var}(z_i + u\iota|X_i, \delta)$ where u is a scalar random variable and ι is vector of ones. This is true only if Ω is unrestricted. If, for example, Ω is diagonal, then adding a scalar random variable will change the covariance structure by adding correlation between choice alternatives and the elements of the diagonal of Ω will be identified. In the case of unrestricted Ω, it is common practice to subtract the

pth equation from each of the first $p-1$ equations to obtain a differenced system. The differenced system can be written

$$w_i = X_i^{\mathrm{d}}\beta + \varepsilon_i, \qquad \varepsilon_i \sim N(0, \Sigma),$$

$$w_{ij} = z_{ij} - z_{ip},\ X_i = \begin{bmatrix} x'_{i1} \\ \vdots \\ x'_{ip} \end{bmatrix},\ X_i^{\mathrm{d}} = \begin{bmatrix} x'_{i1} - x'_{ip} \\ \vdots \\ x'_{i,p-1} - x'_{ip} \end{bmatrix},\ \varepsilon_{ij} = \nu_{ij} - \nu_{ip}, \tag{4.2.2}$$

$$y_i = f(w) = \sum_{j=1}^{p-1} j \times I(\max(w_i) = w_{ij} \text{ and } w_{ij} > 0) + p \times I(w < 0).$$

We also note that if δ contains intercepts, then the intercept corresponding to the pth choice alternative has been set to zero and β contains all of the other elements of δ.

The system in (4.2.2) is still not identified as it is possible to multiply w by a positive scalar and leave the likelihood of the observed data unchanged – in other words, $f(w) = f(cw)$. This identification problem is a normalization problem. We have to fix the scale of the latent variables in order to achieve identification. It is common to set $\sigma_{11} = 1$ to fix the scale of w. However, it is also possible to achieve identification by setting one of the components of the β vector to some fixed value. For example, if price were included as a covariate, we might fix the price coefficient at -1.0. Of course, this requires a prior knowledge of the sign of one of the elements of the β vector. In the case of models with price, we might feel justified in imposing this exact restriction on the sign of the price coefficient, but in other cases we might be reluctant to make such an imposition.

In classical approaches to the MNP, the model in (4.2.2) is reparameterized to the identified parameters $(\tilde{\beta}, \tilde{\Sigma})$ by setting $\sigma_{11} = 1$. In a Bayesian approach, it is not necessary to impose identification restrictions. With proper priors, we can define the posterior in the unidentified space, construct an MCMC method to navigate this space and then simply 'margin down' or report the posterior distribution of the identified quantities. In many cases, it is easier to define an MCMC method on the unrestricted space. Moreover, it turns out that the unrestricted sampler will often have better mixing properties.

In Section 3.8, we saw how data augmentation can be used to propose a Gibbs sampler for the binary probit model. These same ideas can be applied to the MNP model. If β, Σ are a priori independent, the DAG for the MNP model is given by

$$\begin{matrix} \Sigma \searrow & & \\ & w \longrightarrow y & \\ \beta \nearrow & & \end{matrix} \tag{4.2.3}$$

McCulloch and Rossi (1994) propose a Gibbs sampler based on cycling through three conditional distributions:

$$\begin{gathered} w|\beta, \Sigma, y, X_i^{\mathrm{d}} \\ \beta|\Sigma, w \\ \Sigma|\beta, w \end{gathered} \tag{4.2.4}$$

This sampler produces a chain with w, β, $\Sigma|y$ as its stationary distribution. We can simply marginalize out w if the posterior of β, Σ is desired. We note that w is sufficient for β, Σ and this is why it is not necessary to include y as a conditioning argument in the second and third distributions in (4.2.4). Since the underlying latent structure is a normal multivariate regression model, we can use the standard theory developed in Chapter 2 to draw β and Σ. The only difficulty is the draw of w. Given β, Σ, the $\{w_i\}$ are independent, $(p-1)$-dimensional truncated normal random vectors. The regions of truncation are the R_{y_i}, as defined below in (4.2.11). Direct draws from truncated multivariate normal random vectors are difficult to accomplish efficiently. The insight of McCulloch and Rossi (1994) is to recognize that one can define a Gibbs sampler by breaking each draw of w_i into a sequence of $p-1$ univariate truncated normal draws by cycling through the w vector.

That is, we do not draw $w_i|\beta$, Σ directly. Instead, we draw from each of the $p-1$ truncated univariate normal distributions:

$$\begin{aligned} w_{ij}|w_{i,-j}, y_i, \beta, \Sigma &\sim N(m_{ij}, \tau_{jj}^2) \\ &\quad \times [I(j = y_i)I(w_{ij} > \max(w_{i,-j}, 0)) + I(j \neq y_i)I(w_{ij} < \max(w_{i,-j}, 0))], \\ m_{ij} &= x_{ij}^{d'}\beta + F'(w_{i,-j} - X_{i,-j}^{d}\beta), \\ F &= -\sigma^{jj}\gamma_{j,-j}, \\ \tau_{jj}^2 &= 1/\sigma^{jj}, \end{aligned} \tag{4.2.5}$$

where σ^{ij} denotes the (i,j)th element of Σ^{-1} and

$$\Sigma^{-1} = \begin{bmatrix} \gamma_1' \\ \vdots \\ \gamma_{p-1}' \end{bmatrix}. \tag{4.2.6}$$

The univariate distribution in (4.2.5) is a truncated normal distribution. $\gamma_{j,-j}$ refers to the jth row of Σ^{-1} with the jth element deleted. $X_{i,-j}^{d}$ is the matrix X_i^d with the jth column deleted. We start with the first element ($j = 1$) and 'Gibbs through' each observation, replacing elements of w, one by one, until the entire vector is updated. In

R *bayesm*, this is done in C code. To implement this sort of sampler in R, `condMom` can be useful to compute the right moments for each univariate truncated normal draw.

To implement the full Gibbs sampler for the MNP model, we need to specify a prior over the model parameters, β, Σ. The model parameters are not identified without further restrictions due to the scaling problem for the MNP model. The identified parameters are obtained by normalizing with respect to one of the diagonal elements of Σ:

$$\tilde{\beta} = \beta/\sqrt{\sigma_{11}}, \qquad \tilde{\Sigma} = \Sigma/\sigma_{11}. \tag{4.2.7}$$

For ease of interpretation, we will report the correlation matrix and vector of relative variances as the set of identified covariance parameters. One approach would put a prior on the full set of unidentified parameters as in McCulloch and Rossi (1994),

$$\beta \sim N(\bar{\beta}, A^{-1}), \qquad \Sigma \sim \mathrm{IW}(\nu, V_0), \tag{4.2.8}$$

thus inducing a prior on the identified parameters in (4.2.7). The induced prior on the identified parameters is not in standard form.[1] Imai and van Dyk (2005) suggest a very similar prior but with the advantage that the prior induced on the identified regression coefficients is more easily interpretable:

$$\beta|\Sigma \sim N(\sqrt{\sigma_{11}}\,\bar{\beta}, \sigma_{11}A^{-1}), \qquad \Sigma \sim \mathrm{IW}(\nu, V_0). \tag{4.2.9}$$

With the prior in (4.2.9), $\tilde{\beta} \sim N(\bar{\beta}, A^{-1})$. This prior specification makes assessment of the prior more straightforward. It also allows for the use of an improper prior on the identified regression coefficients.

Another approach would be to put a prior directly on the set of identified parameters and define a Gibbs sampler on the space $(w, \tilde{\beta}, \tilde{\Sigma})$ as in McCulloch *et al.* (2000). This requires a prior on the set of covariance matrices with (1,1)th element set equal to 1. This is done via a reparameterization of Σ:

$$\Sigma = \begin{bmatrix} \sigma_{11} & \gamma' \\ \gamma & \Phi + \gamma\gamma' \end{bmatrix}. \tag{4.2.10}$$

This reparameterization is suggested by considering the conditional distribution of $\varepsilon_{-1}|\varepsilon_1 \sim N(\gamma\varepsilon_1, \Phi)$. This is a multivariate regression of ε_{-1} on ε_1 and the standard conjugate prior can be used for γ, Φ (as in Section 2.8.5). σ_{11} can be set to 1 in order to achieve identification. This creates a prior over covariance matrices with the (1,1)th element fixed at 1. However, the Gibbs sampler suggested by McCulloch *et al.* (2000) (hereafter termed the ID MNP sampler) navigates the identified parameter space more slowly than the sampler of McCulloch and Rossi (1994). For this reason, we cover the approach of McCulloch and Rossi (1994) here.

Given the prior in (4.2.8), we can define the McCulloch and Rossi (1994) sampler, which we term the nonidentified (NID) sampler, as follows:

NID MNP GIBBS SAMPLER

Start with initial values, w_0, β_0, Σ_0.

Draw $w_1|\beta_0, \Sigma_0$ using (4.2.5).

Draw $\beta_1|w_1, \Sigma_0 \sim N(\tilde{\beta}, V)$,

$$\begin{aligned} V &= (X^{*\prime}X^* + A)^{-1}, \quad \tilde{\beta} = V(X^{*\prime}w^* + A\bar{\beta}), \\ \Sigma_0^{-1} &= C'C, \\ X_i^* &= C'X_i, \; w_i^* = C'w_i, \\ X &= \begin{bmatrix} X_1 \\ \vdots \\ X_n \end{bmatrix}. \end{aligned}$$

[1] McCulloch *et al.* (2000) derive these distributions. Assessment of these priors for all but the highly diffuse case may be difficult. Under the prior in (4.2.8), $\tilde{\beta}$ and $\tilde{\Sigma}$ are not independent. In addition, the prior on $\tilde{\beta}$ can be skewed for nonzero values of $\bar{\beta}$.

Draw $\Sigma_1|w_1, \beta_1$ using $\Sigma^{-1}|w, \beta \sim W(\nu + n, (V_0 + S)^{-1})$,

$$S = \sum_{i=1}^{n} \varepsilon_i \varepsilon_i',$$

$$\varepsilon_i = w_i - X_i\beta.$$

Repeat as necessary.

R This sampler is implemented in `rmnpGibbs` in *bayesm*.

To illustrate the functioning of the NID sampler, consider a simulated example from McCulloch and Rossi (1994):

$$N = 1600, \qquad p = 6,$$

$$X \sim \text{iid Unif}(-2, 2), \qquad \beta = 2, \qquad \Sigma = \text{diag}(\sigma)(\rho\iota\iota' + (1 - \rho)I_{p-1})\text{diag}(\sigma),$$

$$\rho = 0.5, \qquad \sigma' = (1, 2, 3, 4, 5)^{1/2}.$$

Figure 4.1 shows the MCMC trace plot and ACFs for $\tilde{\beta}$ and ρ_{12}. These are achieved by normalizing the draws of the unidentified parameters appropriately. For example, to construct an MCMC estimate of the posterior distribution of $\tilde{\beta}$, we simply post-process the draws of β by dividing by the (1,1)th element of Σ. Figure 4.1 shows the trace plots for the 'thinned' draw sequence obtained by extracting every 100th draw from a draw sequence of 100,000. However, the ACFs are computed using every draw. β was started at 0 and Σ at the identity matrix. Very diffuse priors were used, $\beta \sim N(0, 100)$ and $\Sigma \sim \text{IW}(\nu = (p - 1) + 2, \nu I_{p-1})$. The MCMC sampler dissipates the initial conditions very rapidly but exhibits very high autocorrelation. One way to gauge the extent of this autocorrelation is to compute the f factor or relative numerical efficiency which is the ratio of the variance of the mean of the MCMC draws to the variance assuming an iid sample. For $\tilde{\beta}, f = 110$ and, for $\rho_{12}, f = 130$. This means that our 'sample' of 100,000 has the information content of approximately 800 ($\approx 100{,}000/f$) iid draws from the posterior. The posterior mean of $\tilde{\beta}$ is estimated to be 2.14 with a numerical standard error of 0.0084, and the posterior mean of ρ_{12} is 0.39 with a numerical standard error of 0.005. These numerical standard errors must be viewed relative to the posterior standard deviations of these quantities. If the posterior standard deviation is large, then it means that we cannot make precise inferences about these parameters and we may be willing to tolerate larger numerical standard errors. In this example, the estimated posterior standard deviations for both $\tilde{\beta}$ and ρ_{12} are about 30 times the size of the numerical standard errors.

4.2.1 Understanding the Autocorrelation Properties of the MNP Gibbs Sampler

The MNP Gibbs sampler exhibits higher autocorrelation than the MCMC examples considered in Chapter 3. As with all Gibbs samplers, high autocorrelation is created by

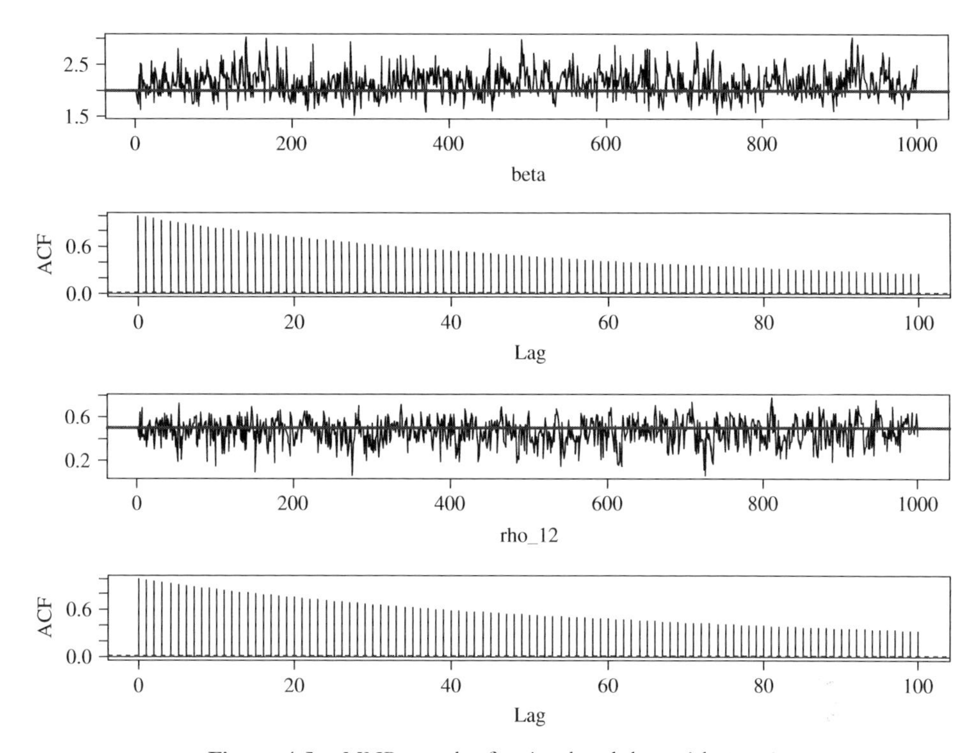

Figure 4.1 MNP sampler for simulated data with $p = 6$

dependence among the random variables being sampled. It appears that the introduction of the latent variables via data augmentation has created the problem. There must be high dependence between w and β and/or Σ. To verify and explore this intuition, consider the binary probit model which is a special case of the MNP model with $p = 2$. For this case, we can easily identify two samplers: the NID sampler,[2] and the ID sampler of Albert and Chib (see also Chapter 3). Both samplers include both w and β, but the Albert and Chib sampler sets $\Sigma = 1$ and navigates in the identified parameter space.

It is easier to develop an intuition for the high dependence between w and the model parameters in the case of the ID sampler as there is one fewer model parameter. Consider a simple case where there is only one X variable and this variable takes on only two values (0,1). In this case, all observations fall into a 2×2 contingency table defined by X and Y. The latent variables are drawn from four different truncated distributions corresponding to the cells of this table.

To see why β and w can be very highly autocorrelated consider the case depicted in Figure 4.2. If $X = 0$, then w will be sampled from the normal distribution centered at zero which is depicted by the dotted density. For $X = 0$ and $Y = 1$, w will be sampled from the positive half normal and, for $X = 0$ and $Y = 0$, from the negative half normal. If β changes, these draws of w will remain unchanged in distribution. However, consider what happens to the draws of w for $X = 1$. These draws are made from truncations of the normal distributions depicted by solid densities in Figure 4.2. If β increases, then

[2] For $p = 2$, our NID sampler is identical to that proposed by Van Dyk and Meng (2001).

	$Y = 0$	$Y = 1$
$X = 0$	w_{00}	w_{01}
$X = 1$	w_{10}	w_{11}

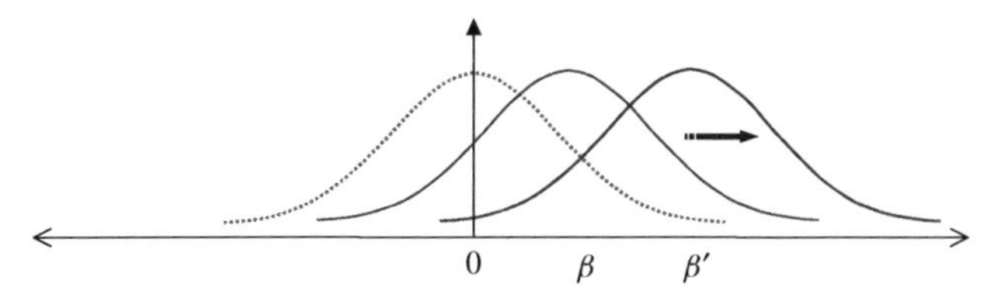

Figure 4.2 Understanding the correlation between latent variable and binary logit parameters

the draws of w will come from a normal distribution centered at a higher mean and truncated either from above by 0 or from below by 0. These draws will be larger in expectation. This sets up a situation in which if β increases, w draws tend to increase for $X = 1$ and remain unchanged for $X = 0$. In turn, the posterior mean of β will increase as the least squares estimate of $\beta|w$ will simply be the difference in mean w for $X = 1$ versus $X = 0$. The only force working against the strong dependence of w and β is the variability in the distribution of $w|X = 1, \beta$.

If $\text{var}(x'\beta)$ is large relative to $\text{var}(\varepsilon)$, then the dependence between β and w will be high. In the ID sampler, $\text{var}(\varepsilon)$ is fixed at 1. However, in the NID sampler the variance of ε is drawn as a parameter. Large values of $\text{var}(\varepsilon)$ will reduce the dependence between w and β. The NID sampler can draw large values of $\text{var}(\varepsilon)$, particularly for very diffuse priors. Van Dyk and Meng (2001) observe that it would be optimal to set an improper prior for Σ in terms of the properties of the NID sampler. However, both the NID and ID samplers can exhibit high dependence between draws. Both samplers will perform poorly when $\text{var}(x'\beta)$ is large. This may be disconcerting to the reader – when the X variables are highly predictive of choice, the performance of both samplers degrades.

To illustrate the dependence between β and w, we consider some simple simulated examples. X is one column which is created by taking draws from a Bernoulli(0.5) distribution, with $N = 100$, $\Sigma = 1$. We will simulate data corresponding to an increasing sequence of β values. The intuition is that, as β increases, we should observe an increase in dependence between w_i draws for $X = 1$, $Y = 1$ and β. Figure 4.3 plots the sample mean of the w draws for $X = 1$, $Y = 1$ against the β draw for the same ID MCMC iteration. Four simulated data sets with $\beta = (1, 2, 3)$ were created. This correlation exceeds 0.99 for $\beta = 3$. Figure 4.4 provides the ACFs for the draw sequence of β for each of the simulated data sets. The dependence between w and β results in extremely high autocorrelation for large values of β. It should be pointed out that a β value of 3 or larger is fairly extreme as this implies that $\Pr(Y = 1|X = 1) > 0.99$.[3]

[3] It should be noted that $\Pr(Y = 0|X = 0) = 0.5$. This means that we are not in the situation in which the observations can be separated and the posterior with flat priors becomes improper.

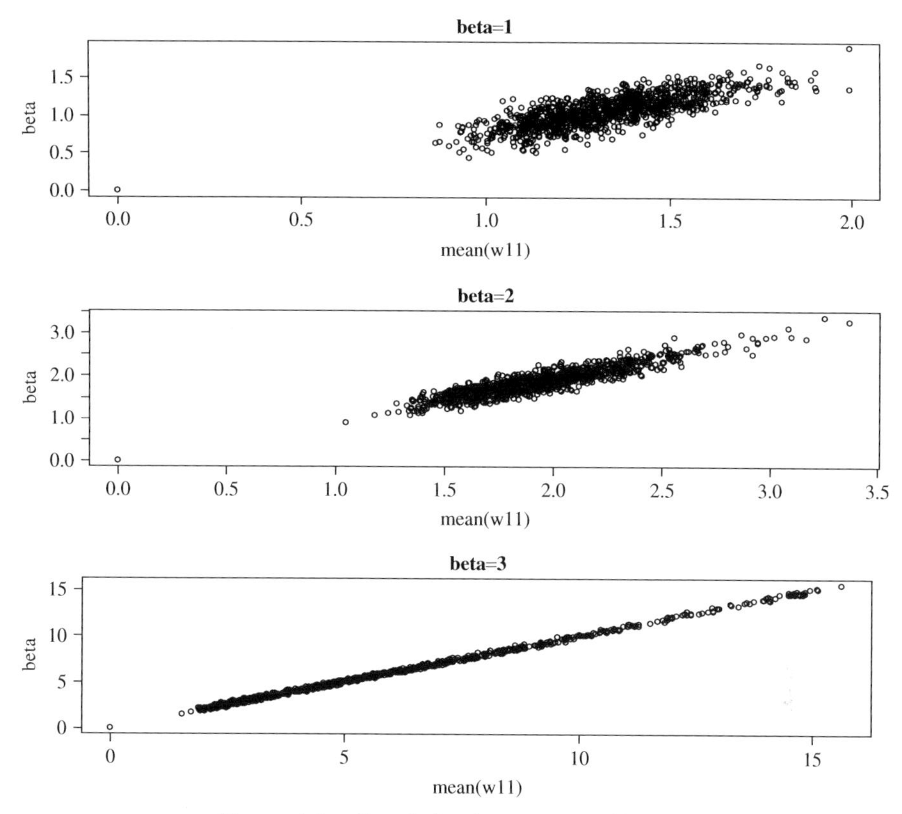

Figure 4.3 Correlation between β and latents

However, the NID sampler performs much better, as illustrated in Figure 4.4. The top two ACFs show the ID and NID samplers for data simulated with $\beta = 2$. The NID sampler exhibits much lower autocorrelation. For data simulated with $\beta = 3$, the autocorrelation for the ID sampler becomes very extreme. In that case, the ID sampler output is some 15 times less informative than an iid sample. The NID sampler is highly autocorrelated but with output approximately twice as informative as the ID sampler. These results for the binary probit model suggest that the ID sampler for the MNP model proposed by McCulloch *et al.* (2000) can exhibit very high autocorrelation. Nobile (2000) and Imai and Van Dyk (2005) point this out. The intuition developed here is that navigating in the unidentified parameter space can result in a chain with better mixing properties in the space of identified parameters. Nobile (1998) proposes a modification of the McCulloch and Rossi (1994) algorithm which includes a Metropolis step that increases the variability of the σ draws. Nobile observes improvement over the McCulloch and Rossi NID sampler under extreme conditions. As Van Dyk and Meng (2001) point out, any latent variable model that margins out to the right likelihood can be used as the basis for a data augmentation MCMC method.[4] It is entirely possible that even better-performing chains can be constructed using different data augmentation schemes.

[4] In the sense that $p(y|\beta, \Sigma) = \int p(y, w|\beta, \Sigma)\mathrm{d}w = \int p(y|w)p(w|\beta, \Sigma)\,\mathrm{d}w$ for many latent models.

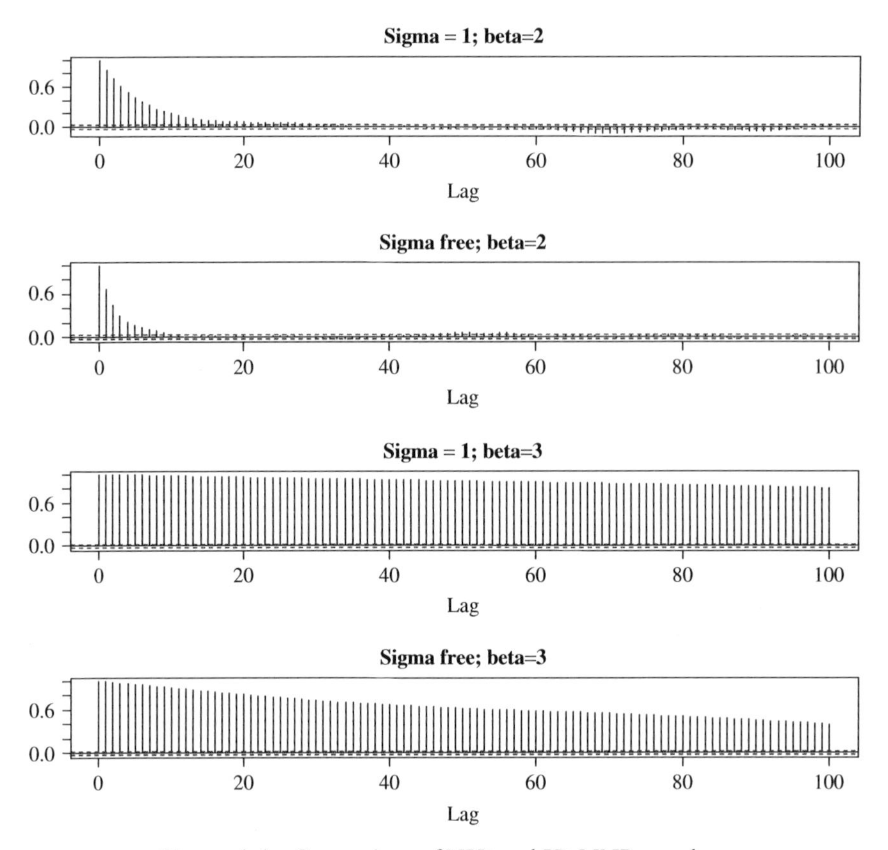

Figure 4.4 Comparison of NID and ID MNP samplers

4.2.2 The Likelihood for the MNP Model

The likelihood for the MNP model is simply the product of the relevant probabilities of the outcomes. Each of these probabilities involves integrating a normal distribution over a $(p-1)$-dimensional cone:

$$\ell(\beta, \Sigma) = \prod_{i=1}^{n} \Pr(y_i|X_i^{\mathrm{d}}, \beta, \Sigma),$$

$$\Pr(y_i|X_i^{\mathrm{d}}, \beta, \Sigma) = \int_{R_{y_i}} \varphi(w|X_i^{\mathrm{d}}\beta, \Sigma)\, \mathrm{d}w, \tag{4.2.11}$$

$$R_{y_i} = \begin{cases} \{w : w_{y_i} > \max(w_{-y_i}, 0)\}, & \text{if } y_i < p, \\ \{w : w < 0\} & \text{if } y_i = p. \end{cases}$$

Here w_{-j} denotes all elements in the w vector except the jth element (a $(p-2)$-dimensional vector) and $\varphi(\bullet)$ is the multivariate normal density function. Figure 4.5 shows this situation for a three-choice example. The contours correspond to a bivariate normal distribution centered at a mean determined by the X values for the ith observation. Three

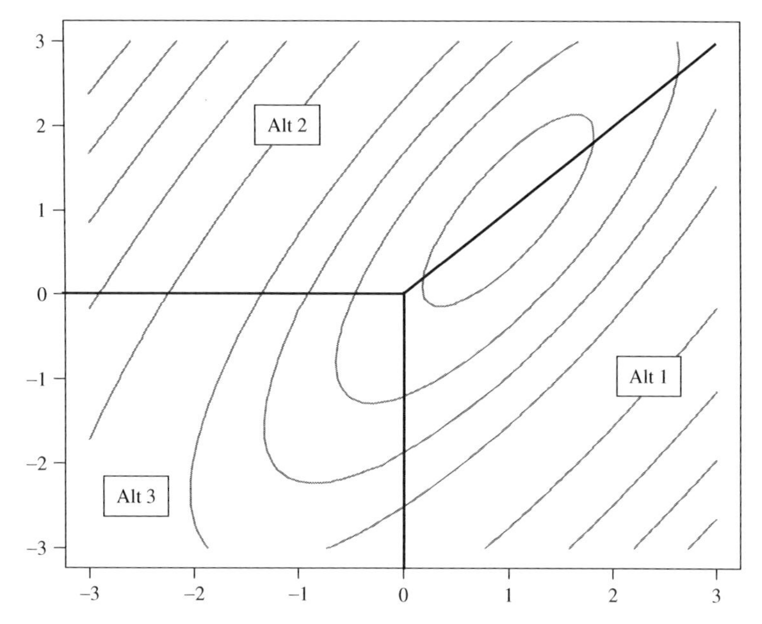

Figure 4.5 Regions of integration for MNP model

regions are shown on the figure, corresponding to each of the three choice alternatives. If $y_i = 1$ then we must have $w_1 > \max(w_2, 0)$. If $y_i = 2$, then $w_2 > \max(w_1, 0)$. If $y_i = 3$, then $w_1, w_2 < 0$. Direct evaluation of the likelihood is only computationally feasible if these integrals can be evaluated accurately. The GHK method discussed in Chapter 2 can be used to evaluate the likelihood in (4.2.11). The GHK method is designed to compute integrals of a normal distribution over a rectangle defined by vectors of lower and upper truncation points. In order to use this method, we have to reexpress the region defined in (4.2.11) as a rectangular region. To do so, we must define a matrix, specific to each choice alternative, that can be used to transform the w vector to a rectangular region.

For $j = 1, \ldots, p-1$, define[5]

$$A_j = \begin{bmatrix} -I_{j-1} & 1 & 0 \\ & \vdots & \\ 0 & 1 & -I_{(p-1)-j} \end{bmatrix}.$$

The condition $w_j > \max(w_{-j})$ and $w_j > 0$ is equivalent to $A_j w > 0$. We can then reexpress the inequalities by applying A_j to both sides of the latent variable equation

$$A_j w = A_j \mu + A_j \varepsilon > 0,$$
$$A_j \varepsilon > -A_j \mu,$$

where $\mu = X^{\mathrm{d}}\beta$.

[5] Note that if $j = 1$ or $j = p - 1$, then we simply omit the requisite I matrix (that is, I_0 means nothing)!

If we define $u = A_j\varepsilon$, then we can express the choice probability as a multivariate normal integral truncated from below:

$$u > -A_j\mu, \qquad u \sim N(0, A_j\Sigma A_j'),$$
$$\Pr(y = j|X, \beta, \Sigma) = \int_{A_jX\beta}^{\infty} \varphi(u|0, A_j\Sigma A_j')\,\mathrm{d}u. \tag{4.2.12}$$

For the case of $j = p$, the region is already defined in terms of ε:

$$\varepsilon < -\mu,$$
$$\Pr(y = p|X, \beta, \Sigma) = \int_{-\infty}^{-X\beta} \varphi(u|0, \Sigma)\,\mathrm{d}u. \tag{4.2.13}$$

R The function `llmnp`, in *bayesm*, implements this approach.

We could use the likelihood evaluated via GHK to implement a Metropolis chain for the MNP without data augmentation. However, this would require a good proposal function for both β, Σ. Experience with Metropolis methods for covariance matrices suggests that these methods are only useful for low-dimensional problems.

4.3 MULTIVARIATE PROBIT MODEL

The multivariate probit model is specified by assuming the same multivariate regression model as for the MNP model but with a different censoring mechanism. We observe the sign of the components of the underlying p-dimensional multivariate regression model:

$$w_i = X_i\beta + \varepsilon_i, \qquad \varepsilon_i \sim N(0, \Sigma),$$
$$y_{ij} = \begin{cases} 1, & \text{if } w_{ij} > 0, \\ 0, & \text{otherwise.} \end{cases} \tag{4.3.1}$$

Here choice alternatives are not mutually exclusive as in the MNP model. The multivariate probit model has been applied to the purchase of products in two different categories (Manchanda *et al.* 1999) or to surveys with pick j of p questions (Edwards and Allenby 2003). In the econometrics literature, the multivariate probit has been applied to a binary phenomenon which is observed over adjacent time periods (for example, labor force participation observed for individual workers).

The identification problem in the multivariate probit can be different from the identification problem for the MNP depending on the structure of the X array. Consider the general case which includes intercepts for each of the p choice alternatives and covariates that are allowed to have different coefficients for the p choices:

$$X_i = (z_i' \otimes I_p). \tag{4.3.2}$$

Here z is a $d \times 1$ vector of observations on covariates. Thus, X is a $p \times k$ matrix with $k = p \times d$. Also

$$\beta = \begin{bmatrix} \beta_1 \\ \vdots \\ \beta_d \end{bmatrix}, \tag{4.3.3}$$

where the β_i are p-dimensional coefficient vectors. The identification problem arises from the fact that we can scale each of the p means for w with a different scaling constant without changing the observed data. This implies that only the correlation matrix of Σ is identified and that transformation from the unidentified to the identified parameters $((\beta, \Sigma) \rightarrow (\tilde{\beta}, R))$ is defined by

$$\begin{aligned} \tilde{B} &= \Lambda B, \\ \tilde{\beta} &= \text{vec}(\tilde{B}), \\ R &= \Lambda \Sigma \Lambda, \end{aligned} \tag{4.3.4}$$

where

$$B = [\beta_1, \ldots, \beta_d],$$

$$\Lambda = \begin{bmatrix} 1/\sqrt{\sigma_{11}} & & \\ & \ddots & \\ & & 1/\sqrt{\sigma_{pp}} \end{bmatrix}.$$

However, if the coefficients on a given covariate are restricted to be equal across all p choices, then there are fewer unidentified parameters. Then we cannot scale each equation by a ***different*** positive constant. This brings us back in to the same situation as in the MNP model where we must normalize by one of the diagonal elements of Σ. For example, we might have an attribute like price of the p alternatives under consideration. We might want to restrict the price attribute to have the same effect on w for each alternative. This amounts to the restriction that $\beta_{j1} = \beta_{j2} = \ldots = \beta_{jp}$ for covariate j.

To construct an MCMC algorithm for the multivariate probit model, we can use data augmentation just as in the MNP model by adding w to the parameter space. We simply 'Gibbs through' the w vector using the appropriate conditional univariate normal distribution but with an upper (lower) truncation of 0 depending on the value of y:

$$\begin{aligned} w_{ij} | w_{i,-j}, y_i, \beta, \Sigma &\sim N(m_{ij}, \tau_{jj}^2) \times [I(y_{ij} = 1)I(w_{ij} > 0) + I(y_{ij} = 0)I(w_{ij} < 0)], \\ m_{ij} &= x_{ij}'\beta + F'(w_{i,-j} - X_{i,-j}\beta), \\ F &= -\sigma^{jj}\gamma_{j,-j}, \\ \tau_{jj}^2 &= 1/\sigma^{jj}. \end{aligned} \tag{4.3.5}$$

Here y is an np-vector of indicator variables.

We must choose whether to navigate in the unidentified (β, Σ) space or the identified $(\tilde{\beta}, R)$ space. The unidentified parameter space is larger by p dimensions than the identified space. The intuition developed from the MNP model and generalized by Van Dyk and Meng is that navigating in the higher-dimensional unidentified parameter space with diffuse priors will produce a chain with superior mixing properties to a chain defined on the identified space. An additional complication will be the method for drawing valid R matrices. The algorithm of Chib and Greenberg (1998) or Barnard *et al.* (2000) can be used to draw R. However, given the additional complication and computational cost of these methods and the fact that we expect these chains to have inferior mixing properties, we recommend using a more straightforward Gibbs sampler on the unidentified parameter space; for details, see Edwards and Allenby (2003).

NID MULTIVARIATE PROBIT GIBBS SAMPLER

Start with initial values w_0, β_0, Σ_0.

Draw $w^1|\beta_0, \Sigma_0, y$ using (4.3.5).

Draw $\beta_1|w_1, \Sigma_0 \sim N(\tilde{\beta}, V)$,

$$
\begin{aligned}
&V = (X^{*\prime}X^* + A)^{-1}, \quad \tilde{\beta} = V(X^{*\prime}w^* + A\bar{\beta}),\\
&\Sigma_0^{-1} = C'C,\\
&X_i^* = C'X_i; \qquad w_i^* = C'w_i,\\
&X = \begin{bmatrix} X_1 \\ \vdots \\ X_n \end{bmatrix}.
\end{aligned}
$$

Draw $\Sigma_1|w_1, \beta_1$ using $\Sigma^{-1}|w, \beta \sim W(\nu + n, (V_0 + S)^{-1})$,

$$
\begin{aligned}
S &= \sum_{i=1}^{n} \varepsilon_i \varepsilon_i',\\
\varepsilon_i &= w_i - X_i\beta.
\end{aligned}
$$

Repeat as necessary.

R This algorithm has been implemented in the function `rmvpGibbs`, in *bayesm*.

To illustrate this sampler, we consider a data example from Edwards and Allenby (2003). This data set is available in *bayesm* and can be loaded using the R command

R `data(Scotch)`. 2218 respondents were given a list of 21 Scotch whisky brands and asked to indicate whether or not they drink these brands on a regular basis. The interest in this example is in understanding the correlation structure underlying brand choice. Correlation in the latent variable can be viewed as a measure of similarity of two brands. In this example, $X_i = I_{21}$ so that the β vector is simply a vector of intercepts or means for the latent variable w.

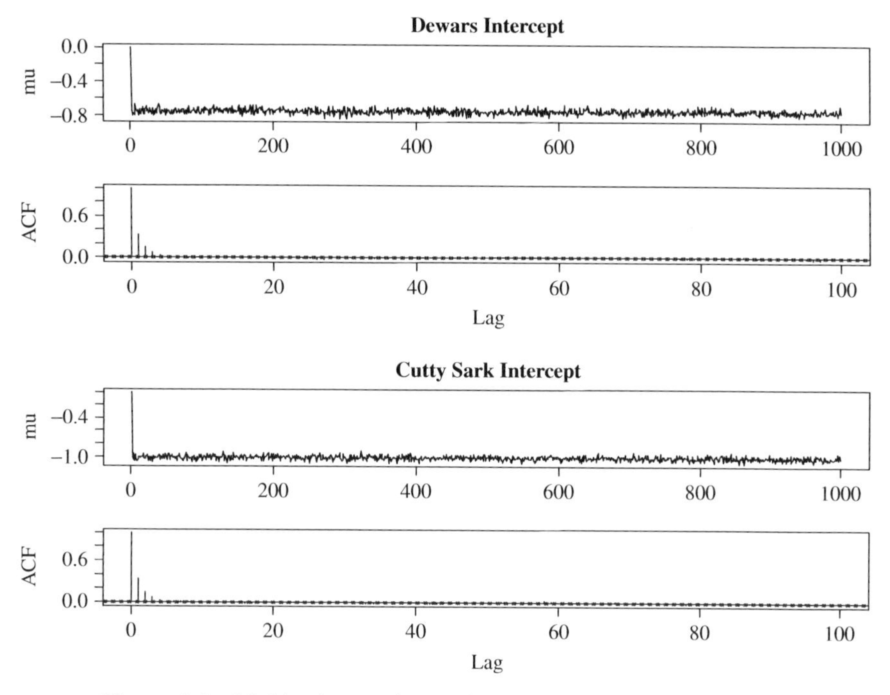

Figure 4.6 Multivariate probit model intercepts: Scotch survey data

Figure 4.6 shows MCMC traces (every 20th draw from a sequence of 20,000) and ACFs for two of the elements of β corresponding to popular blended whiskies (note: the brand chosen most often is Chivas Regal, so all other brands have smaller intercepts). These plots show quick dissipation of the initial condition ($\beta_0 = 0$ and $\Sigma_0 = I$). The autocorrelations of the intercept draws are small with a numerical efficiency roughly half of that of an iid sample. Figure 4.7 shows MCMC traces (again every 20th draw from a sequence of 20,000) and ACFS for two of the correlations. These correlations are between two single malts and between the most popular blended whisky (Chivas Regal) and a single malt. The single malts exhibit high correlation, showing similarity in product branding and taste. There is a negative correlation between the single malt and the blended whisky, showing some divergence in consumer preferences. The autocorrelations for the correlation parameters are higher than the intercept or mean parameters, but still less than for some of the MNP examples considered in Section 4.2 (numerical efficiency here is one-ninth of a random sample as compared to one-eleventh for the MNP examples). The intuition for the better performance of the NID sampler for the multivariate probit (in comparison to the NID MNP sampler) is that there is a higher-dimensional unidentified parameter space improving the mixing characteristics of the sampler in the identified parameter space.

In our analysis of the Scotch data, we used a relatively diffuse but proper prior: $\beta \sim N(0, 100I_p)$ and $\Sigma \sim \text{IW}(p+2, (p+2)I_p)$. Our analysis of the binary probit model developed an intuition that the diffusion of the prior on Σ would affect the performance of the sampler. In particular, as emphasized by Meng and Van Dyk (1999) and Van Dyk and Meng (2001), mixing should be maximized by allowing the prior on Σ to

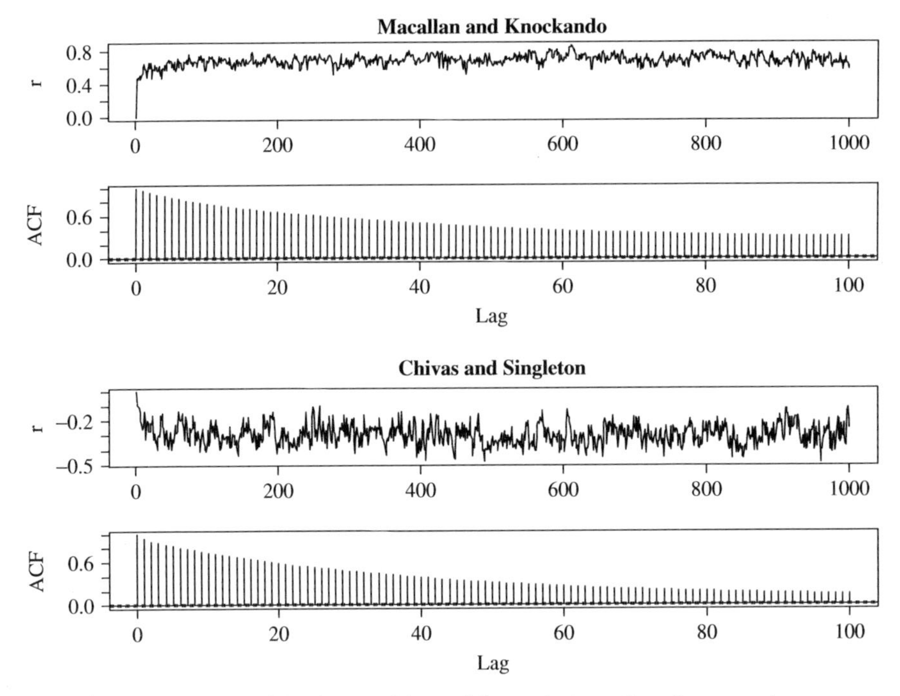

Figure 4.7 Multivariate probit model correlations: Scotch survey data

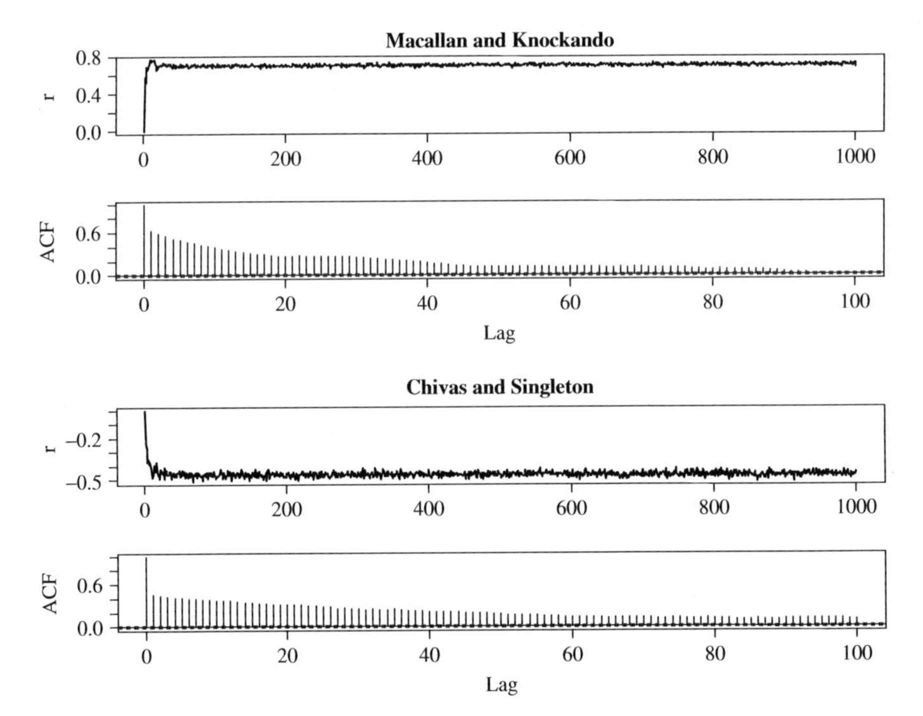

Figure 4.8 Multivariate probit model correlations: Scotch survey data with improper prior on Σ

be improper. However, as McCulloch *et al.* (2000) have pointed out, improper priors on Σ can be extraordinarily informative on functions of Σ such as correlations. The implied marginal prior on each correlation will be U-shaped, with substantial mass near -1 and 1.

We reran the analysis with an improper prior on Σ and show the results in Figure 4.8. As expected, the improper prior produces better mixing, reducing the autocorrelations substantially. Numerical efficiency is now at one-seventh of an iid sample. However, the improper prior on Σ is extraordinarily informative on the correlations changing both the location and tightness of the posterior distributions of the correlations. Figure 4.9 shows the prior distribution of a correlation for the barely proper case, $\Sigma \sim \text{IW}(p, pI_p)$. This 'U-shaped' distribution puts high mass near high positive or negative correlations. The improper prior can be viewed as the limit of proper priors as the diffusion increases and, therefore, will be even more informative than the barely proper case, putting high mass on extreme values of correlation.

Thus, improper priors are very dangerous for this analysis even though they might receive attention due to mixing considerations. The problem here is that there is a conflict between prior settings which promote maximum mixing and substantive informative considerations. It would be desirable to produce a MCMC sampler with a prior that separates 'tuning' from substantive informativeness.

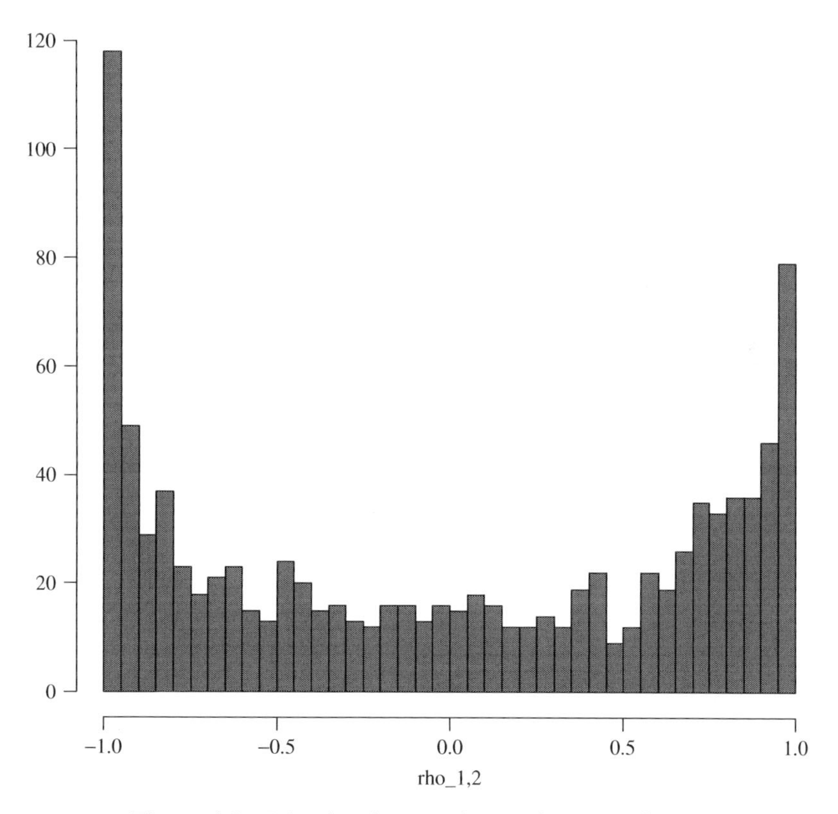

Figure 4.9 Prior distribution of a correlation coefficient

4.4 DEMAND THEORY AND MODELS INVOLVING DISCRETE CHOICE

The multinomial and multivariate choice models developed in Sections 4.2 and 4.3 are simply statistical models that use a latent variable device to link discrete outcomes to covariate information. Since the outcomes in many marketing situations are the result of customer decisions, we should relate these statistical models to models of optimizing behavior and consumer demand theory. If we assume that consumers are maximizing utility subject to a budget constraint, then our modeling efforts focus on choice of the utility function and assumptions regarding the distribution of unobservables.

In much of standard demand theory (c.f. Deaton and Muellbauer 1980), utility functions are chosen so that only interior solutions are obtained. That is, consumers faced with a set of goods would purchase all goods in various amounts. For aggregate data or for highly aggregated goods (such as food and housing), these utility functional forms are appropriate. However, at the level of the individual customer, few of the possible alternative goods are chosen. When faced with a set of similar products, customers are often observed to only purchase one product. For these situations, utility functions must be chosen which allow for corner solutions. This means that marginal utility must remain finite as consumption approaches zero (at least for some of the goods). In other words, the indifference curves cannot be tangent to the axes.

For situations in which consumers are only observed to choose one alternative, a linear utility function is a reasonable place to start.

$$\max_x U(x) = \psi' x \qquad \text{such that } p'x = E. \tag{4.4.1}$$

For this functional form, marginal utility is constant and the optimal solution is to choose the alternative with the highest ratio of marginal utility to price (ψ_j / p_j) and consume $x = E/p_j$ units. In this situation, if we observe p and E, then consumer choice will be deterministic. The standard random utility approach is to assume that there are factors affecting the marginal utility which are not observed by the researcher but are known to the consumer. These factors could be time-varying (consumer inventory or promotional activities) or represent taste differences between consumers. Since marginal utility must be positive, it is natural to consider a multiplicative specification for inclusion of these omitted unobservables:

$$\psi = \overline{\psi}\, e^{\varepsilon}. \tag{4.4.2}$$

If we take logs, allow $\overline{\psi}$ to be a function of observable covariates such as consumer characteristics and product attributes, and subtract log price, we obtain a model similar to the standard multinomial models:

$$\ln \psi - \ln p = \overline{\psi} - \ln p + \varepsilon = \tilde{X}\beta - \ln p + \varepsilon. \tag{4.4.3}$$

$\tilde{X}$ is a $p \times k$ matrix containing the revelevant covariates. Letting $z = \ln \psi - \ln p$ and $X = [\,\tilde{X} \quad \ln p\,]$, we have the latent variable formulation of a multinomial choice model in (4.2.1). Alternative j is chosen if $z_j = \max(z)$.

To complete the model, we must specify the distribution of the latent component of marginal utility. First, consider iid errors with a given scale, σ. In (4.4.3), σ is identified as the price coefficient is set to -1. Alternatively, we could introduce a price coefficient. If we allow a free price coefficient, we must fix the scale of the marginal utility errors to avoid the scale invariance identification problem. This means that we can interpret a price coefficient as $1/\sigma$. As the variance of the marginal utility errors grows, the influence of price on the probabilities of choice declines since the factors other than price are 'larger' in scale. If we allow ε to have a multivariate normal distribution, we obtain the standard MNP model. The standard multinomial logit model can be derived by assuming that errors are iid extreme value type I with zero mean and scale parameter of 1 (denoted EV(0,1)). Demand theory provides some guidance as to the sort of covariates which might be included as well as the restriction that the price coefficient is negative and the same across choice alternatives.

4.4.1 A Nonhomothetic Choice Model

The restrictive properties of the multinomial logit model can be relaxed either by specifying a richer class of error distributions (as in the MNP model which avoids the independence of irrelevant alternatives property) or by specifying a more flexible utility structure. The linear structure assumes constant marginal utility and no interactions (for example, the marginal utility of alternative i is not affected by the consumption of other alternatives). One could argue that, for choice data, constant marginal utility is appropriate in that we would need quantity information to estimate the curvature of the utility function. However, even with only access to choice information, the constant marginal utility assumption implies that price changes can only have substitution and not income effects (that is, the utility function is homothetic). In the homothetic world, as greater amounts are allocated to expenditure on the products under study, we continue to purchase the same brand. In many marketing applications, there are products of differing quality levels. We observe consumers trading up to high-quality models as expenditure increases or during sales. The opposite phenomena of 'trading down' to lower-quality brands as expenditure decreases or as lower-quality brands are discounted does not occur as frequently. This phenomenon has been dubbed 'asymmetric switching'. The standard homothetic logit or probit models cannot exhibit asymmetric switching.

A nonhomothetic choice model can be specified by retaining the linear structure of utility but assuming that marginal utility is a function of the overall level of attainable utility. This retains the assumption of no interactions in utility but allows for a nontrivial income effect which occurs either across consumers as consumers allocate a greater expenditure to the product category or as the category becomes 'cheaper' due to price reductions on some of the items. One convenient specification which defines a valid utility function is given by

$$\psi_j(u) = \exp\{\alpha_j - k_j u\}. \tag{4.4.4}$$

Here the marginal utility of the ith product is a function of the maximum attainable level of utility. This defines the utility function implicitly. k_j governs the rate at which marginal utility changes as the level of attainable utility increases. If $k_1 < k_2$, then, as the budget increases, consumers will tend to purchase product 1 more than product 2. The utility function for this case has a set of linear but rotating indifference curves. In this sense, we can regard product 1 as a superior product that provides a higher marginal utility for consumers willing to allocate a greater real expenditure to the product category.

Since the indifference curves are linear, consumers will choose only one of the products in the category. As in the standard linear utility model, consumers choose the alternative which provides the highest level of attainable utility conditional on the level of expenditure for the category, E, and vector of prices, p. That is, consumers find the maximum of $\{u^1, u^2, \ldots, u^p\}$. u^i solves

$$u^i = \psi_i(u^i)E/p_i. \tag{4.4.5}$$

Taking logs of both sides, we can write this as the implicit solution to an equation of the form $\ln(y) = C_1 - C_2 y$. If C_2 is positive, this equation has an implicit solution which can be easily calculated by Newton's method. To complete this model, we must specify a random utility error as well as an expenditure function. In Allenby and Rossi (1991), an expenditure function was specified as

$$\ln\ E = \gamma' z, \tag{4.4.6}$$

where z is a vector of covariates. In Allenby and Rossi, z includes a price index for the product category. This is a somewhat ad hoc solution which avoids the specification of a bivariate utility function over all other goods and this product category. In addition, one might be tempted to make the α and k parameters a function of demographic variables. However, since marginal utility is specified as an implicit nonlinear function of α and k, we would not be able to write the nonhomothetic model in form of (4.4.3).

If we add a standard extreme value type I error to the implicitly defined marginal utility, we obtain a logit model in which the probability of choice is given by

$$\Pr(i) = \frac{\exp(\tau \nu_i)}{\sum_j \exp(\tau \nu_j)}, \tag{4.4.7}$$

where $\nu_j = \alpha_j - k_j u^j - \ln p_j$. τ is the scale parameter of the extreme value distribution. Note that u^j is also a function of E, p_j, α_j, and k_j. Thus, we can write $\nu_j = f(\alpha_j, k_j | E, p_j)$.

R `llnhlogit` in *bayesm* valuates the log-likelihood for this model.

4.4.2 Demand for Discrete Quantities

The application of discrete choice models to packaged goods requires researchers to adjust their models to accommodate demand quantities. Packaged goods are often available in multiple sizes, and it is not appropriate to treat the various package sizes as independent alternatives since the same good is contained in each of the packages. While it is possible to estimate discrete choice models that allow for dependence among the alternatives, additional restrictions on the coefficients are needed so that parameter estimates from the model conform to economic theory. For example, a discrete choice model calibrated on soft drink purchases would need to impose ordinal restrictions on the intercepts so that the utility of 6-pack, 12-pack, and 18-pack offerings would reflect diminishing marginal returns to quantity.

Quantity can be incorporated into models of consumer demand by embedding a utility function for a discrete choice model into a utility function that relates the product class to an outside good. For example, consider the Cobb–Douglas utility function:

$$\ln U(x, z) = \alpha_0 + \alpha_x \ln U(x) + \alpha_z \ln (z), \tag{4.4.8}$$

where $x = (x_1, \ldots, x_K)$ is the vector of the amount of each alternative (that is, brand) purchased, K represents the number of brands in the product class, z represents the amount of the outside good purchased, and $U(x)$ denotes a subutility function. The subutility function can be specified as the linear function in equation (4.4.1) or the nonhomothetic function in equation (4.4.4).

Maximizing (4.4.8) subject to a budget constraint leads to a vector of demand (x,z) that, in general, is a mixture of corner and interior solutions. However, as discussed in Case Study 1, the utility-maximizing solution will always be a corner solution in x when per-unit price schedules are concave, for example, the cost per ounce of a 6-pack of is greater than the cost per ounce of a 12-pack. When this occurs, the choice probability of observing quantity x_i is:

$$\Pr(x_i) = \frac{\exp[\ln(\psi_i) + \ln(x_i) + (\alpha_z/\alpha_x)\ln(E - p_i(x_i))]}{\sum_{k=1}^{K} \exp[\ln(\psi_k) + \ln(x_k) + (\alpha_z/\alpha_x)\ln(E - p_k(x_k))]}, \tag{4.4.9}$$

where ψ_i is the marginal utility of brand i, E is the budgetary allotment, $p_i(x_i)$ is the price of x_i units of brand i, and x_k is the quantity of brand k that maximizes equation (4.4.8).

4.4.3 Demand for Variety

In some product categories, consumers are observed to purchase a subset of products in the category. For example, consumers purchase multiple varieties of soft drinks or yogurts. The standard multinomial models have zero likelihood for this sort of consumer behavior as the choice options are regarded as mutually exclusive. On the other hand, many common utility specifications are designed to give rise to strictly interior solutions in which all products in the category are purchased. What is needed is a demand system which can give rise to a mixture of corner and interior solutions. This can be achieved by translating a utility function so that its indifference curves intersect the axes with finite slope. One simple additive structure is given by

$$\overline{U}(x) = \sum_j \psi_j (x_j + \gamma_j)^{\alpha_j}. \tag{4.4.10}$$

The $\{\gamma_j\}$ parameters serve to translate an additive power utility to admit the possibility of corner solutions. The utility function also exhibits curvature or diminishing marginal utility which allows for the possibility of 'wear-out' in the consumption of a particular variety. The utility in (4.4.10) is an additive, but nonlinear utility function. Equation (4.4.10) defines a valid utility function under the restrictions that $\psi_j > 0$ and $0 < \alpha_j \leq 1$.

This utility specification can accommodate a wide variety of situations, including the purchase of a large number of different varieties as well as purchases where only one variety is selected. If a particular variety has a high value of ψ_j and a value of α_j close to one, then we would expect to see purchases of large quantities of only one variety (high baseline preference and low satiation). On the other hand, small values of α imply a high satiation rate, and we would expect to see multiple varieties purchased if the ψs are not too different.

To develop a statistical specification, we follow a standard random utility approach and introduce a multiplicative normal error into the marginal utility:

$$\ln(U_j) = \ln(\overline{U}_j) + \varepsilon_j, \quad \varepsilon \sim N(0, \Sigma), \tag{4.4.11}$$

where $\overline{U}_j$ is the derivative of the utility function in (4.4.10) with respect to x_j. We use a lognormal error term to enforce the positivity of the marginal utility. We specify a full covariance matrix for the random marginal utility errors. In some applications, it may be difficult to identify this covariance matrix. Further restrictions may be necessary. Even the assumption that Σ is the identity matrix is not necessarily too restrictive as we have specified a log-normal distribution of marginal utility errors which exhibits heteroskedasticity of a reasonable form. However, this is largely an empirical matter.

We derive the demand system for the set of goods under study conditional on the expenditure allocation to this set of goods. In the random utility approach, it is assumed that the consumer knows the value of ε and that this represents omitted factors which influence marginal utility but are not observable to the data analyst. If we derive the optimal demand by maximizing utility subject to the budget constraint and conditional on the random utility error, we define a mapping from p, E and ε to demand. Assuming a distribution for ε provides a basis for deriving the distribution of optimal demand, denoted x^*. There are two technical issues in deriving the distribution of demand. First, optimal demand is a nonlinear function of ε and requires use of change-of-variable calculus. Second, the possibility of corner solutions means that there are point masses in the distribution of demand and, thus, the distribution of demand will be a mixed discrete–continuous distribution. Computing the size of these point masses involves integrating the normal distribution of ε over rectangular regions.

To solve for optimal demand, we form the Lagrangian for the problem and derive the standard Kuhn–Tucker first-order conditions. In the utility function specified in (4.4.11), $\overline{U}$, is the deterministic (observable) part of utility and that the consumer maximizes U, which includes the realization of the random utility errors. The Lagrangian is given by

$$U(x) - \lambda(p'x - E).$$

Differentiating the Lagrangian gives the standard Kuhn–Tucker first-order conditions:

$$\begin{aligned} \overline{U}_j e^{\varepsilon_j} - \lambda p_j &= 0, \qquad x_j^* > 0, \\ \overline{U}_j e^{\varepsilon_j} - \lambda p_j &< 0, \qquad x_j^* = 0. \end{aligned}$$

x^* is the vector of optimal demands for each of the m goods under consideration. Dividing by price and taking logs, the Kuhn–Tucker conditions can be rewritten as:

$$\begin{aligned} V_j(x_j^*|p) + \varepsilon_j &= \ln \lambda, \qquad \text{if } x_j^* > 0, \\ V_j(x_j^*|p) + \varepsilon_j &< \ln \lambda, \qquad \text{if } x_j^* = 0, \end{aligned} \tag{4.4.12}$$

where λ is the Lagrange multiplier and

$$V_j(x_j^*|p) = \ln\left(\psi_j\alpha_j(x_j^* + \gamma_j)^{\alpha_j - 1}\right) - \ln(p_j), \qquad j = 1, \ldots, m.$$

Optimal demand satisfies the Kuhn–Tucker conditions in (4.4.12) as well as the 'adding-up' constraint that $p'x^* = E$. The 'adding-up' constraint induces a singularity in the distribution of x^*. To handle this singularity, we use the standard device of differencing the first-order conditions with respect to one of the goods. Without loss of generality, we assume that the first good is always purchased (one of the m goods must be purchased since we assume that $E > 0$) and subtract condition (4.4.12) for good 1 from the others. This reduces the dimensionality of the system of equations by one. Now (4.4.12) is equivalent to

$$\begin{aligned} \nu_j &= h_j(x^*, p), \qquad \text{if } x_j^* > 0, \\ \nu_j &< h_j(x^*, p), \qquad \text{if } x_j^* = 0, \end{aligned} \tag{4.4.13}$$

where $\nu_j = \varepsilon_j - \varepsilon_1$ and $h_j(x^*, p) = V_1 - V_j, j = 2, \ldots, m$.

The likelihood for $x^* = (x_1^*, \ldots, x_m^*)'$ can be constructed by utilizing the density of $\nu = (\nu_2, \ldots, \nu_m)'$, the Kuhn–Tucker conditions in (4.4.13), and the adding-up constraint $p'x^* = E$. ν is distributed as $N(0, \Omega)$, $\Omega = A\Sigma A'$, where $A = (-\iota \quad I_{m-1})$. Given that corner solutions will occur with nonzero probability, optimal demand will have a mixed discrete–continuous distribution with lumps of probability corresponding to regions of ε which imply corner solutions. Thus, the likelihood function will have a density component corresponding to the goods with nonzero quantities and a mass function corresponding to the corners in which some of the goods will have zero optimal demand. The probability that n of the m goods are selected is equal to:

$$\begin{aligned} &\Pr(x_i^* > 0 \text{ and } x_j^* = 0; i = 2, \ldots, n \text{ and } j = n+1, \ldots m) \\ &= \int_{-\infty}^{h_m} \cdots \int_{-\infty}^{h_{n+1}} \varphi(h_2, \ldots, h_n, \nu_{n+1}, \ldots, \nu_m|0, \Omega)|J|\, d\nu_{n+1} \cdots d\nu_m, \end{aligned} \tag{4.4.14}$$

where $\varphi(\bullet)$ is the normal density, $h_j = h_j(x^*, p)$, and J is the Jacobian,

$$J_{ij} = \frac{\partial h_{i+1}(x^*; p)}{\partial x_{j+1}^*}, \qquad i, j = 1, \ldots, n-1.$$

We should note that the adding-up constraint, $p'x = E$, makes this Jacobian non-diagonal as we can always express the demand for the 'first' good with nonzero demand as a function of the other demands.

The intuition behind the likelihood function in (4.4.14) can be obtained from the Kuhn–Tucker conditions in (4.4.13). For goods with nonzero demand, the first condition in (4.4.13) means that optimal demand is an implicitly defined nonlinear function of ε given by h. We use the change-of-variable theorem to derive the density of x^* (this generates the Jacobian term). For goods not purchased, the second Kuhn–Tucker condition defines a region of possible values of ν which are consistent with this specific corner solution. The probability that these goods have zero demand is calculated by integrating the normal distribution of ν over the appropriate region.

If there are only corner solutions with one good chosen, our model collapses to a standard choice model. The probability that only good one is chosen is given by

$$\Pr(x_j^* = 0, j = 2, \ldots, m) = \int_{-\infty}^{h_m} \cdots \int_{-\infty}^{h_2} \varphi(\nu_2, \ldots, \nu_m)\, d\nu_2 \cdots d\nu_m.$$

Similarly, we can derive the distribution of demand for the case in which all goods are at an interior solution:

$$\Pr(x_i^* > 0; i = 2, \ldots, m) = \varphi(h_2, \ldots, h_m | 0, \Omega)|J|.$$

The joint distribution of $(x_2^*, \ldots, x_m^*)'$ in (4.4.14) can be evaluated by noting that it can be factored into discrete and continuous parts. In evaluating the likelihood, we write (4.4.14) as the product of two factors as follows. By partitioning $\nu = (\nu_2, \ldots, \nu_m)'$ into $\nu_a = (\nu_2, \ldots, \nu_n)'$ and $\nu_b = (\nu_{n+1}, \ldots, \nu_m)'$ such that

$$\begin{bmatrix} \nu_a \\ \nu_b \end{bmatrix} \sim N\left(\begin{bmatrix} 0 \\ 0 \end{bmatrix}, \begin{bmatrix} \Omega_{aa} & \Omega_{ab} \\ \Omega_{ba} & \Omega_{bb} \end{bmatrix}\right),$$

ν_a and $\nu_b|\nu_a$ are normally distributed, then $\nu_a \sim N(0, \Omega_{aa})$ and $\nu_b|\nu_a = h_a \sim N(\mu, \Sigma)$ where $\mu = \Omega_{ba}\Omega_{aa}^{-1}h_a$, $\Sigma = \Omega_{bb} - \Omega_{ba}\Omega_{aa}^{-1}\Omega_{ab}$, and $h_a = (h_2, \ldots, h_n)'$. Then, (4.4.14) can be rewritten as the product of two factors:

$$\begin{aligned} &\Pr(x_i^* > 0 \text{ and } x_j^* = 0; i = 2, \ldots, n \text{ and } j = n+1, \ldots, m) \\ &= \varphi_{\nu_a}(h_2, \ldots, h_n | 0, \Omega_{aa})|J| \int_{-\infty}^{h_m} \cdots \int_{-\infty}^{h_{n+1}} \varphi_{\nu_b|\nu_a}(\nu_{n+1}, \ldots, \nu_m | \mu, \Sigma)\, d\nu_{n+1} \cdots d\nu_m. \end{aligned} \tag{4.4.15}$$

We use the GHK simulator (Section 2.10; see also Keane 1994; Hajivassiliou *et al.* 1996) to evaluate the multivariate normal integral in (4.4.15).

In Case Study 5 we will apply a heterogeneous version of this model to data on purchase of yogurt varieties. We will also discuss the R implementation of this model.

The additive utility model used does not include any interactions in the utility function. That is, the marginal utility of consumption of good i does not depend on the consumption level of other goods. In particular, additive utility specifications impose the restriction that all goods are substitutes, ruling out complementarity. Gentzkow (2005) includes utility interaction in a choice model which allows for the possibility that two goods are complements.[6] He applies this formulation to purchase data on the print and on-line versions of newspapers. His results suggest that there are complementarities between the print and on-line versions of newspapers.

[6] We note that with only two goods there can only be substitution in demand. However, Gentzkow includes the usual outside alternative.

5

Hierarchical Models for Heterogeneous Units

Using this Chapter

This chapter provides a comprehensive treatment of hierarchical models. Hierarchical models are designed to measure differences between units using a particular prior structure. Choice of the form of the hierarchical model (that is, the form of the prior) as well as the MCMC algorithm to conduct inference is an important issue. We explore a new class of hybrid MCMC algorithms that are customized or tuned to the posteriors for individual units. We also implement a mixture of normals prior for the distribution of model coefficients across units. We illustrate these methods in the context of a panel of household purchase data and a base or unit-level multinomial logit model. Those interested in the main points without technical details are urged to concentrate on Sections 5.1, 5.2, 5.4, and 5.5.3.

One of the greatest challenges in marketing is to understand the diversity of preferences and sensitivities that exists in the market. Heterogeneity in preferences gives rise to differentiated product offerings, market segments and market niches. Differing sensitivities are the basis for targeted communication programs and promotions. As consumer preferences and sensitivities become more diverse, it becomes less and less efficient to consider the market in the aggregate. Marketing practices which are designed to respond to consumer differences require an inference method and model capable of producing individual or unit-level parameter estimates. Optimal decision-making requires not only point estimates of unit-level parameters but also a characterization of the uncertainty in these estimates. In this chapter, we will show how Bayesian hierarchical approaches are ideal for these problems as it is possible to produce posterior distributions for a large number of unit-level parameters.

In contrast to this emphasis on individual differences, economists are often more interested in aggregate effects and regard heterogeneity as a statistical nuisance parameter problem. Econometricians frequently employ methods which do not allow for the estimation of individual-level parameters. For example, random coefficient models are often implemented through an unconditional likelihood approach in which only

Bayesian Statistics and Marketing P. E. Rossi, G. M. Allenby and R. McCulloch

hyperparameters are estimated. Furthermore, the models of heterogeneity considered in the econometrics literature often restrict heterogeneity to subsets of parameters such as model intercepts. In the marketing context, there is no reason to believe that differences should be confined to the intercepts and, as indicated above, differences in slope coefficients are critically important. Finally, economic policy evaluation is frequently based on estimated hyperparameters which are measured with much greater certainty than individual-level parameters. This is in contrast to marketing policies which often attempt to respond to individual differences that are measured less precisely.

This new literature emphasizing unit-level parameters is made possible by the explosion in the availability of disaggregate data. Scanner data at the household and store level is now commonplace. In the pharmaceutical industry, physician-level prescription data is also available. This raises modeling challenges as well as opportunities for improved profitability through decentralized marketing decisions that exploit heterogeneity. This new data comes in panel structure in which N, the number of units, is large relative to T, the length of the panel. Thus, we may have a large amount of data obtained by observing a large number of decision units. For a variety of reasons, it is unlikely that we will ever have a very large amount of information about any one decision unit. Data describing consumer preferences and sensitivities to variables such as price is typically obtained through surveys or household purchase histories which yield very limited individual-level information. For example, household purchases in most product categories often total less than 12 per year. Similarly, survey respondents become fatigued and irritable when questioned for more than 20 or 30 minutes. As a result, the amount of data available for drawing inferences about any specific consumer is very small, although there may exist many consumers in a particular study.

The classical fixed effects approach to heterogeneity has some appeal since it delivers the individual unit-level parameter estimates and does not require the specification of any particular probability distribution of heterogeneity. However, the sparseness of individual-level data renders this approach impractical. In many situations, incomplete household-level data causes a lack of identification at the unit level. In other cases, the parameters are identified in the unit-level likelihood but the fixed effects estimates are measured with huge uncertainty which is difficult to quantify using standard asymptotic methods.

From a Bayesian perspective, modeling panel data is about the choice of a prior over a high-dimensional parameter space. The hierarchical approach is one convenient way of specifying the joint prior over unit-level parameters. Clearly, this prior will be informative and must be so in order to produce reasonable inferences. However, it is reasonable to ask for flexibility in the form of this prior distribution. In this chapter, we will introduce hierarchical models for general unit-level models and apply these ideas to a hierarchical logit setting. Recognizing the need for flexibility in the prior, we will expand the set of priors to include mixtures of normal distributions.

5.1 HETEROGENEITY AND PRIORS

A useful general structure for disaggregate data is a panel structure in which the units are regarded as independent conditional on unit-level parameters (see Case Study 2 for an example which relaxes this assumption).We monitor m units each for a possibly different number of observations. Given a joint prior on the collection of unit-level parameters, the posterior distribution can be written as

$$p(\theta_1, \ldots, \theta_m | y_1, \ldots, y_m) \propto \left[\prod_i p(y_i|\theta_i) \right] \times p(\theta_1, \ldots, \theta_m|\tau). \tag{5.1.1}$$

The term in brackets is the conditional likelihood and the rightmost term is the joint prior with hyperparameter τ. In many instances, the amount of information available for many of the units is small. This means that the specification of the functional form and hyperparameter for the prior may be important in determining the inferences made for any one unit. A good example of this can be found in choice data sets in which consumers are observed to be choosing from a set of products. Many consumers ('units') do not choose all of the alternatives available during the course of observation. In this situation, most standard choice models do not have a bounded maximum likelihood estimate (the likelihood will asymptote in a certain direction in the parameter space). For these consumers, the prior is, in large part, determining the inferences made.

Assessment of the joint prior for $(\theta_1, \ldots, \theta_m)$ is difficult due to the high dimension of the parameter space and, therefore, some sort of simplification of the form of the prior is required. One frequently employed simplification is to assume that, conditional on the hyperparameter, $(\theta_1, \ldots, \theta_m)$ are a priori independent:

$$p(\theta_1, \ldots, \theta_m | y_1, \ldots, y_m) \propto \prod_i p(y_i|\theta_i)p(\theta_i|\tau). \tag{5.1.2}$$

This means that inference for each unit can be conducted independently of all other units *conditional* on τ. This is the Bayesian analogue of fixed effects approaches in classical statistics.

The specification of the conditionally independent prior can be very important due to the scarcity of data for many of the units. Both the form of the prior and the values of the hyperparameters are important and can have pronounced effects on the unit-level inferences. For example, it is common to specify a normal prior, $\theta_i \sim N(\bar{\theta}, V_\theta)$. The normal form of this prior means that influence of the likelihood for each unit may be attenuated for likelihoods centered far away from the prior. That is, the thin tails of the normal distribution diminish the influence of outlying observations. In this sense, the specification of a normal form for the prior, whatever the values of the hyperparameters, is far from innocuous.

Assessment of the prior hyperparameters can also be challenging in any applied situation. For the case of the normal prior, a relatively diffuse prior may be a reasonable default choice. Allenby and Rossi (1993) use a prior based on a scaled version of the pooled model information matrix. The prior covariance is scaled back to represent the expected information in one observation to ensure a relatively diffuse prior. Use of this sort of normal prior will induce a phenomenon of 'shrinkage' in which the Bayes estimates (posterior means) $\{\tilde{\theta}_i = \mathrm{E}[\theta_i|\text{data}_i, \text{prior}]\}$ will be clustered more closely to the prior mean than the unit-level maximum likelihood estimates $\{\hat{\theta}_i\}$. For diffuse prior settings, the normal form of the prior will be responsible for the shrinkage effects. In particular, outliers will be 'shrunk' dramatically toward the prior mean. For many applications, this is a very desirable feature of the normal form prior. We will 'shrink' the outliers in toward the rest of the parameter estimates and leave the rest pretty much alone.

5.2 HIERARCHICAL MODELS

In general, however, it may be desirable to have the amount of shrinkage induced by the priors driven by information in the data. That is, we should 'adapt' the level of shrinkage to the information in the data regarding the dispersion in $\{\theta_i\}$. If, for example, we observe that the $\{\theta_i\}$ are tightly distributed about some location or that there is very little information in each unit-level likelihood, then we might want to increase the tightness of the prior so that the shrinkage effects are larger. This feature of 'adaptive shrinkage' was the original motivation for work by Efron and Morris (1975) and others on empirical Bayes approaches in which prior parameters were estimated. These empirical Bayes approaches are an approximation to a full Bayes approach in which we specify a second-stage prior on the hyperparameters of the conditional independent prior. This specification is called a hierarchical Bayes model and consists of the unit-level likelihood and two stages of priors:

likelihood prior	$p(y_i\|\theta_i)$
first-stage prior	$p(\theta_i\|\tau)$
second-stage prior	$p(\tau\|h)$

The joint posterior for the hierarchical model is given by

$$p(\theta_1, \ldots, \theta_m, \tau|y_1, \ldots, y_m, h) \propto \left[\prod_i p(y_i|\theta_i)p(\theta_i|\tau)\right] \times p(\tau|h).$$

In the hierarchical model, the prior induced on the unit-level parameters is not an independent prior. The unit-level parameters are conditionally, but not unconditionally, a priori independent:

$$p(\theta_1, \ldots, \theta_m|h) = \int \prod_i p(\theta_i|\tau)p(\tau|h)\,\mathrm{d}\tau.$$

If, for example, the second-stage prior on τ is very diffuse, the marginal priors on the unit-level parameters, θ_i, will be highly dependent as each parameter has a large common component. Improper priors on the hyperparameters are extremely dangerous not only because of their extreme implications for some marginals of interest, as we have seen in Chapter 4, but also because the posterior may not be proper. As Hobert and Casella (1996) point out, it is possible to define an MCMC method for a hierarchical model which does not have any posterior as its invariant distribution in the case of improper priors.

The first-stage prior (or random effect distribution) is often taken to be a normal prior. Obviously, the normal distribution is a flexible distribution with easily interpretable parameters. In addition, we can increase the flexibility of this distribution using a mixture of normals approach, as outlined in Section 5.5. We can easily incorporate observable features of each unit by using a multivariate regression specification:

$$\theta_i = \Delta' z_i + u_i, \qquad u_i \sim N(0, V_\theta) \quad \text{or} \quad \Theta = Z\Delta + U. \tag{5.2.1}$$

Θ is an $m \times k$ matrix whose rows contain each of the unit-level parameter vectors. Z is an $m \times n_z$ matrix of observations on the n_z covariates which describe differences between units. $\theta_i \sim N(\bar{\theta}, V_\theta)$ is a special case of (5.2.1) where Z is a vector of ones with length equal to the number of units. Given the Θ array, draws of Δ and V_θ can be accomplished using either a Gibbs sampler or direct draws for the multivariate regression model as

R outlined in Section 2.12 and implemented in the *bayesm* function `rmultireg`.

The hierarchical model specifies that both prior and sample information will be used to make inferences about the common parameter, τ. For example, in a normal prior, $\theta_i \sim N(\bar{\theta}, V_\theta)$, the common parameters provide the location and the spread of the distribution of θ_i. Thus, the posterior for the θ_i will reflect a level of shrinkage inferred from the data. It is important to remember, however, that the normal functional form will induce a great deal of shrinkage for outlying units even if the posterior of V_θ is centered on large values.

In classical approaches to these models, the first-stage prior is called a random effects model and is considered part of the likelihood. The random effects model is used to average the conditional likelihood to produce an unconditional likelihood which is a function of the common parameters alone:

$$\ell(\tau) = \prod_i \int p(y_i|\theta_i)p(\theta_i|\tau)\,\mathrm{d}\theta_i.$$

In the classic econometrics literature, much is made of the distinction between random coefficient models and fixed effect models. Fixed effect models are considered 'nonparametric' in the sense that there is no specified distribution for the θ_i parameters.[1] Random coefficient models are often considered more efficient but subject to specification error in the assumed random effects distribution, $p(\theta_i|\tau)$. In a Bayesian treatment, we see that the distinction between these two approaches is in the formulation of the joint prior on $\{\theta_1, \ldots, \theta_m\}$. A Bayesian 'fixed effects' approach specifies independent priors over each of the unit-level parameters while the "random effects" approach specifies a highly dependent joint prior.

The use of a hierarchical model for prediction also highlights the distinction between various priors. A hierarchical model assumes that each unit is a draw from a 'superpopulation' or that the units are exchangeable (conditional, perhaps, on some vector of covariates). This means that if we want to make a prediction regarding a new unit we can regard this new unit as drawn from the same population. Without the hierarchical structure, all we know is that this new unit is different and have little guidance as to how to proceed.

5.3 INFERENCE FOR HIERARCHICAL MODELS

Hierarchical models for panel data structures are ideally suited for MCMC methods. In particular, a 'Gibbs'-style Markov chain can often be constructed by considering the

[1] Classical inference for fixed effects models faces a fundamental conundrum: more time series observations are required for application of asymptotic theory (which is needed for nonlinear models). However, we invariably have a short panel. Various experiments in which both the number of cross-sectional units and the time dimension increase are unpersuasive. While we might accept asymptotics that allow only N to increase to infinity, we are unlikely ever to see T increase as well. But most importantly, we avoid this altogether in the Bayesian approach.

basic two sets of conditionals:

(i) $\theta_i|\tau, y_i$

(ii) $\tau|\{\theta_i\}$

The first set of conditionals exploit the fact that the θ_i are conditionally independent. The second set exploit the fact that $\{\theta_i\}$ are sufficient for τ. That is, once the $\{\theta_i\}$ are drawn from (i), these serve as 'data' to the inferences regarding τ. If, for example, the first-stage prior is normal, then standard natural conjugate priors can be used, and all draws can be done one-for-one and in logical blocks. This normal prior model is also the building block for other more complicated priors. The normal model is given by

$$\theta_i \sim N(\bar{\theta}, V_\theta)$$

$$\bar{\theta} \sim N(\bar{\bar{\theta}}, A^{-1})$$

$$V_\theta \sim \mathrm{IW}(\nu, V)$$

In the normal model, the $\{\theta_i\}$ drawn from (i) are treated as a multivariate normal sample and standard conditionally conjugate priors are used. It is worth noting that in many applications the second-stage priors are set to be very diffuse ($A^{-1} = 100I$ or larger) and the Wishart is set to have expectation I with very small degrees of freedom such as $\dim(\theta) + 3$. As we often have a larger number of units in the analysis, the data seems to overwhelm these priors and we learn a great deal about τ or, in the case of the normal prior, $(\bar{\theta}, V_\theta)$.

Drawing the $\{\theta_i\}$ given the unit-level data and τ is dependent on the unit-level model. For linear models, as illustrated in Chapter 3, we can implement a Gibbs sampler by making direct draws from the appropriate conjugate distributions. However, in most marketing applications, there is no convenient conjugate prior or a convenient way of sampling from the conditional posteriors. For this reason, most rely on some sort of Metropolis algorithm to draw θ_i. As discussed in Chapter 3, there are two very useful variants of the Metropolis algorithm – independence and random walk. Both could be used to develop a general-purpose drawing method for hierarchical models. In either case, the candidate draws require a candidate sampling density (as in the case of the independence Metropolis) or an increment density (in the case of the RW Metropolis). The performance of these algorithms will depend critically on the selection of these densities.

In both the independence and RW cases, the densities should be selected to capture the curvature and tail behavior of the conditional posterior,

$$p(\theta_i|y_i, \tau) \propto p(y_i|\theta_i)p(\theta_i|\tau). \tag{5.3.1}$$

This suggests that the Metropolis algorithm used to draw each θ_i should be customized for each cross-sectional unit. In the Metropolis literature, there is also a tradition of experimentation with the scaling of the covariance matrix of either the RW increments or the independence chain candidate density. Clearly, it is not practical to experiment with scaling factors which are customized to each individual unit. In order to develop a practical Metropolis algorithm for hierarchical models, we must provide a method of customization to the unit level which does not require experimentation.

A further complication for a practical implementation is that individual-level likelihoods may not have a maximum. For example, suppose the unit-level model is a logit model and the unit does not choose all alternatives in the sample. Then there is no maximum likelihood estimator for this unit, if we include intercepts for each choice alternative. The unit-level likelihood is increasing in any direction which moves the intercepts for alternatives never chosen to $-\infty$. Most common proposals for Metropolis sampling densities involve use of maximum likelihood estimators. One could argue that the prior in (5.3.1) avoids this problem. Proper priors as well as normal tails will usually ensure that a maximum exists – Allenby and Rossi (1993) suggest an approximate Bayes estimator which uses this posterior mode. However, using the posterior mode (and associated Hessian) in a Metropolis step would require computation of the mode and Hessian at every MCMC step and for each cross-sectional unit.[2] This could render the Metropolis algorithm computationally impractical.

There is a folk literature on the choices of Metropolis proposal densities for hierarchical models. Some advocate using the same proposal for all units and base this proposal on the asymptotic normal approximation to the pooled likelihood. Obviously, the pooled likelihood is a mixture of unit-level likelihoods so that it is possible that this proposal (even if scaled for the relative number of unit-level and total observations) has a location and curvature that is different from any single unit-level likelihood. Another popular proposal is to use the current draw of the prior as the basis of an RW chain. That is, if the first-stage prior is normal, we use the current draw of the variance of the θ as the variance of the proposal distribution. This is clearly not a good idea as it does not adapt to the information content available for a specific unit. If all units have very little information, then the prior will dominate and this idea may work acceptably. However, when there are units with a moderate to large amount of information, this RW chain will exhibit very high autocorrelation due to rejected draws which come from a prior which is much less tight than the unit-level likelihood.

We propose a class of Metropolis algorithms which use candidate sampling distributions which are customized to the unit-level likelihoods but are computationally practical in the sense that they do not require order R (the number of MCMC draws) optimizations but only require an initial set of optimizations. These candidate sampling distributions can be used as the basis of either an independence or RW Metropolis chain. In addition, our proposal does not require that each unit-level likelihood have a maximum. To handle the problem of nonexistence of maxima, we use a 'fractional' likelihood approach in which we modify the individual-level likelihood (but only for the purpose of a Metropolis proposal density) by multiplying it by a likelihood with a defined maximum:

$$\ell_i^*(\theta) = \ell_i(\theta)^{(1-w)}\bar{\ell}(\theta)^{w\beta}. \qquad (5.3.2)$$

$\bar{\ell}$ can be the pooled likelihood which almost certainly has a maximum. The β weight is designed to scale the pooled likelihood to the appropriate order so that it does not

[2] Note that the parameters of the prior in (5.3.1) will vary from MCMC step to MCMC step.

dominate the unit-level likelihood:

$$\beta = \frac{n_i}{N}. \tag{5.3.3}$$

n_i is the number of observations for the ith unit and N is the total number of observations in all units. w is a tuning constant which represents the weight of the scaled pooled likelihood relative to the individual likelihood. We only bring in the pooled likelihood for the purpose of 'regularizing' the problem, so we would typically set w to a small value such as 0.1. The pseudo-likelihood[3] in (5.3.2) can be maximized to obtain a location and scale, $\hat{\theta}_i$ and H_i, $H_i = -\left.\frac{\partial^2 \log(\ell_i^*)}{\partial\theta\partial\theta'}\right|_{\theta=\hat{\theta}_i}$. These quantities can then be combined with the prior to form a Metropolis proposal distribution. In many cases, the prior will be in a normal form so that we can combine the prior and normal approximation to the pseudo unit-level likelihood using standard theory for the Bayes linear model. This provides us with a proposal that is customized to the curvature and possible location of each unit-level likelihood. That is, if the prior is $N(\bar{\theta}, V_\theta)$, then our proposal will be based on a normal density with mean and variance

$$\begin{aligned} \theta_i^* &= (H_i + V_\theta^{-1})^{-1}(H_i\hat{\theta}_i + V_\theta^{-1}\bar{\theta}), \\ &(H_i + V_\theta^{-1})^{-1}, \end{aligned} \tag{5.3.4}$$

respectively.

For an independence Metropolis, we will use both the customized location and the curvature estimate for each unit. We note that these will be updated from draw to draw of the prior τ parameters as the chain progresses. However, each update will only use the current draw of τ and the proposal location and scale parameters. An RW chain will use only the scale parameters combined with τ. We will adopt the scaling proposal of Roberts and Rosenthal (2001) and set scaling to $2.93/\sqrt{\dim(\theta)}$. This provides us with two Metropolis algorithms which are automatically tuned. The independence chain might be regarded as somewhat higher risk than the RW chain as we require that both the curvature and the location obtained by the approximate pseudo-likelihood procedure be correct. If the location is off, the independence chain can be highly autocorrelated as it rejects 'out-of-place' candidates. The RW chain will adapt to the location of the unit-level likelihoods, but this could be at the price of higher autocorrelation. Thus, it is the risk-averse alternative.

5.4 A HIERARCHICAL MULTINOMIAL LOGIT EXAMPLE

To examine the performance of various proposed chains, we consider first the case of a hierarchical logit model. Each of the units is assumed to have an MNL likelihood and we specify a normal distribution of the logit parameters over units with mean $\Delta' z_i$ as

[3] Computation of the pseudo-likelihood estimates need only be performed once prior to initiation of the Metropolis algorithm. It should also be noted that for models without lower-dimensional sufficient statistics, the evaluation of the pseudo-likelihood in (5.3.2) requires evaluation of the pooled likelihood as well. To reduce this computational burden we can use the asymptotic normal approximation to the pooled likelihood. As we are using this likelihood only for the purpose of 'regularizing' our unit-level likelihood, the quality of this approximation is not crucial.

in (5.2.1). The hierarchical logit model takes the form:

$$
\begin{aligned}
&\ell(\beta_i|y_i, X_i) \text{ [MNL]}, \\
&B = Z\Delta + U, \qquad u_i \sim N(0, V_\beta), \\
&\text{vec}(\Delta|V_\beta) \sim N(\text{vec}(\bar{\Delta}), V_\beta \otimes A^{-1}), \\
&V_\beta \sim \text{IW}(\nu, V).
\end{aligned}
\tag{5.4.1}
$$

u_i and β_i are the ith rows of B and U. The DAG for the model in (5.4.1) is given by

$$
\begin{array}{ccc}
V_\beta & \searrow & \\
\downarrow & & \beta_i \longrightarrow y_i \\
\Delta & \nearrow &
\end{array}
\tag{5.4.2}
$$

Given a draw of B, draws of Δ, V_β can be made using standard conjugate theory for the MRM. We can define three possible chains for drawing the β_i:

(i) An independence Metropolis with draws from a multivariate Student t with location and scale given by (5.3.4). Note that both the location and scale will be influenced by the current draw of both Δ and V_β. Candidates will be drawn from a multivariate Student distribution with mean $\beta^* = (H_i + (V_\beta^r)^{-1})^{-1}(H_i\hat{\beta}_i + (V_\beta^r)^{-1}(\Delta^r)'z_i)$ and covariance proportional to $(H_i + (V_\beta^r)^{-1})^{-1}$.

(ii) An RW Metropolis with increments having covariance $s^2 V_\beta^r$, where s is a scaling constant and V_β^r is the current draw of V_β.

(iii) An improved RW Metropolis with increments having covariance $s^2(H_i + (V_\beta^r)^{-1})^{-1}$ where H_i is the Hessian of the ith unit likelihood evaluated at the MLE for the fraction likelihood defined by multiplying the MNL unit likelihood by the pooled likelihood raised to the β power.

We will choose an 'automatic' tuning scheme in which $s = 2.93/\sqrt{\dim(\beta_i)}$.

We might expect the chain defined by (i) to perform well if our location estimates for the posteriors of each MNL given the current draw of Δ and V_β are good following the intuition developed for the single-logit model in Chapter 3. However, we must recognize that the normal approximation to the logit likelihood may break down for likelihoods with no defined maximum as we have if a unit does not choose from all alternatives available. The independence Metropolis chain will not adapt to the proper location, unlike the RW chains.

The RW chains offer adaptability in location at the expense of possibly slower navigation. The RW chain defined by (ii), which simply uses V_β for the covariance of increments, is not expected to perform well for cases in which some units have a good deal of information and others very little. For units with little information, the unit-level conditional posteriors are dominated by the prior term (the unit likelihood is relatively flat) and the RW defined by (ii) may have increments of approximately the right scale. However, if a unit has a more sharply defined likelihood, the increments proposed by the chain in (ii) will be too large on average. This could create high autocorrelation due to the rejection of candidates and the consequent 'stickiness' of the chain. The RW sampler defined by (iii) does not suffer from this problem.

To investigate the properties of these three chains, we simulated data from a five-choice hierarchical logit model with four intercepts and one X variable drawn as Unif$(-1.5, 0)$ which is meant to approximate a log-price variable with a good deal of variation. One hundred units were created: 50 with only 5 observations and 50 with 50 observations. β_i is distributed as $N(\mu, V_\beta)$, and $\mu' = (1, -1, 0, 0, -3)$; V_β has diagonal elements all equal to 3 and (4,5)th and (5,4)th elements set to 1.5. Diffuse priors were used, $A = 0.01, \nu = 5 + 3, V = \nu I_5$.

Figure 5.1 shows draw sequences (every 20th draw) from the independence chain (i) and the improved RW chain (iii) for the (5,5)th element of V_β. The dark horizontal line is the 'true' parameter value. The independence chain takes an extraordinary number of draws to dissipate the initial conditions. It appears to take at least 15000 draws to reach the stationary distribution. On the other hand, the improved RW chain mixes well and dissipates the initial condition in fewer than 500 iterations. Figure 5.2 compares the two RW chains for a unit with 50 observations. The RW chain with increments based on V_β alone exhibits very poor mixing. Long runs of repeated values are shown in the figure as the chain rejects draws too far away from the mass of the posterior. The improved RW chain shows much better mixing. The numerical efficiency measure (see Section 3.9) for the improved RW chain is 3.83, compared to 8.66 for the RW chain proposed in (ii). For a unit with a small number of observations, both chains have comparable numerical efficiency (5 for the improved chain and 5.94 for the chain proposed in (ii)). The improved RW chain is implemented in the *bayesm* function `rhierMnlRwMixture`.

R

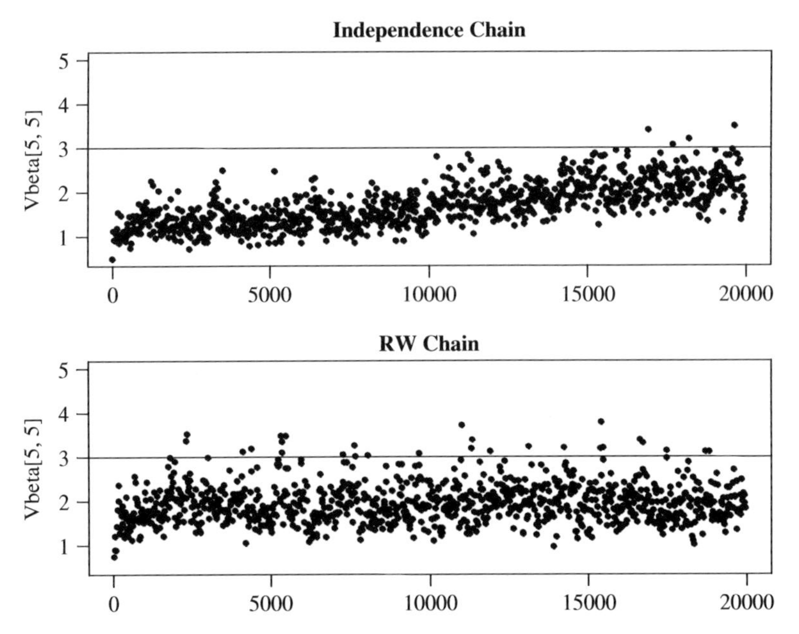

Figure 5.1 Comparison of draw sequences: independence and random-walk chains

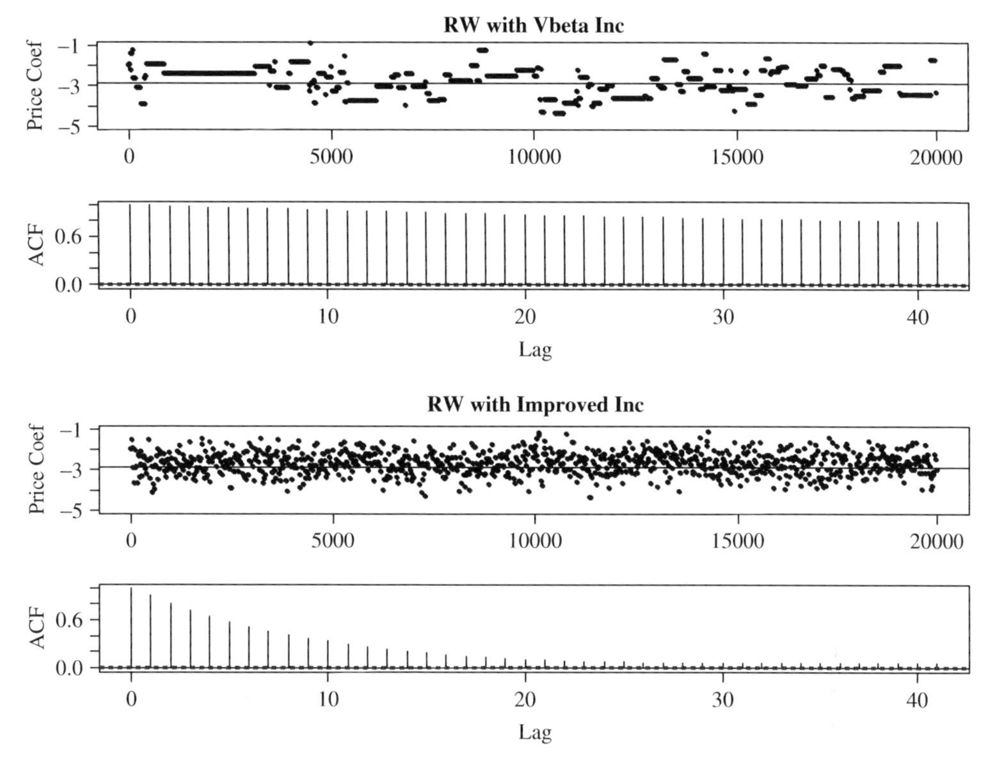

Figure 5.2 Comparison of two RW chains: draws of unit-level parameters

We also consider an example using scanner panel data on purchases of margarine. This data set was originally analyzed in Allenby and Rossi (1991) and contains purchases on ten brands of margarine and some 500 panelists. This data set is available in `bayesm` and can be loaded with the command, `data(margarine)`. Several of the brands have only very small shares of purchases and, thus, for the purposes of illustration, we consider a subset of data on purchases of six brands: (1) Parkay stick, (2) Blue Bonnett stick, (3) Fleischmanns stick, (4) House brand stick, (5) Generic stick, and (6) Shed Spread tub. We also restricted attention to those households with five or more purchases. This gives us a data set with 313 households making a total of 3405 purchases. We have brand-specific intercepts and a log-price variable in the hierarchy for a total of six unit-level logit coefficients. We also have information on various demographic characteristics of each household, including household income and family size, which form the Z matrix of observable determinants in the hierarchy. We use 'standard' diffuse prior settings of $A = 0.01$ I, $\nu = 6 + 3$, $V = \nu I$, and run the improved RW chain for 20000 iterations. Figure 5.3 shows the posterior distributions of the price coefficient for selected households. In the top row, the posterior is displayed for two households with a relatively small amount of information. It should be emphasized that these households do not have defined maxima for their unit-level likelihoods. This does not mean that we cannot learn something about their price sensitivity. The household-level data plus

R

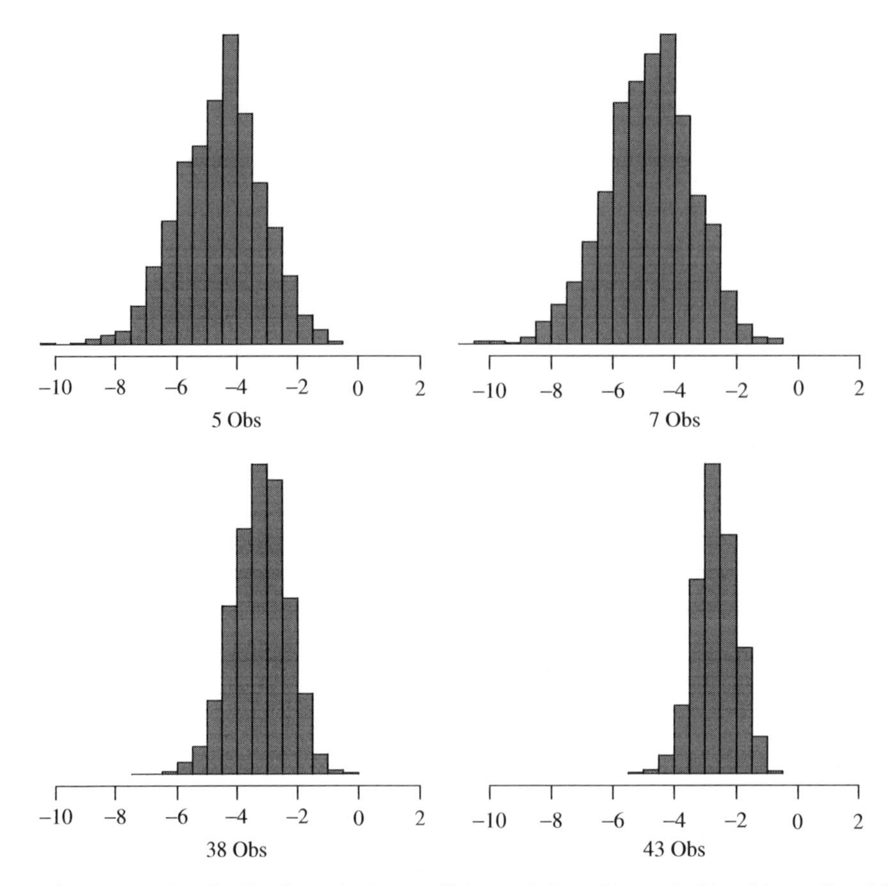

Figure 5.3 Posterior distribution of price coefficients: Selected households with small and large number of observations

the first-stage prior provide some limited information. In the bottom row of Figure 5.3 we display marginal posteriors for households with a larger number of observations. As might be expected, the posteriors sharpen up considerably.

Figure 5.4 shows the marginal posteriors for various functions of V_β. The top histogram shows the marginal posterior of the correlation between the house and generic brand intercepts. This is centered tightly over rather large values, suggesting that household preferences for house and generic brands are highly correlated, as has been suggested in the literature on private label brands. In the bottom panel of Figure 5.4 the posterior distribution of the standard deviation of the price coefficient is displayed. We note that both quantities are nonlinear functions of V_β; it is inappropriate to apply these functions to the posterior mean, $\mathrm{E}[V_\beta|\text{data}]$. Both posterior distributions exhibit substantial skewness and show, yet again, that asymptotic normal approximations to the posterior distribution of key parameters can be poor.

Table 5.1 shows the posterior means (standard deviations) of all correlations in the off-diagonal and the standard deviations of each β on the diagonal. The posterior standard deviations of the households βs are very large. This shows tremendous heterogeneity between households in brand preference and price sensitivity. Table 5.2 shows that

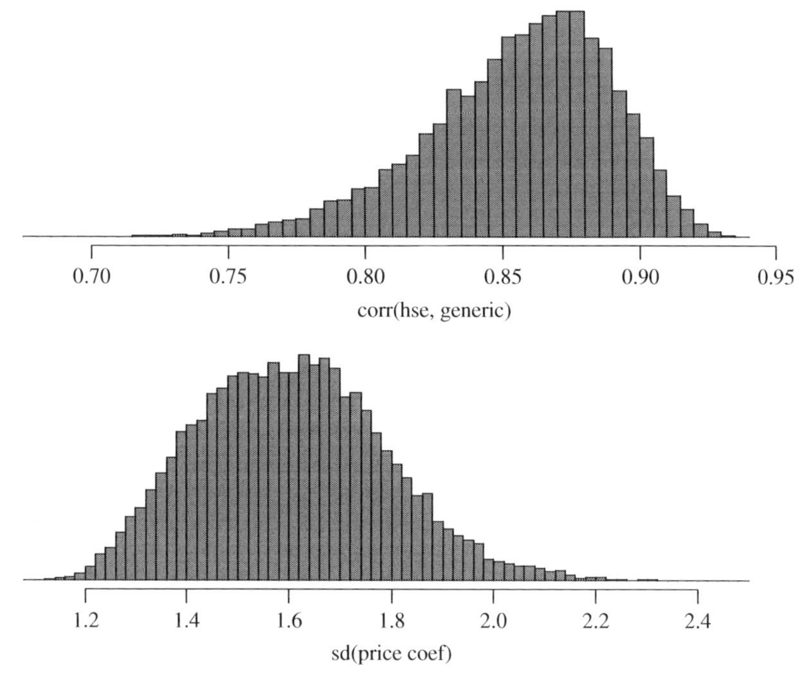

Figure 5.4 Posterior distribution of selected variance parameters

Table 5.1 Correlations and standard deviations of betas

Blue Bonnett	1.53 (0.13)	0.39 (0.13)	0.43 (0.10)	0.46 (0.10)	0.27 (0.13)	−0.07 (0.14)
Fleischmanns		3.44 (0.55)	0.31 (0.15)	0.28 (0.18)	0.09 (0.17)	0.48 (0.15)
House			2.5 (0.19)	0.86 (0.03)	0.49 (0.10)	−0.05 (0.14)
Generic				3.0 (0.27)	0.55 (0.10)	−0.08 (0.14)
Shed Spread Tub					3.0 (0.33)	0.05 (0.15)
Price						1.6 (0.18)

Diagonal contains standard deviations; off-diagonal the correlations.

very little of this measured heterogeneity can be attributed to the household demographic attributes, log income and family size. Most of the elements of Δ displayed in Table 5.2 are very imprecisely measured, particularly for the effects of income. Larger families show some preference toward the house and generic brands and shy away from Fleischmanns. However, the general impression is of a weak relationship with these demographic variables.

Table 5.2 Posterior distribution of Δ

	Blue Bonnet Intercept	Fleischmanns intercept	House intercept	Generic intercept	Shed Spread intercept	log(price)
Intercept	−1.27	−3.37	−3.31	−4.96	0.03	−3.48
	(0.64)	(1.8)	(0.99)	(1.2)	(1.2)	(0.85)
log(Income)	0.07	0.80	0.02	−0.51	−0.62	−0.26
	(0.21)	(0.59)	(0.32)	(0.40)	(0.42)	(0.28)
Family size	−0.03	−0.70	0.24	0.55	0.06	0.08
	(0.10)	(0.28)	(0.14)	(0.18)	(0.20)	(0.12)

Figure 5.5 displays the distributions of posterior means of coefficients across the 313 households. These distributions exhibit a good deal of skewness. In Section 5.7 we develop a diagnostic for our normal model of household heterogeneity which is based on comparing these distributions to the predictive distribution from our model. The predictive distribution will not be normal as we will integrate out the parameters of the first-stage prior. However, for the settings of the hyperparameters in this data analysis, the predictive distribution will be symmetric, albeit fatter-tailed than the normal. This informal evidence suggests that the normal first-stage prior may not be adequate. In the next section we will allow for a more flexible family of priors based on mixtures of normals.

5.5 USING MIXTURES OF NORMALS

Much of the work in both marketing and in the general statistics literature has used the normal prior for the first stage of the hierarchical model. The normal prior offers a great deal of flexibility and fits conveniently with a large Bayesian regression/multivariate analysis literature. The standard normal model can easily handle analysis of many units (Steenburgh *et al.* 2003), and can include observable determinants of heterogeneity (see Allenby and Ginter 1995; Rossi *et al.* 1996). Typically, we might postulate that various demographic or market characteristics might explain differences in intercepts (brand preference) or slopes (marketing mix sensitivities). In linear models, this normal prior specification amounts to specifying a set of interactions between the explanatory variables in the model explaining y; see McCulloch and Rossi (1994) for further discussion of this point.

While the normal model is flexible, there are several drawbacks for marketing applications. As discussed above, the thin tails of the normal model tend to shrink outlying units greatly toward the center of the data. While this may be desirable in many applications, it is a drawback in discovering new structure in the data. For example, if the distribution of the unit-level parameters is bimodal (something to be expected in models with brand intercepts) then a normal first-stage prior may shrink the unit-level

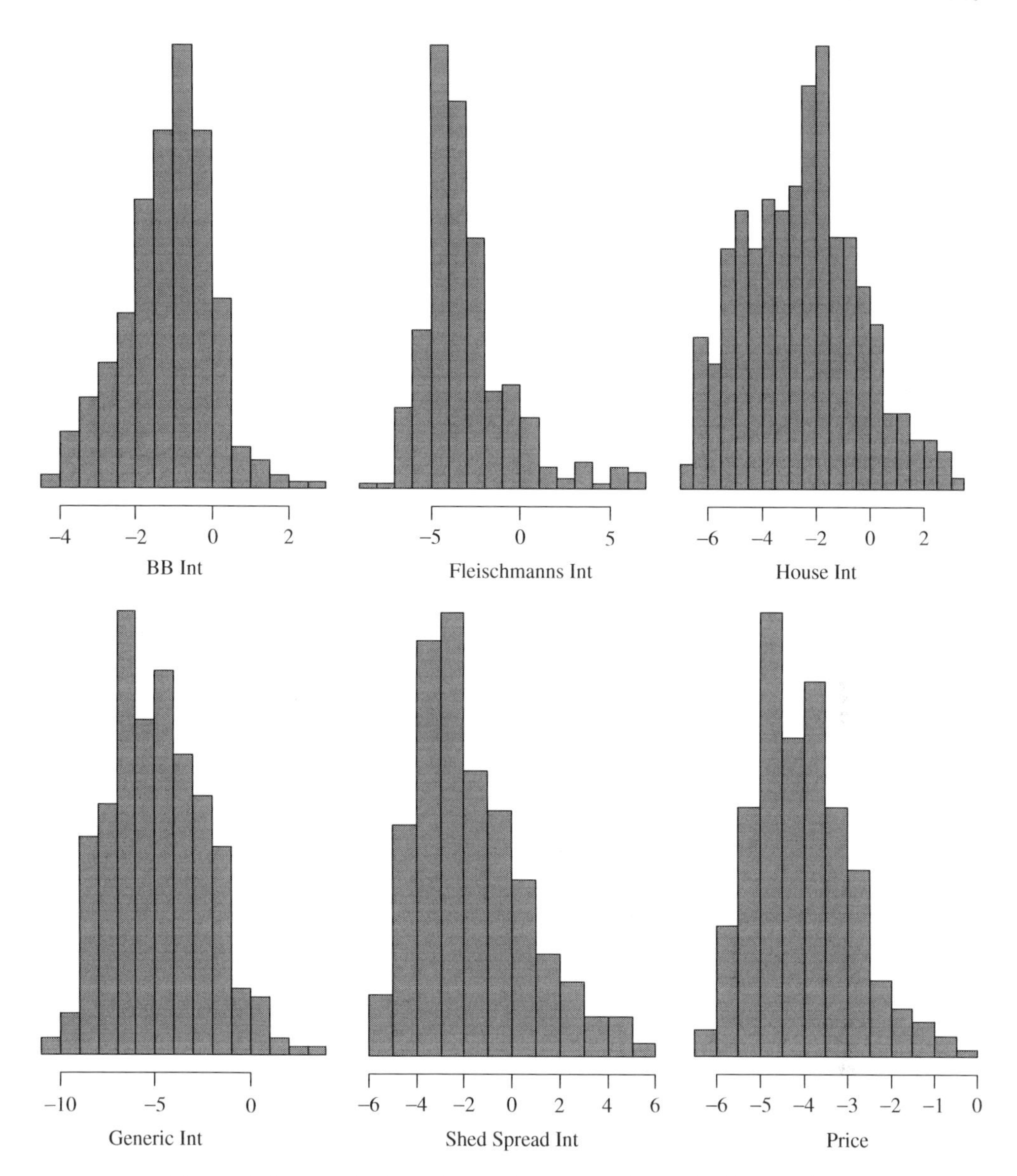

Figure 5.5 Distribution of posterior means of household coefficients

estimates to such a degree as to mask the multimodality (see below for further discussions of diagnostics). Fortunately, the normal model provides a building block for a mixture of normals extension of the first-stage prior. Mixtures of normal models provide a great deal of added flexibility. In particular, multiple modes are possible. Fatter tails than the normal can also be accommodated by mixing in normal components with large variance. It is well known that the mixture of normals model provides a great deal of flexibility and that, with enough components, virtually any multivariate density can be approximated. However, as a practical matter, we may not be able to identify significant deviations from a normal model of heterogeneity as we only observe the unit-level parameters with considerable error. Intuition developed by direct application of the mixture of normals approach to estimation of densities for directly observed data may not carry over well to the use of mixture of normals in a hierarchical setting.

The mixture of normals model can also be viewed as a generalization of the popular finite-mixture model. The finite-mixture model views the prior as a discrete distribution with a set of mass points. This approach has been very popular in marketing due to the interpretation of each mixture point as representing a 'segment' and to the ease of estimation. In addition, the finite-mixture approach can be given the interpretation of a nonparametric method as in Heckman and Singer (1984). Critics of the finite-mixture approach have pointed to the implausibility of the existence of a small number of homogeneous segments as well as the fact that the finite-mixture approach does not allow for extreme units whose parameters lie outside the convex hull of the support points. The mixture of normals approach avoids the drawbacks of the finite-mixture model while incorporating many of the more desirable features.

The mixture of K multivariate normal models can be used as the basis of the heterogeneity distribution as follows:

$$\begin{aligned} \theta_i &= \Delta' z_i + u_i, \\ u_i &\sim N(\mu_{\text{ind}_i}, \Sigma_{\text{ind}_i}), \\ \text{ind}_i &\sim \text{Multinomial}_K(\text{pvec}). \end{aligned} \tag{5.5.1}$$

ind_i is an indicator latent variable for which component observation i is from. ind takes on values $1, \ldots, K$. pvec is a vector of mixture probabilities of length K. In (5.5.1), the z vector of observable characteristics of the population does not include an intercept and has n_z elements. For this reason, we advise that z be centered so that the mean of θ given average z values will be entirely determined by the normal mixture component means. The moments of θ are as follows:[4]

$$\begin{aligned} E[\theta_i | z_i = \bar{z}, p, \{\mu_k\}] &= \bar{\mu} = \sum_{k=1}^{K} \text{pvec}_k \mu_k, \\ \text{var}(\theta_i | z_i, p, \{\mu_k\}, \{\Sigma_k\}) &= \sum_{k=1}^{K} \text{pvec}_k \Sigma_k + \sum_{k=1}^{K} \text{pvec}_k (\mu_k - \bar{\mu})(\mu_k - \bar{\mu})'. \end{aligned} \tag{5.5.2}$$

Of course, the variance loses much of its meaning and interpretability as we move farther from an elliptically symmetric distribution.

[4] The variance can be derived by using the identity $\text{var}(\theta_i) = E[\text{var}(\theta_i|\text{ind})] + \text{var}(E[\theta_i|\text{ind}])$.

As in Section 3.9, priors for the mixture of normals model can be chosen in convenient conditionally conjugate forms:

$$\begin{aligned}
&\operatorname{vec}(\Delta) = \delta \sim N(\bar{\delta}, A_{\delta}^{-1}),\\
&\text{pvec} \sim \text{Dirichlet}(\alpha)\\
&\mu_k \sim N(\bar{\mu}, \Sigma_k \otimes a_{\mu}^{-1}),\\
&\Sigma_k \sim \text{IW}(\nu, V).
\end{aligned} \tag{5.5.3}$$

The DAG for this model can be written as

$$\text{pvec} \rightarrow \text{ind} \; \begin{matrix} \nearrow \Sigma_k \searrow \\ \downarrow \\ \searrow \mu_k \nearrow \end{matrix} \; \theta_i \rightarrow y_i, \quad \Delta \uparrow \theta_i \tag{5.5.4}$$

The K parameters in α determine the tightness of the prior on the mixture component probabilities as they are from a natural conjugate prior in which they can be interpreted as cell counts for the components from a previous sample of size, $n^* = \sum_k \alpha_k$. The priors on the mixture components are iid across components in (5.5.3). Much of the statistics literature on mixtures of normals has considered only mixtures of normals for univariate or for low-dimensional multivariate distributions. Unless there is only a small amount of data or a very large number of components, the priors for the mixture component parameters may not be very influential in this setting and, thus, may not require too careful consideration. However, in the case of hierarchical models and marketing applications, the mixture model may be applied to parameter vectors of relatively high dimension (such as multinomial choice model parameters) and the priors will matter as the dimension of parameter space of the normal components may easily exceed 200 or 300. This means that more attention should be paid to prior settings in our applications.

5.5.1 A Hybrid Sampler

We can easily define an MCMC chain of Gibbs style by alternating between the draws of individual unit-level parameters and mixture components:

$$\theta_i | \text{ind}_i, \Delta' z_i, \mu_{\text{ind}_i}, \Sigma_{\text{ind}_i}, \tag{5.5.5}$$

$$\text{pvec}, \text{ind}, \Delta, \{\mu_k\}, \{\Sigma_k\} | \{\Theta\}, \tag{5.5.6}$$

$$\Theta = \begin{bmatrix} \theta_1' \\ \vdots \\ \theta_m' \end{bmatrix}; \text{ind} = \begin{bmatrix} \text{ind}_1 \\ \vdots \\ \text{ind}_m \end{bmatrix}.$$

Some advocate margining out the indicators of the mixture components and using a direct Metropolis step for (5.5.6). The argument here is that removal of these latent variables may improve the mixing of the chain. The likelihood function for the mixture of normals can be evaluated at very low cost. This is certainly possible for mixtures of univariate normals in which one could stack up the means and log variances into a vector to be drawn either piecemeal or in one shot by an RW Metropolis. However, in the case of mixtures of multivariate normals, this would require using a Metropolis algorithm to navigate in a very high-dimensional space of positive definite matrices. Experience with Metropolis algorithms for covariance structures has shown they are very difficult to tune for satisfactory performance for 5 and higher dimensions. For a mixture of normals, we can easily require a parameter space with as many as 10 covariance matrices, each one of which might have 20 or more parameters.

The draw of the hierarchical parameters in (5.5.6) can be broken down into a succession of conditional draws (note that the prior parameters are suppressed in (5.5.7) to focus discussion on the nature of the draws):

$$\begin{aligned} &\text{ind}|\text{pvec}, Z, \Delta, \{\mu_k, \Sigma_k\}, \Theta \\ &\text{pvec}|\text{ind} \\ &\{\mu_k, \Sigma_k\}|\text{ind}, \Theta \\ &\Delta|\text{ind}, Z, \{\mu_k, \Sigma_k\}, \Theta \end{aligned} \tag{5.5.7}$$

Here we view the $\{\theta_i\}$ or Θ as the 'data' generated by a mixture of normals with mean driven by a multivariate regression with explanatory variables in the $m \times n_z$ matrix Z. First, we draw the indicators for each component, which provides a classification of the 'observations' into one of each of the K components. Given the indicators, there are essentially K independent multivariate normal samples on which conjugate draws can be performed to update the $\{\mu_k, \Sigma_k\}$ parameters. We also must update our views on the mixture probabilities. Since the observable variables in Z affect the means of all components, the draw of Δ must be done by pooling across all observations, adjusting for heteroskedasticity.

The role of the z variables is to shift the mean of the normal mixture on the basis of observations. All of the normal mixture parameters should, therefore, be drawn on the 'data' with this component of the mean removed:

$$\Theta^* = \Theta - Z\Delta. \tag{5.5.8}$$

As in Section 3.9, the draw of the indicators is a multinomial draw based on the likelihood ratios with pvec_k as the prior probability of membership in each component:

$$\begin{aligned} &\text{ind}_i \sim \text{multinomial}(\pi_i); \qquad \pi' = (\pi_{i,1}, \ldots, \pi_{i,K}), \\ &\pi_{i,k} = \frac{\text{pvec}_k \varphi(\theta_i^* | \mu_k, \Sigma_k)}{\sum_m \text{pvec}_m \varphi(\theta_i^* | \mu_m, \Sigma_m)}. \end{aligned} \tag{5.5.9}$$

Here $\varphi(\bullet)$ is the multivariate normal density.

The draw of pvec given the indicators is a Dirichlet draw:

$$
\begin{aligned}
&\text{pvec} \sim \text{Dirichlet}(\tilde{\alpha}), \\
&\tilde{\alpha}_k = n_k + \alpha_k, \\
&n_k = \sum_{i=1}^{n} I(\text{ind}_i = k).
\end{aligned} \tag{5.5.10}
$$

The draw of each (μ_k, Σ_k) can be made using the algorithm to draw from the multivariate regression model as detailed in Section 2.8.5. For each subgroup of observations, we have an MRM model of the form

$$
\Theta_k^* = \iota\mu_k' + U, \qquad U = \begin{bmatrix} u_1' \\ \vdots \\ u_{n_k}' \end{bmatrix}, \qquad u_i \sim N(0, \Sigma_k). \tag{5.5.11}
$$

Here Θ_k^* is the submatrix of Θ^* that consists of the n_k rows where $\text{ind}_i = k$. We can use

R the function `rmultireg` in *bayesm*, to achieve these draws.

The draw of Δ requires that we pool data from all K components into one regression model. Since we are proceeding conditionally on the component means and variances, we can appropriately standardize the 'data' and perform one draw from a standard Bayesian regression model. To motivate the final draw result, let us first consider the kth component. We subset both the Θ and Z matrices to consider only those observations from the kth component and subtract the mean. If nvar is the dimension of the parameter vectors $\{\theta_i\}$, let Θ_k, Z_k be $n_k \times \text{nvar}$ and $n_k \times n_z$ arrays consisting of only those observations for which $\text{ind}_i = k$:

$$
Y_k = \Theta_k - \iota\mu_k. \tag{5.5.12}
$$

We can write the model for these observations in the form

$$
Y_k = Z_k\Delta + U_k \quad \text{or} \quad Y_k' = \Delta' Z_k' + U_k'. \tag{5.5.13}
$$

We will stack these nvar equations up to see how to standardize:

$$
\text{vec}(Y_k') = (Z_k \otimes I_{\text{nvar}})\text{vec}(\Delta') + \text{vec}(U_k'), \tag{5.5.14}
$$

where $\text{var}(\text{vec}(U_k')) = I_{n_k} \otimes \Sigma_k$ and $\Sigma_k = R_k' R_k$. Therefore, if we multiply through by $I_k \otimes (R_k^{-1})'$, this will standardize the error variances in (5.5.14) to give an identity covariance structure:

$$
\begin{aligned}
&\left(I_{n_k} + (R_k^{-1})'\right)\ \text{vec}(Y_k') = \left(Z_k \otimes (R_k^{-1})'\right)\ \text{vec}(\Delta') + z_k, \\
&\text{var}(z_k) = I_{n_k \times \text{nvar}}.
\end{aligned} \tag{5.5.15}
$$

We can stack up the K equations of the form of (5.5.15):

$$y = X\delta + z,$$

$$y = \begin{bmatrix} I_{n_1} \otimes (R_1^{-1})' \operatorname{vec}(Y_1') \\ \vdots \\ I_{n_K} \otimes (R_K^{-1})' \operatorname{vec}(Y_K') \end{bmatrix},$$

$$X = \begin{bmatrix} Z_1' \otimes (R_1^{-1})' \\ \vdots \\ Z_K' \otimes (R_K^{-1})' \end{bmatrix}. \tag{5.5.16}$$

Here $\delta = \operatorname{vec}(\Delta')$. Given our prior, $\delta \sim N\left(\bar{\delta}, (A_\delta)^{-1}\right)$, we can combine with (5.5.16) to compute the conditional posterior in the standard normal form:

$$\delta | y, X, \bar{\delta}, A_\delta \sim N\left((X'X + A_\delta)^{-1}(X'y + A_\delta\bar{\delta}), (X'X + A_\delta)^{-1}\right). \tag{5.5.17}$$

The moments needed for (5.5.17) can be calculated efficiently as follows:

$$X'X = \sum_{k=1}^{K} \left(Z_k'Z_k \otimes R_k^{-1}(R_k^{-1})'\right) = \sum_{k=1}^{K} (Z_k'Z_k \otimes \Sigma_k)$$

$$X'y = \operatorname{vec}\left(\sum_{k=1}^{K} \Sigma_j^{-1} Y_k' Z_k\right) \tag{5.5.18}$$

5.5.2 *Identification of the Number of Mixture Components*

Given that it is possible to undertake posterior simulation of models with 10 or more components, there is some interest in determining the number of components from the data and priors. For mixtures of univariate normals, Richardson and Green (1997) propose an application of the reversible jump sampler that, in principle, allows for MCMC navigation of different size mixture models. The Richardson and Green sampler can 'jump' up or down to mixture models of different sizes. In theory, one might be able to use the frequency with which the chain visits a given size component model as an estimate of the posterior probability of that size model. The reversible jump sampler requires a mapping from a lower-dimensional mixture component model to a higher-dimensional mixture component model. Dellaportas and Papageorgiou (2004) propose a method for extending the Richardson and Green ideas to multivariate mixtures. It remains to be seen how well this will work for the case of a mixture model embedded within a hierarchical setting.

The other approach to determining the number of mixture components is to attempt to compute the posterior probability of models with a fixed number of components on the basis of simulation output. That is to say, we run models with 1, 5, and 10 components and attempt to compute the Bayes factors for each model. Some have used asymptotic approximations to the Bayes factors such as the Schwarz approximation. DiCiccio *et al.*

(1997) provide a review of various methods which use simulation output and various asymptotic approximations. All asymptotic methods require finding the posterior mode either by simulation or numerical optimization. This may be particularly challenging in the case of the mixture of multivariate normals in which the likelihood exhibits multiple modes and the parameter space can be extremely high-dimensional. In Chapter 6, we will review a number of these methods and return to the problem of computing Bayes factors for high-dimensional models with nonconjugate set-ups. Lenk and DeSarbo (2000) compute Bayes factors for the number of mixture components in hierarchical generalized linear models.

Given that the normal mixture model is an approximation to some underlying joint density, the main goal in exploring models with different numbers of components is to ensure the adequacy of the approximation. That is to say, we want to ensure that we include enough components to capture the essential features of the data. The danger of including too many components is that our estimated densities may 'overfit' the data. For example, the mixture approximation may build many small lobes on the joint density in an attempt to mimic the empirical distribution of the data in much the same way as kernel smoothing procedures produce lumpy or multimodal density estimates with a too small bandwidth selection. In a hierarchical setting, this is made all the more difficult by the fact that the 'data' consist of unknown parameters and we are unable to inspect the empirical distribution. This means that prior views regarding the smoothness of the density are extremely important in obtaining sensible and useful density estimates. The fact that we are in a hierarchical setting where the parameters are not observed directly may help us obtain smoother density estimates as the normal mixture will not be able to fit particular noise in the empirical distribution of the 'data' as this distribution is only known with error. Thus, devoting a mixture component to accommodating a few outlying data points will not occur unless these outliers are determined very precisely.

'Testing' for the number of components or computing Bayes factors is of greater interest to those who wish to attach substantive importance to each component, for example, to proclaim there are X number of subpopulations in the data. It is also possible to use posterior probabilities for model averaging. The purpose of model averaging, in this context, is to ensure smoothness by averaging over models of varying numbers of components. Our view is that individual components are not very interpretable and are only included for flexibility. Adequate smoothness can be built in via the priors on mixture component parameters. Thus, we take a more informal approach where we investigate fits from models with varying numbers of components. With informative priors that ensure adequate smoothness, addition of components that are not necessary to fit the patterns in the data will not change the fitted density.

5.5.3 Application to Hierarchical Models

The normal mixture model provides a natural generalization to the standard normal model of the distribution of heterogeneity. In this section, we will apply this model to a hierarchical MNL model. In the literature on mixtures of normals, investigators typically use very diffuse informative priors on the mixture component parameters $\{\mu_k, \Sigma_k\}$ and pvec. Improper priors or even very diffuse proper priors are dangerous in any Bayesian context, and especially so in the case of marketing data. Typically, panel data on the choice of products includes a subset of panelists who do not purchase from the complete set of products. Prior beliefs about logit model parameters will be very important for this set of panelists. Diffuse but proper priors applied to these panelists will result in the inference that these panelists are essentially never willing to purchase these products under any setting of the model covariates if product or brand intercepts are included in the model. We will simply set the intercepts for products not purchased to large negative numbers. Given the logistic probability locus, this will result in zero probability of purchase for all intents and purposes. We do not find this plausible. The probability of purchase of these products may be low but is not zero. In a one-component normal mixture, the other households inform the first-stage prior so that we never obtain extreme estimates of intercepts for these panelists with incomplete purchase histories. The thin tails of the normal density as well as reasonable values of the covariance matrix keep us from making extreme inferences. However, in the case of more than one normal component, this can change. If, for example, there is a group of panelists who do not purchase product A, then the mixture model can assign these panelists to one component. Once this assignment is made, the mean product A intercept values for this component will drift off to very large negative numbers. This problem will be particularly acute when a reasonably large number of components are considered.

There are two ways to deal with this problem (note that the option of deleting panelists with incomplete purchase records is not defensible!): either use models with very small numbers of components or use informative priors on the means of each component. Given that we center the Z variables, the prior on $\{\mu_k\}$ reflects our views regarding intercepts. Recall that we use an $N(\bar{\mu}, \Sigma_k \otimes A_{\mu}^{-1})$ prior. In much of the work with mixtures, A_{μ} is set to very small values (say, 0.01). Our view is that this admits implausible intercept values of -20 or $+20$. This, of course, is only meaningful if all X variables are on the same scale and location. For this reason, we advocate standardizing the X variables. We then set A_{μ} to $1/16$ or so rather than $1/100$.

The prior on Σ_k is also important. If we set a tight prior around a small value, then we may force the normal mixture to use a large number of components to fit the data. In addition, the natural conjugate prior links the location and scale so that tight priors over Σ will influence the range of plausible μ values. We will set the prior on Σ to be relatively diffuse by setting ν to nvar $+3$ and $V = \nu I$.

We return to the margarine example discussed in Section 5.4 There are five brand intercepts and one price sensitivity parameter so that nvar $= 6$ and we are fitting a six-dimensional distribution of β_i over households. We can combine the Gibbs sampler for normal mixtures with an RW Metropolis chain defined along the lines of Section 5.3 to draw the household-level parameters. We should emphasize that we do have some 300 households, but this is not the same as 300 direct observations on six-dimensional data. With 300 direct observations, the normal mixture sampler works very well, recovering components with relative ease. However, in the hierarchical logit example, there is only a small amount of information about each household parameter vector. It will be much more difficult to recover complex structure with this effective sample size.

Figure 5.6 presents posterior means of the marginal densities, contrasting one- and five-component mixture models. For each MCMC draw, we have one fitted multivariate density and we can average these densities over the R draws. To obtain the posterior

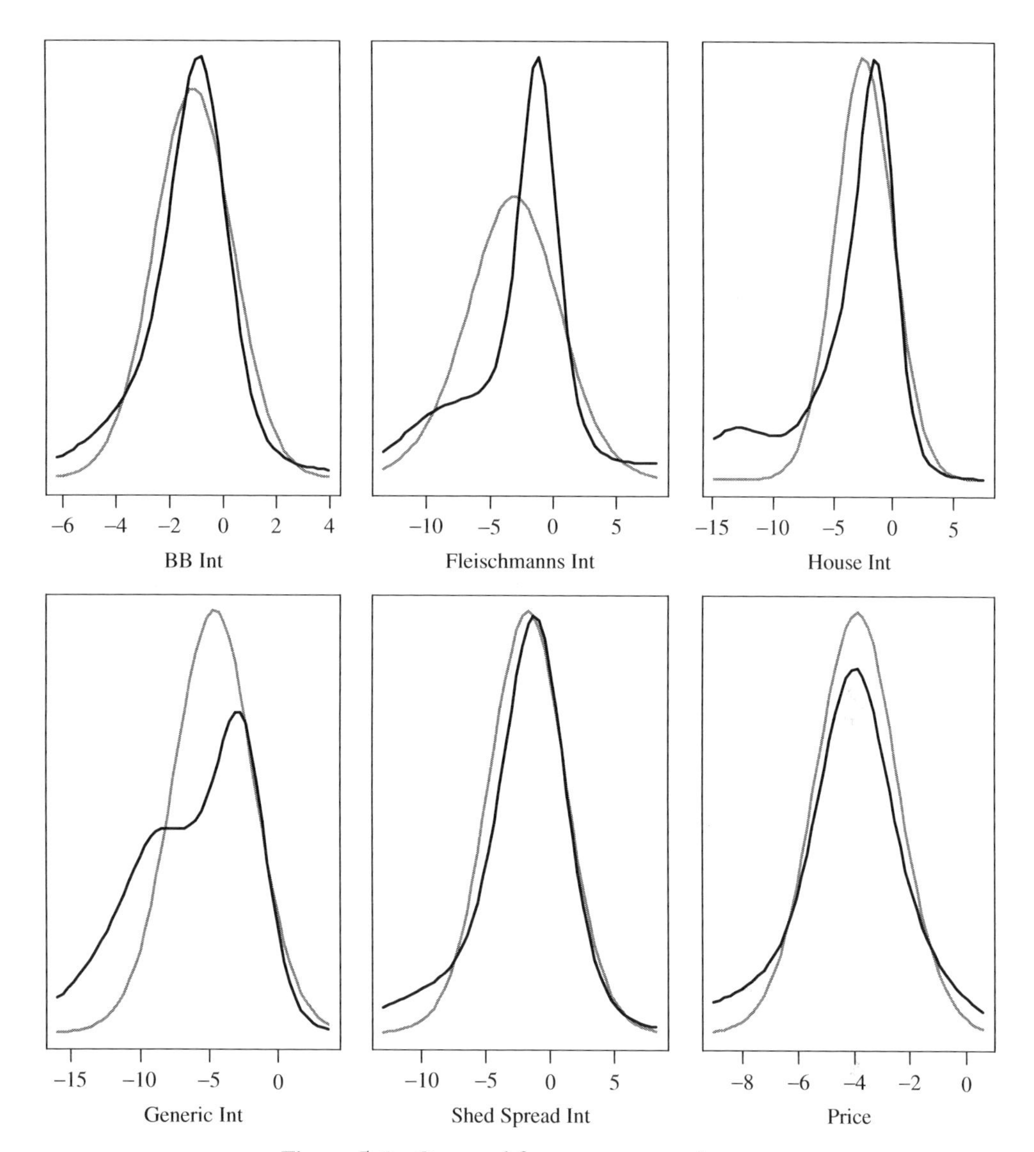

Figure 5.6 One- and five-component mixtures

mean of a marginal density for a specific element of β, we average the marginal densities:

$$\bar{d}_j(\beta_j) = \frac{1}{R}\sum_{r=1}^{R}\sum_{k=1}^{K} \text{pvec}_k^r \phi_j(\beta_j|\mu_k^r, \Sigma_k^r). \tag{5.5.19}$$

We set down a grid of possible values for each element of β and then evaluate the posterior mean of the density in (5.5.19). These densities are shown in the figure for $K = 1$ and $K = 5$. For at least four of the six elements of β, we see pronounced deviations from normality. For Fleischmanns, house and generic intercepts, we obtain highly left-skewed distributions with some left lobes. Recall that these are intercepts with the base brand set to Parkay stick. This means that these are relative preferences, holding price constant. For Fleischmanns and the house brands there is a mass centered close to zero with a very thick left tail. We can interpret this as saying that there are a number of households who view the house and Fleischmanns brands as equivalent in quality to Parkay but that there are a number of other households who regard these brands as decidedly inferior.

The nonnormality of the estimated first stage prior also has a strong influence on the estimates of household posterior means, as illustrated in Figure 5.7. The fat tails of the five-component normal mixture allow for more extreme estimates of both brand intercepts and price sensitivity. Thus, Figure 5.7 demonstrates that the observed differences in Figure 5.6 make a material difference even if one is only concerned with developing household estimates. However, one should be cautious before using the household estimates based on the five component model. The one-component model provides very strong shrinkage of extreme estimates and should, therefore, be regarded as somewhat conservative.

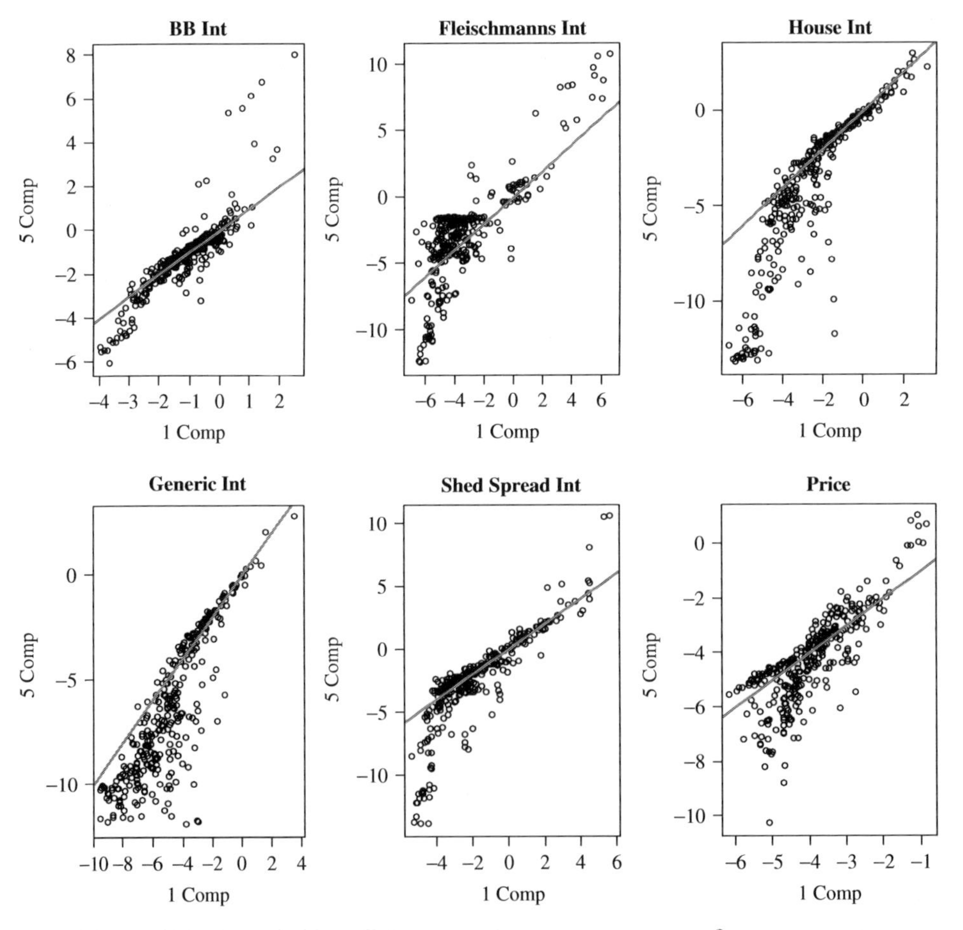

Figure 5.7 Household coefficient posterior means: one versus five components

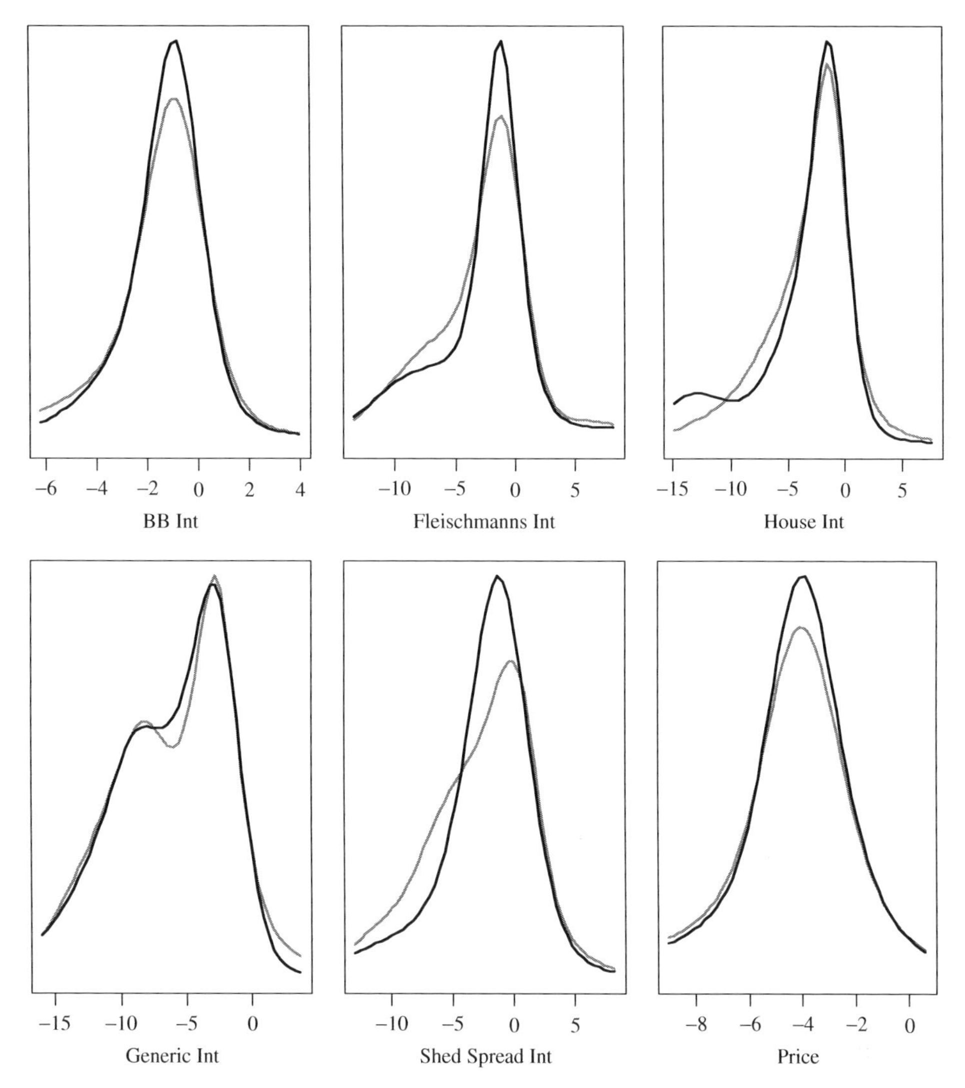

Figure 5.8 Five-component (black) and ten-component (gray) mixtures

Our fitting strategy for normal mixture models also included adding a large number of components to see if this makes a material difference in the estimated distribution of household-level parameters. It should be pointed out that the marginal computation cost of increasing the number of normal components is rather trivial compared to the cost of the Metropolis draws of the individual household parameters. From a computational point of view, the mixture model is basically free and, therefore, can be used routinely. In Figure 5.8, we consider a 10-component mixture and compare this to the five-component one. Our view is that the differences in fitted densities in Figure 5.8 are rather small. Figure 5.9 compares the household posterior means for five- and ten-component models. There is quite close agreement between the estimates derived from five and ten component models.

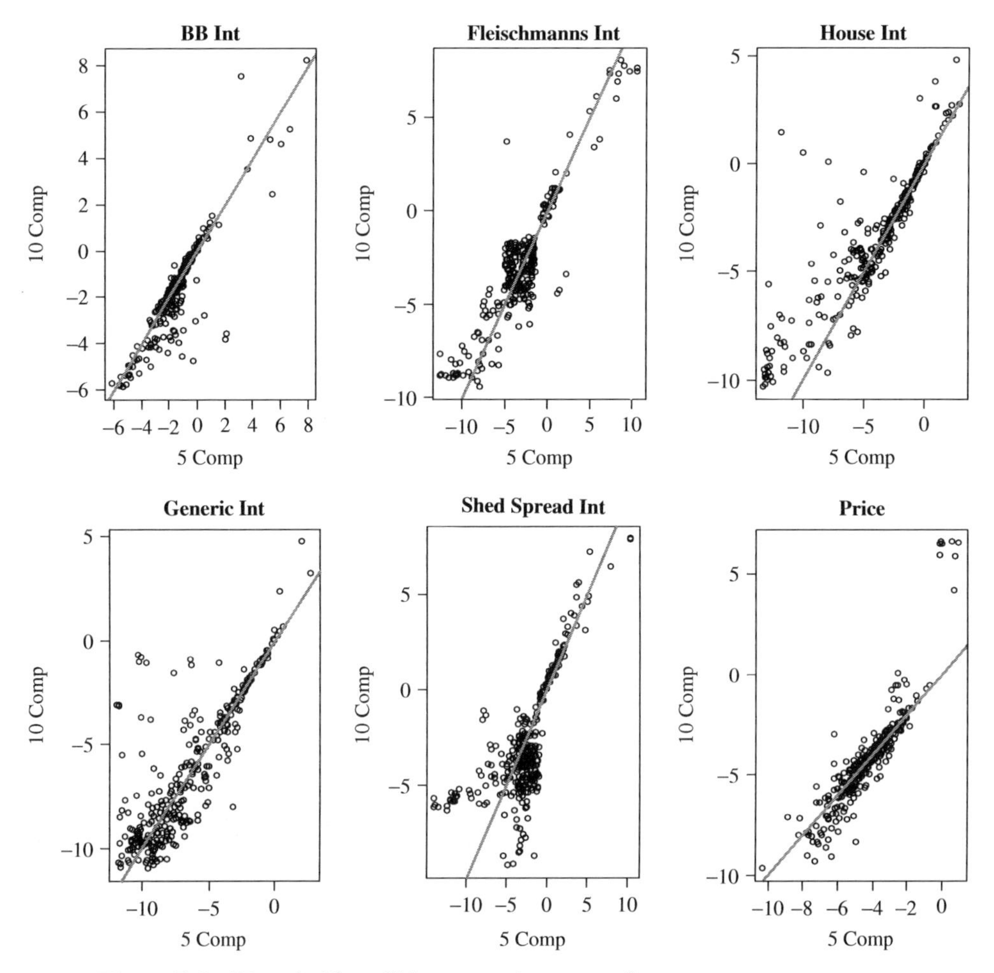

Figure 5.9 Household coefficient posterior means: five versus ten components

5.6 FURTHER ELABORATIONS OF THE NORMAL MODEL OF HETEROGENEITY

In many situations, we have prior information on the signs of various coefficients in the base model. For example, price parameters are negative and advertising effects are positive. In a Bayesian approach, this sort of prior information can be included by modifying the first-stage prior. We replace the normal distribution with a distribution with restricted support, corresponding to the appropriate sign restrictions. For example, we can use a log-normal distribution for a parameter which is restricted via sign by the reparameterization $\theta' = \ln(\theta)$. However, note that this change in the form of the prior can destroy some of the conjugate relationships which are exploited in the Gibbs sampler. However, if Metropolis-style methods are used to generate draws in the Markov chain, it is a simple matter to directly reparameterize the likelihood function, by substituting

$\exp(\theta')$ for θ, rather than rely on the heterogeneity distribution to impose the range restriction.

What is more important is to ask whether the log-normal prior is appropriate. The left tail of the log-normal distribution declines to zero, ensuring a mode for the log-normal distribution at a strictly positive value. For situations in which we want to admit zero as a possible value for the parameter, this prior may not be appropriate. Boatwright *et al.* (1999) explore the use of truncated normal priors as an alternative to the log-normal reparameterization approach. Truncated normal priors are much more flexible, allowing for mass to be piled up near zero.

Bayesian models can also accommodate structural heterogeneity, or changes in the likelihood specification for a unit of analysis. The likelihood is specified as a mixture of likelihoods:

$$p(y_{it}|\{\theta_{ik}\}) = r_1 p_1(y_{it}|\theta_{i1}) + \cdots + r_K p_K(y_{it}|\theta_{iK}),$$

and estimation proceeds by appending indicator variables for the mixture component to the state space. Conditional on the indicator variables, the datum y_{it} is assigned to one of K likelihoods. The indicator variables, conditional on all other parameters, have a multinomial distribution with probabilities proportional to the number of observations assigned to the component and the probability that the datum arises from the likelihood. Models of structural heterogeneity have been used to investigate intra-individual change in the decision process due to environmental changes (Yang and Allenby 2000) and fatigue (Otter *et al.* 2003).

Finally, Bayesian methods have recently been used to relax the commonly made assumption that the unit parameters θ_i are iid draws from the distribution of heterogeneity. Ter Hofstede *et al.* (2002) employ a conditional Gaussian field specification to study spatial patterns in response coefficients:

$$p(\theta_i|\tau) = p(\theta_i|\{\theta_j : j \in S_i\}, V_\theta),$$

where S_i denote units that are spatially adjacent to unit i. Since the MCMC estimation algorithm employs full conditional distributions of the model parameters, the draw of θ_i involves using a local average for the mean of the mixing distribution. Yang and Allenby (2003) employ a simultaneous specification of the unit parameters to reflect the possible presence of interdependent effects due to the presence of social and information networks:

$$\theta = \rho W\theta + u, \qquad u \sim N(0, \sigma^2 I),$$

where W is a matrix that specifies the network, ρ is a coefficient that measures the influence of the network, and u is an innovation. We discuss this model at length in Case Study 2.

5.7 DIAGNOSTIC CHECKS OF THE FIRST-STAGE PRIOR

In the standard normal hierarchical model, the prior is specified in a two-stage process:

$$\theta \sim N(\bar{\theta}, V_\theta),$$

$$p(\bar{\theta}, V_\theta).$$

In the classical literature, the normal distribution of θ would be called the random effects model and would be considered part of the likelihood rather than part of the prior. Typically, very diffuse priors are used for the second stage. Thus, it is the first-stage prior which is important and will always remain important as long as there are only a few observations available per household. Since the parameters of the first-stage prior are inferred from the data, the main focus of concern should be on the form of this distribution.

In the econometrics literature, the use of parametric distributions of heterogeneity (for example, normal distributions) is often criticized on the grounds that their misspecification leads to inconsistent estimates of the common model parameters (see Heckman and Singer 1984). For example, if the true distribution of household parameters were skewed or bimodal, our inferences based on a symmetric, unimodal normal prior could be misleading. One simple approach would be to plot the distribution of the posterior household means and compare this to the implied normal distribution evaluated at the Bayes estimates of the hyperparameters, $N(\mathrm{E}[\bar{\theta}|\text{data}], \mathrm{E}[V_\theta])$. The posterior means are not constrained to follow the normal distribution since the normal distribution is only part of the prior and the posterior is influenced by the unit-level data. This simple approach is in the right spirit but could be misleading since we do not properly account for uncertainty in the unit-level parameter estimates.

Allenby and Rossi (1999) provide a diagnostic check of the assumption of normality in the first stage of the prior distribution that properly accounts for parameter uncertainty. To handle uncertainty in our knowledge of the common parameters of the normal distribution, we compute the predictive distribution of $\theta_{i'}$ for unit i' selected at random from the population of households with the random effects distribution. Using our data and model, we can define the predictive distribution of $\theta_{i'}$ as follows:

$$\theta_{i'}|\text{data} = \iint \phi(\theta_{i'}|\bar{\theta}, V_\theta) p(\bar{\theta}, V_\theta|\text{data})\, \mathrm{d}\bar{\theta}\, \mathrm{d}V_\theta.$$

Here $\phi(\theta_{i'}|\bar{\theta}, V_\theta)$ is the normal prior distribution. We can use our MCMC draws of $\bar{\theta}$, V_θ, coupled with draws from the normal prior, to construct an estimate of this distribution. The diagnostic check is constructed by comparing the distribution of the unit-level posterior means to the predictive distribution based on the model, given above.

5.8 FINDINGS AND INFLUENCE ON MARKETING PRACTICE

The last decade of work on heterogeneity in marketing has yielded several important findings. Researchers have explored a rather large set of first-stage models with a normal distribution of heterogeneity across units. In particular, investigators have considered a first-stage normal linear regression (Blattberg and George 1991), a first-stage logit model (Allenby and Lenk 1994, 1995), a first-stage probit (McCulloch and Rossi 1994), a first-stage Poisson (Neelamegham and Chintagunta 1999), and a first-stage generalized gamma distribution model (Allenby *et al.* 1999). The major conclusion is that there is a substantial degree of heterogeneity across units in various marketing data sets. This finding of a large degree of heterogeneity holds out substantial promise for the study of preferences, in terms of both substantive and practical significance (Ansari *et al.* 2000). There may be substantial heterogeneity bias in models that do not properly account for heterogeneity (Chang *et al.* 1999), and there is great value in customizing marketing decisions to the unit level (see Rossi *et al.* 1996).

Yang *et al.* (2002) investigate the source of brand preference, and find evidence that variation in the consumption environment, and resulting motivations, leads to changes in a unit's preference for a product offering. Motivating conditions are an interesting domain for research as they preexist the marketplace, offering a measure of demand that is independent of marketplace offerings. Other research has documented evidence that the decision process employed by a unit is not necessarily constant throughout a unit's purchase history (Yang and Allenby 2000). This evidence indicates that the appropriate unit of analysis for marketing is at the level that is less aggregate than a person or respondent, although there is evidence that household sensitivity to marketing variables (Ainslie and Rossi 1998) and state dependence (Seetharaman *et al.* 1999) is constant across categories.

The normal continuous model of heterogeneity appears to do reasonably well in characterizing this heterogeneity, but there has not yet been sufficient experimentation with alternative models such as the mixture of normals to draw any definitive conclusions. With the relatively short panels typically found in marketing applications, it may be difficult to identify much more detailed structure beyond that afforded by the normal model. In addition, relatively short panels may produce a confounding of the finding of heterogeneity with various model misspecifications in the first stage. If only one observation is available for each unit, then the probability model for the unit level is the mixture of the first-stage model with the second-stage prior:

$$p(y|\tau) = \int p(y|\theta)p(\theta|\tau)\,d\theta.$$

This mixing can provide a more flexible probability model. In the one-observation situation, we can never determine whether it is 'heterogeneity' or lack of flexibility that causes the Bayesian hierarchical model to fit the data well. Obviously, with more than one observation per unit, this changes and it is possible to separately diagnose first-stage model problems and deficiencies in the assumed heterogeneity distribution. However, with short panels there is unlikely to be a clean separation between these problems and it may be the case that some of the heterogeneity detected in marketing data is really due to lack of flexibility in the base model.

There have been some comparisons of the normal continuous model with the discrete approximation approach of a finite-mixture model. It is our view that it is conceptually inappropriate to view any population of units of being comprised of only a small number of homogeneous groups and, therefore, the appropriate interpretation of the finite-mixture approach is as an approximation method. Allenby *et al.* (1998), Allenby and Rossi (1999) and Lenk *et al.* (1996) show some of the shortcomings of the finite-mixture model and provide some evidence that it does not recover reasonable unit-level parameter estimates.

At the same time that the Bayesian work in the academic literature has shown the ability to produce unit-level estimates, there has been increased interest on the part of practitioners in unit-level analysis. Conjoint researchers have always had an interest in respondent-level part-worths and had various ad hoc schemes for producing these estimates. Recently, the Bayesian hierarchical approach to the logit model has been implemented in the popular Sawtooth conjoint software. Experience with this software and simulation studies have led Rich Johnson, Sawtooth Software's founder, to conclude that Bayesian methods are superior to others considered in the conjoint literature (Sawtooth Software 2001).

Retailers are amassing volumes of store-level scanner data. Not normally available to academic researchers, this store-level data is potentially useful for informing the basic retail decisions such as pricing and merchandizing. Attempts to develop reliable models for pricing and promotion have been frustrated by the inability to produce reliable promotion and price response parameters. Thus, the promise of store-level pricing has gone unrealized. Recently, a number of firms have appeared in this space, offering data-based pricing and promotion services to retail customers. At the heart of some of these firms' approach is a Bayesian shrinkage model applied to store–item–week data obtained directly from the retail client. The Bayesian shrinkage methods produce reasonable and relatively stable store-level parameter estimates. This approach builds directly on the work of Montgomery (1997).

6
Model Choice and Decision Theory

Using this Chapter

This chapter discusses Bayesian model choice and decision theory. Bayesian model choice involves various approaches to computing the posterior probability of a model. Posterior model probabilities are useful in comparing two or more competing models or in the choice from a class of models. While there are some methods which can use standard MCMC output to approximate these probabilities, most problems require additional computations for accurate evaluation of posterior probabilities. Sections 6.1–6.9 introduce various methods for computation of model probabilities and compare some of the most useful methods in the context of a model comparison motivated by the multinomial probit model. Many marketing problems suggest a natural decision problem (such as profit maximization) so that there is more interest in nontrivial applications of decision theory. Sections 6.10 and 6.11 introduce a Bayesian decision-theoretic approach to marketing problems and provide an example by considering the valuation of disaggregate sample information.

Most of the recent Bayesian literature in marketing emphasizes the value of the Bayesian approach to inference, particularly in situations with limited information. Bayesian inference is only a special case of the more general Bayesian decision-theoretic approach. Bayesian decision theory has two critical and separate components: a loss function and the posterior distribution. The loss function associates a loss with a state of nature and an action, $\ell(a, \theta)$, where a is the action and θ is the state of nature. The optimal decision-maker chooses the action so as to minimize expected loss, where the expectation is taken with respect to the posterior distribution:

$$\min_a \; \overline{\ell}(a) = \int \ell(a, \theta) p(\theta|\text{Data}) \, d\theta.$$

As indicated in Chapter 2, inference about the model parameters can be viewed as a special case of decision theory where the 'action' is to choose an estimate based on the data. Quadratic loss yields the posterior mean as the optimal (minimum expected loss) estimator.

Bayesian Statistics and Marketing P. E. Rossi, G. M. Allenby and R. McCulloch

Model choice can also be thought of as a special case of decision theory where there is a zero–one loss function. If loss is 0 when the correct model is chosen and 1 if not, then the optimal action is to choose the model with highest posterior probability. In Sections 6.1–6.9 we develop methods for computing posterior probabilities for a set of models.

In marketing applications, a natural loss function is the profits of a firm. The firm seeks to determine marketing actions so as to maximize the expected profits which arise from these actions. In Section 6.10 we develop this loss framework and apply this idea to the valuation of various information sets. An example of targeted couponing is introduced to make these ideas concrete in Section 6.11.

6.1 MODEL SELECTION

In many scientific settings, the action is a choice between competing models. In the Bayesian approach, it is possible to define a set of models $M_1, \ldots, M_k$ and calculate the posterior probability of each of the models. The posterior probability of a model can be obtained from the data likelihood and the prior model probability in the usual manner[1].

$$p(M_i|y) = \frac{p(y|M_i)p(M_i)}{p(y)}. \tag{6.1.1}$$

If the set of models $\{M_1, \ldots, M_k\}$ is exhaustive, we can compute the posterior probability of model i as

$$p(M_i|y) = \frac{p(y|M_i)p(M_i)}{\sum_j p(y|M_j)p(M_j)}. \tag{6.1.2}$$

In many instances, we might wish to compare two models and choose the model with higher posterior probability. For these cases, the ratio of the posterior probabilities (called the posterior odds ratio) is relevant. The posterior odds ratio is the ratio of relative likelihood times the prior odds ratio:

$$\begin{aligned} \frac{p(M_1|y)}{p(M_2|y)} &= \frac{p(y|M_1)}{p(y|M_2)} \times \frac{p(M_1)}{p(M_2)}. \\ &= \text{BF} \times \text{Prior odds}, \end{aligned} \tag{6.1.3}$$

where BF is the Bayes factor.

In a parametric setting, the posterior probability of a model requires that we integrate out the parameters by averaging the density of the data over the prior density of the parameters,

$$p(y|M_i) = \int p(y|\theta, M_i)p(\theta|M_i)\,d\theta. \tag{6.1.4}$$

---

[1] Throughout this chapter, we will use the notation y to refer to the observed data. This is for ease of reference to the Bayesian model choice literature.

Some write this as the expectation of the likelihood with respect to the prior distribution and, thus, call it the 'marginal likelihood':

$$\ell^*(y|M_i) = E_{\theta|M_i}[\ell(\theta|y, M_i)]. \tag{6.1.5}$$

However, it should be noted that the likelihood is any function proportional to the data density so that this interpretation is somewhat imprecise. Henceforth in this chapter, when we use the notation $\ell(\bullet)$, we mean the density of the data including all normalizing constants. The intuition is that if the likelihood of the model is high where our prior puts most probability mass then the model has high posterior probability.

The posterior odds ratio for parametric models can be written

$$\frac{p(M_1|y)}{p(M_2|y)} = \frac{\int \ell_1(\theta_1)p_1(\theta_1)\,d\theta_1}{\int \ell_2(\theta_2)p_2(\theta_2)\,d\theta_2} \times \frac{p(M_1)}{p(M_2)} = \text{BF} \times \text{Prior odds}. \tag{6.1.6}$$

BF stands for 'Bayes factor', the ratio of marginal likelihoods for the two models.

In the Bayesian approach, the posterior probability only requires specification of the class of models and the priors. There is no distinction between nested and non-nested models as in the classical hypothesis-testing literature. However, we do require specification of the class of models under consideration; there is no omnibus measure of the plausibility of a given model or group of models versus some unspecified and possibly unknown set of alternative models.

In the classical testing literature, there is an important distinction made between the non-nested and nested cases. The Neyman–Pearson approach to hypothesis testing requires the specification of a specific null hypothesis. In the case of non-nested models, there is no natural 'null' model and classical methods for hypothesis testing can lead to contradictory results in which model 1 is rejected in favor of model 2 and vice versa. The Bayesian approach uses the predictive density of the data under each model in the Bayes factor (6.1.4). The predictive density is the density of the data averaged over the prior distribution of the model parameters. This can be defined for any set of models, nested or non-nested.

Equation (6.1.4) also reveals the sensitivity of the Bayes factor to the prior. This is not a limitation or weakness of the Bayesian approach but simply a recognition that model comparison depends critically on the assessment of a prior. From a practical point of view, this means that the researcher must carefully select the prior for each model to ensure that it does indeed reflect his views regarding the possible set of parameter values. In particular, 'standard' diffuse prior settings can be deceptive. As the diffuseness of the prior increases for a particular model, the value of the predictive density in (6.1.4) will decline, at least in the limit. This means that relative diffuseness of the priors is important in the computation of the Bayes factor for comparison of two models. If one prior is very diffuse relative to the other, then the Bayes factor will tend to favor the model with the less diffuse prior. Improper priors are the limiting case of proper but diffuse priors, and must be avoided in the computation of Bayes factors.

We also can interpret the marginal density of the data given model i as the normalizing constant of the posterior:

$$p(\theta|y, M_i) = \frac{\ell(\theta|y, M_i)p(\theta|M_i)}{p(y|M_i)} \tag{6.1.7}$$

If we let $\tilde{p}(\theta|y, M_i)$ denote the 'unnormalized' posterior, then the marginal density of the data can be written

$$p(y|M_i) = \int \tilde{p}(\theta|y, M_i)\,d\theta = \frac{\tilde{p}(\theta|y, M_i)}{p(\theta|y, M_i)}. \tag{6.1.8}$$

6.2 BAYES FACTORS IN THE CONJUGATE SETTING

For conjugate models, (6.1.8) can be used to compute Bayes factors. That is, the product of the data density and the prior can be divided by the full density form for the posterior to obtain the marginal likelihood of the data. However, this requires fully conjugate priors, not just conditionally conjugate priors. Care must be taken to include all appropriate normalizing constants of the posterior, data density and priors:

$$p(y|M_i) = \frac{\tilde{p}(\theta|y, M_i)}{p(\theta|y, M_i)} = \frac{p(y|\theta, M_i)p(\theta|M_i)}{p(\theta|y, M_i)}.$$

For nested hypotheses, a simplification of the Bayes factor, called the Savage–Dickey density ratio, can be used (see Allenby (1990) for an application of the Savage-Dickey approach to scanner data). Consider the case of comparison of model M_0 and M_1, where M_0 is a restricted version of M_1. Transform[2] θ to ϕ so that the restriction amounts to setting a subvector of ϕ to some specified value (often 0):

$$M_0 : \phi_1 = \phi_1^h, \qquad M_1 : \text{unrestricted},$$

where $\phi' = (\phi_1', \phi_2')$. In this case, the Bayes factor for comparison of M_0 to M_1 is given by

$$\frac{p(y|M_0)}{p(y|M_1)} = \frac{\int \ell(\phi_2|y)p(\phi_2)\,d\phi_2}{\iint \ell(\phi_1, \phi_2|y)p(\phi_1, \phi_2)\,d\phi_1\,d\phi_2}, \tag{6.2.1}$$

where $\ell(\phi_2|y) = \ell(\phi_1, \phi_2|y)|_{\phi_1=\phi_1^h}$.

One 'natural' choice for the prior on the unrestricted component, ϕ_2, is the conditional prior derived from the joint prior under M_1:

$$p(\phi_2|\phi_1 = \phi_1^h) = \left.\frac{p(\phi_1, \phi_2)}{\int p(\phi_1, \phi_2)\,d\phi_2}\right|_{\phi_1=\phi_1^h}. \tag{6.2.2}$$

Using the prior in (6.2.2), the Bayes factor can be written

$$\text{BF} = \frac{\int \ell(\phi_1, \phi_2|y)(p(\phi_1, \phi_2)/p(\phi_1))\,d\phi_2|_{\phi_1=\phi_1^h}}{\iint \ell(\phi_1, \phi_2)p(\phi_1, \phi_2)\,d\phi_1\,d\phi_2}$$

[2] In most cases, this must be a linear transformation to take full advantage of the Savage–Dickey simplication.

$$= \left. \frac{\int \ell(\phi_1, \phi_2|y)p(\phi_1, \phi_2)\,d\phi_2}{p(\phi_1)\iint \ell(\phi_1, \phi_2|y)p(\phi_1, \phi_2)\,d\phi_1\,d\phi_2} \right|_{\phi_1=\phi_1^h} \tag{6.2.3}$$

$$= \left. \frac{\int p(\phi_1, \phi_2|y)\,d\phi_2}{p(\phi_1)} \right|_{\phi_1=\phi_1^h}.$$

Thus, the Bayes factor can be written as the ratio of the marginal posterior of ϕ_1 to the marginal prior of ϕ_1 evaluated at $\phi_1 = \phi_1^h$. We also note that if the conditional posterior of $\phi_1|\phi_2$ is of known form, then we can write the marginal posterior as the average of the conditional. We can estimate this with MCMC output from the marginal posterior of ϕ_2:

$$\mathrm{BF} = \left. \frac{\int p(\phi_1|\phi_2, y)p(\phi_2|y)\,d\phi_2}{p(\phi_1)} \right|_{\phi_1=\phi_1^h},$$
$$\widehat{\mathrm{BF}} = \left. \frac{R^{-1}\sum_r p(\phi_1|\phi_2^r, y)}{p(\phi_1)} \right|_{\phi_1=\phi_1^h}. \tag{6.2.4}$$

Outside of the conjugate setting, the computation of Bayes factors must rely on various numerical methods for computing the requisite integrals unless asymptotic methods are used. For most problems, asymptotic methods which rely on the approximate normality of the posterior are not reliable.[3] We briefly review the asymptotic approach to computation of Bayes factors.

6.3 ASYMPTOTIC METHODS FOR COMPUTING BAYES FACTORS

Asymptotic methods can be used to approximate model probabilities. The idea is that the posterior converges to a normal distribution and then we can use results for the multivariate normal distribution to approximate the marginal likelihood and, therefore, the posterior model probability:

$$p(y|M_i) = \int p(y|\theta_i, M_i)p(\theta_i|M_i)\,d\theta_i. \tag{6.3.1}$$

We can approximate the integral in (6.3.1) using the normal approximation to the posterior. This is achieved by expanding the log of the unnormalized posterior around its mode in a Taylor series:

$$\begin{aligned} p(y|M_i) &= \int \exp(\Gamma(\theta))\,d\theta \\ &\approx \int \exp\left(\Gamma(\tilde\theta) - \tfrac{1}{2}(\theta-\tilde\theta)'H(\tilde\theta)(\bullet)\right) d\theta = \exp(\Gamma(\tilde\theta))(2\pi)^{p/2}|H(\tilde\theta)|^{-1/2} \\ &= p(y|\tilde\theta, M_i)p(\tilde\theta|M_i)(2\pi)^{p/2}|H(\tilde\theta)|^{-1/2} \end{aligned} \tag{6.3.2}$$

[3] One notable exception is the multinomial logit model whose likelihood, we have already seen, closely resembles a normal density.

where $\tilde{\theta}$ is the posterior mode and

$$H(\tilde{\theta}) = -\left.\frac{\partial^2 \tilde{p}(\theta|y)}{\partial\theta\partial\theta'}\right|_{\theta=\tilde{\theta}},$$

is the negative of the Hessian of the unnormalized posterior, with $\tilde{p}(\theta|y) = \exp(\Gamma(\theta))$. The approximate Bayes factor for comparison of two models will depend on the ratio of the likelihoods as well as the ratio of the prior densities evaluated at the posterior mode:

$$\mathrm{BF} = \frac{p(y|M_1)}{p(y|M_2)} \approx \frac{p(\tilde{\theta}_1|M_1)}{p(\tilde{\theta}_2|M_2)} \times \frac{p(y|\tilde{\theta}_1, M_1)|H_1(\tilde{\theta}_1)|^{-1/2}}{p(y|\tilde{\theta}_2, M_2)|H_2(\tilde{\theta}_2)|^{-1/2}} \times (2\pi)^{(p_1-p_2)/2}. \tag{6.3.3}$$

If we make one of the priors in (6.3.3) more diffuse, then the prior density evaluated at the posterior mode will decline and the approximate Bayes factor will move in favor of the other model.

Of course, it is possible to base the asymptotic approximation by expanding about the MLE rather than the posterior mode:

$$\mathrm{BF} = \frac{p(y|M_1)}{p(y|M_2)} \approx \frac{p(\hat{\theta}_{\mathrm{MLE},1}|M_1)}{p(\hat{\theta}_{\mathrm{MLE},2}|M_2)} \times \frac{p(y|\hat{\theta}_{\mathrm{MLE},1}, M_1)|\mathrm{Inf}_1(\hat{\theta}_{\mathrm{MLE},1})|^{-1/2}}{p(y|\hat{\theta}_{\mathrm{MLE},2}, M_2)|\mathrm{Inf}_2(\hat{\theta}_{\mathrm{MLE},2})|^{-1/2}} \times (2\pi)^{(p_1-p_2)/2}, \tag{6.3.4}$$

where $\mathrm{Inf}_i(\theta) = [\partial^2 \log \ell_i/\partial\theta\partial\theta']$ is the observed information matrix. Note that this is the observed information in the sample of size N. For vague priors, there will be little difference. However, for informative priors, maximum accuracy can be obtained by expanding about the posterior mode. Computation of the posterior mode is no more difficult or time-intensive than computation of the MLE. Moreover, the posterior may be a more regular surface to maximize over than the likelihood. An extreme example of this occurs when the maximum of the likelihood fails to exist or in cases of nonidentified parameters.

The approximate Bayes factors in (6.3.3) and (6.3.4) differ only by constants which do not depend on n. While these two approximations are asymptotically equivalent, we have reason to believe that the Bayes factor based on the posterior mode may be more accurate. Both expressions are dependent on the ordinate of the prior, while the expression in (6.3.3) also depends on the curvature of the prior (the curvature of the log posterior is the sum of the curvature of the prior and the likelihood). It is possible to define an asymptotic approximation to the marginal density of the data which depends only on the dimension of the model (see Schwarz 1978). If we expand about the MLE and rewrite the information matrix in terms of the average information, the posterior probability of model i can be written as

$$\begin{aligned} p(M_i|y) &\propto p(y|M_i) \\ &\approx k_i p(\hat{\theta}_{\mathrm{MLE},i}|M_i) p(y|\hat{\theta}_{\mathrm{MLE},i}, M_i)(2\pi)^{p_i/2} n^{-p_i/2} \left|\mathrm{Inf}_i(\hat{\theta}_{\mathrm{MLE},i})/n\right|^{-1/2}, \end{aligned} \tag{6.3.5}$$

where p_i is the dimension of the parameter space for model i. Asymptotically, the average information converges to the expected information in one observation. If we drop everything that is not of order n, then (6.3.5) simplifies to

$$p(M_i|y) \approx p(y|\hat{\theta}_{\text{MLE},i}, M_i)n^{-p_i/2}. \tag{6.3.6}$$

Expression (6.3.6) is often computed in log form to select a model from a group of models by picking the model with highest approximate log-posterior probability:

$$\log(p(M_i|y)) \approx \log(\ell_i(\hat{\theta}_{\text{MLE},i})) - \frac{p_i}{2}\log(n). \tag{6.3.7}$$

This form is often called the Bayesian information criterion (BIC) or Schwarz criterion. The prior has no influence on the BIC expression. In some situations, the BIC is used to compute an approximation to the Bayes factor:

$$\begin{aligned} \text{BF} &\approx \log(\ell_1(\hat{\theta}_1)) - \log(\ell_2(\hat{\theta}_2)) - \left(\frac{p_1 - p_2}{2}\right)\log(n) \\ &= \log(\text{LR}_{1,2}) - \frac{\Delta p}{2}\log(n). \end{aligned} \tag{6.3.8}$$

The BIC can be extremely inaccurate and should be avoided whenever possible. However, (6.3.8) is useful to illustrate a fundamental intuition for Bayes factors. The posterior model probability includes an 'automatic' or implicit penalty for models with a larger number of parameters. The Bayes factor recognizes that adding parameters can simply 'overfit' the data and this is automatically accounted for without resort to ad hoc procedures such as out-of-sample validation.

6.4 COMPUTING BAYES FACTORS USING IMPORTANCE SAMPLING

The marginal density of the data for each model can be written as the integral of the unnormalized posterior over the parameter space as in (6.1.8). We can apply importance sampling techniques to this problem (as in Gelfand and Dey 1994, eqn (23)):

$$p(y|M_i) = \int \tilde{p}(\theta|y)\,d\theta = \int \frac{\tilde{p}(\theta|y)}{q(\theta)}q(\theta)\,d\theta. \tag{6.4.1}$$

Here $q(\bullet)$ is the importance sampling density. We note that, unlike applications of importance sampling to computing posterior moments, we require the full importance sampling density form, including normalizing constants. Using draws from the importance sampling density, we can estimate the Bayes factor as a ratio of integral estimates. We note that a separate importance density will be required for both models:

$$\begin{aligned} \widehat{\text{BF}} &= \frac{R^{-1}\sum_r w_r(M_1)}{R^{-1}\sum_r w_r(M_2)}, \\ w_r(M_i) &= \frac{\tilde{p}(\theta_i^r|y, M_i)}{q_i(\theta_i^r)}, \qquad \theta_i^r \sim q_i. \end{aligned} \tag{6.4.2}$$

Choice and calibration of the importance sampling density are critical to the accuracy of the importance sampling approach. Two suggestions can be helpful. First, we should transform to a parameterization that is unrestricted in order to use an elliptically symmetric importance density. For example, we should transform variance/scale parameters to an unrestricted parameterization. For the scalar case, we can write $\theta = \exp(\gamma)$ and use the associated Jacobian, $\exp(\gamma)$. For covariance matrices we can transform to the non-diagonal elements of the Cholesky root and exp() of the diagonal elements.

$$\Sigma = U'U,$$

$$U = \begin{bmatrix} e^{\gamma_{11}} & \gamma_{12} & \cdots & \gamma_{1p} \\ 0 & e^{\gamma_{22}} & \ddots & \vdots \\ \vdots & \ddots & \ddots & \gamma_{p-1,p} \\ 0 & \cdots & 0 & e^{\gamma_{pp}} \end{bmatrix}, \tag{6.4.3}$$

$$J(\Gamma) = 2^p \prod_{i=1}^{p} e^{\gamma_{ii}(p-i+1)} \prod_{i=1}^{p} e^{\gamma_{ii}} = 2^p \prod_{i=1}^{p} e^{\gamma_{ii}(p+2-i)}. \tag{6.4.4}$$

Typically, we would use a multivariate Student t importance density with moderate degrees of freedom (to ensure fatter tails than the unnormalized posterior) and use the MCMC draws from the posterior to assess a mean and covariance matrix in the transformed parameters:

$$\gamma \sim \text{MSt}(\nu, \overline{\gamma}, s^2 \hat{V}),$$

$$\overline{\gamma} = \frac{1}{R} \sum_r \gamma_{\text{mcmc}}^r,$$

$$\hat{V} = \frac{1}{R} \sum_r (\gamma_{\text{mcmc}}^r - \overline{\gamma})(\gamma_{\text{mcmc}}^r - \overline{\gamma})'. \tag{6.4.5}$$

The importance density would be tuned by selecting the constant s to ensure that the distribution of the importance weights is reasonable (not driven by a small number of outliers). The importance sampling estimate of the model probability would now be expressed as

$$\hat{p}(y) = \frac{1}{R} \sum_r \frac{\ell(\theta(\gamma^r)) p(\theta(\gamma^r)) J(\gamma^r)}{q(\gamma^r)}. \tag{6.4.6}$$

6.5 BAYES FACTORS USING MCMC DRAWS

Typically, we have an MCMC method implemented for each model under consideration. This gives us the ability to simulate from the posterior distribution of each model's parameters. Therefore, there is a natural interest in methods which can express Bayes factors as the expectation of quantities with respect to the posterior distribution. These identities allow for the 'reuse' of already existing posterior draws for the purpose of estimating the posterior expectation.

Gelfand and Dey (1994) provide one such basic identity,

$$\int \frac{q(\theta)}{\tilde{p}(\theta|y, M_i)} p(\theta|y, M_i)\, d\theta = \frac{1}{p(y|M_i)}, \tag{6.5.1}$$

which can be verified as follows:

$$\begin{aligned}\int \frac{q(\theta)}{\tilde{p}(\theta|y, M_i)} p(\theta|y, M_i)\, d\theta &= \int \frac{q(\theta)}{p(\theta|M_i)p(y|\theta, M_i)} p(\theta|y, M_i)\, d\theta \\ &= \frac{1}{p(y|M_i)} \int q(\theta)\, d\theta = \frac{1}{p(y|M_i)}.\end{aligned}$$

This derivation makes it clear that q must be a proper density. Equation (6.5.1) can be used to express the Bayes factor as a ratio of posterior expectations:

$$\mathrm{BF}(M_1 \text{ vs. } M_2) = \frac{p(y|M_1)}{p(y|M_2)} = \frac{\mathrm{E}_{\theta|y,M_2}\left[q_2(\theta)/\tilde{p}(\theta|y, M_2)\right]}{\mathrm{E}_{\theta|y,M_1}\left[q_1(\theta)/\tilde{p}(\theta|y, M_1)\right]} \tag{6.5.2}$$

We can estimate each of the marginal densities of the data by

$$\hat{p}(y|M_i) = \left(\frac{1}{R}\sum_{r=1}^{R} \frac{q_i(\theta^r)}{\ell(\theta^r|M_i)p(\theta^r|M_i)}\right)^{-1}. \tag{6.5.3}$$

It should be noted that for some models evaluation of the likelihood can be computationally demanding and (6.5.3) will require many thousands of likelihood evaluations.

The q function above plays a role analogous to the reciprocal of an importance function. As with an importance function, it is important that the q function 'match' or mimic the posterior as closely as possible. This will minimize the variance of the 'weights'

$$\begin{aligned}\hat{p}(y|M_i) &= \frac{1}{R^{-1}\sum_{r=1}^{R} w^r(M_i)}, \\ w^r(M_i) &= \frac{q_i(\theta^r)}{\ell(\theta^r|M_i)p(\theta^r|M_i)}.\end{aligned}$$

However, the desirable tail behavior for the q function is exactly the opposite of that of an importance density. The tails of the q function serve to attenuate the influence of small values of the posterior density on the estimator. Because of the reciprocal formula, small values of the posterior density can create an estimator with infinite variance. On a practical level, a few 'outliers' in the $\{\theta^r\}$ can dominate all other draws in the estimate of the marginal data density in (6.5.3). For this reason it is important to choose a q function with thin tails relative to the posterior. For problems in high dimensions, it may be difficult to select a q function that works well in the sense of matching the posterior while still having thin tails.

A special case of (6.5.2)–(6.5.3) is the estimator of Newton and Raftery (1994) where $q(\theta) = p(\theta|M_i)$:

$$\hat{p}(y|M_i) = \left(\frac{1}{R}\sum_{r=1}^{R}\frac{1}{\ell(\theta^r|M_i)}\right)^{-1}. \tag{6.5.4}$$

This is the harmonic mean of the likelihood values evaluated at the posterior draws. **R** The function `logMargDenNR` in *bayesm* computes this estimator. Thus, only the likelihood must be evaluated to compute the Newton–Raftery (NR) estimate. Many researchers examine the sequence plot of the log-likelihood over the MCMC draws as an informal check on the model fit and convergence. The NR estimate uses these same likelihood evaluations. In our experience, the sequence plots of log-likelihood values for two or more competing models can be more informative than the computation of Bayes factors via the NR method.

The NR estimate has been criticized as having undesirable sampling properties. In many applications, only a handful of draws determine the value of the NR estimate. If the data is not very informative about the parameters and vague or relative diffuse priors are used, then some of the posterior draws can give rise to very small values of the likelihood and make the NR estimate unstable. More carefully assessed informative priors can improve the performance of the NR and Gelfand–Dey estimates.

Jacquier and Polson (2002) provide another useful identity for the computation of Bayes factors from MCMC output. They consider the case of nested models:

$$\begin{aligned}\theta' &= (\theta_1', \theta_2'),\\ M_0 &: \theta_1 = \theta_1^h,\\ M_1 &: \text{unrestricted}.\end{aligned}$$

We can write the Bayes factor as

$$\mathrm{BF} = \frac{p(y|M_0)}{p(y|M_1)} = \frac{\int p(y|\theta_2, M_0)p(\theta_2|M_0)\,d\theta_2}{p(y|M_1)}. \tag{6.5.5}$$

Integrating this expression over the marginal prior distribution of θ_1 under M_1, we will still obtain the Bayes factor as it is a constant not dependent on model parameters:

$$\mathrm{BF} = \int\left[\frac{\int p(y|\theta_2, M_0)p(\theta_2|M_0)\,d\theta_2}{p(y|M_1)}\right]p(\theta_1|M_1)\,d\theta_1. \tag{6.5.6}$$

Using the relationship

$$p(y|M_1) = \frac{p(y|\theta_1, \theta_2, M_1)p(\theta_1, \theta_2|M_1)}{p(\theta_1, \theta_2|M_1)},$$

we can write (6.5.6) as

$$\mathrm{BF} = \iint p(y|\theta_2, M_0)p(\theta_2|M_0)p(\theta_1|M_1)\frac{p(\theta_1, \theta_2|y, M_1)}{p(y|\theta_1, \theta_2, M_1)p(\theta_1, \theta_2|M_1)}\,d\theta_1\,d\theta_2. \tag{6.5.7}$$

Recognizing that

$$\frac{p(\theta_1|M_1)}{p(\theta_1, \theta_2|M_1)} = \frac{1}{p(\theta_2|\theta_1, M_1)},$$

Equation (6.5.7) becomes

$$\mathrm{BF} = \iint \frac{p(y|\theta_2, M_0)p(\theta_2|M_0)}{p(y|\theta_1, \theta_2, M_1)p(\theta_2|\theta_1, M_1)} p(\theta_1, \theta_2|y, M_1)\, \mathrm{d}\theta_1\, \mathrm{d}\theta_2. \tag{6.5.8}$$

We note that

$$p(y|\theta_2, M_0) = p(y|\theta_1, \theta_2, M_1)|_{\theta_1 = \theta_1^b}.$$

Equation (6.5.8) suggests that we can use the posterior draws to form an estimated Bayes factor of the form

$$\widehat{\mathrm{BF}} = \frac{1}{R} \sum_r \frac{p(y|\theta_1, \theta_2^r)|_{\theta_1 = \theta_1^b} p(\theta_2^r|M_0)}{p(y|\theta_1^r, \theta_2^r)p(\theta_2^r|\theta_1^r, M_1)}, \tag{6.5.9}$$

where $\{\theta^r\}$ are MCMC draws from the posterior under the unrestricted model. We note that, unlike the Savage–Dickey set-up, the priors under M_0 and M_1 need not be linked via conditioning. What is required, however, is that the conditional prior for $\theta_2|\theta_1$ under M_1 must be available as a normalized density.

6.6 BRIDGE SAMPLING METHODS

Meng and Wong (1996) provide an identity which links together methods that rely on expectations with respect to the posterior with importance sampling methods. A hybrid procedure which relies on both is termed 'bridge sampling'. The bridge sampling identity starts with a pair of functions $\alpha(\theta)$ and $q(\theta)$ such that $\int \alpha(\theta)p(\theta|y)q(\theta)\, \mathrm{d}\theta > 0$:

$$1 = \frac{\int \alpha(\theta)p(\theta|y)q(\theta)\, \mathrm{d}\theta}{\int \alpha(\theta)q(\theta)p(\theta|y)\, \mathrm{d}\theta} = \frac{\mathrm{E}_q[\alpha(\theta)p(\theta|y)]}{\mathrm{E}_p[\alpha(\theta)q(\theta)]}. \tag{6.6.1}$$

Using the relationship between the marginal density of the data and the unnormalized posterior, we can establish the following identity:

$$p(y) = \frac{\mathrm{E}_q[\alpha(\theta)\tilde{p}(\theta|y)]}{\mathrm{E}_p[\alpha(\theta)\mathrm{q}(\theta)]}. \tag{6.6.2}$$

We can estimate (6.6.2) by approximating both expectations in the numerator and denominator by iid draws from q and MCMC draws from p:

$$\hat{p}(y) = \frac{R_q^{-1} \sum \alpha(\theta_q^r)\tilde{p}(\theta_q^r|y)}{R_p^{-1} \sum \alpha(\theta_p^r)q(\theta_p^r)}. \tag{6.6.3}$$

Meng and Wong (1996) point out that the Gelfand–Dey estimator and the importance sampling estimator are special cases of bridge sampling with $\alpha(\bullet)$ chosen to be either the importance density or the unnormalized posterior. They consider the question of an 'optimal' choice for $\alpha(\bullet)$ (in the sense of smallest mean squared error in estimation) and provide an iterative scheme for constructing $\alpha(\bullet)$ as a weighted combination of Gelfand–Dey and importance sampling. Frühwirth-Schnatter (2004) applies this iterative scheme to the construction of Bayes factors for mixtures of normals and finds the Meng and Wong estimator to be an improvement over standard procedures.

6.7 POSTERIOR MODEL PROBABILITIES WITH UNIDENTIFIED PARAMETERS

We have considered a number of models with unidentified parameters. We have also seen that, in some cases, it is desirable to navigate in the unidentified parameter space and then margin down or 'post-process' the MCMC draws to make inferences regarding the identified parameters. A reasonable question to ask is whether draws of the unidentified parameters can be used to estimate posterior model probabilities. From a purely theoretical point of view, this can be justified. Let us assume that θ is not identified but $\tau(\theta)$ is.

We can then define a transformation, $g(\theta)$, which partitions the transformed parameters into those which are identified and those which are not:

$$\delta = g(\theta) = \begin{bmatrix} \nu(\theta) \\ \tau(\theta) \end{bmatrix} = \begin{bmatrix} \delta_1 \\ \delta_2 \end{bmatrix}. \tag{6.7.1}$$

δ_1 is a subvector of length $k - k_1$ containing the unidentified parameters. We start with a prior over the full vector of parameters, $p_\theta(\bullet)$, and then compute the induced prior over the identified parameters:

$$\begin{aligned} p_{\delta_2}(\delta_2) &= \int p_\delta(\delta_1, \delta_2)\, d\delta_1, \\ p_\delta(\delta_1, \delta_2) &= p_\theta(g^{-1}(\delta)) J_{\theta \to \delta}. \end{aligned} \tag{6.7.2}$$

We can compute the marginal density of the data in two ways. We can compute the density in the space of unidentified parameters or directly on the identified parameters using the induced prior in (6.7.2).

Working in the identified parameter space,

$$p'(y) = \int p(y|\delta_2) p_{\delta_2}(\delta_2)\, d\delta_2. \tag{6.7.3}$$

Now consider the marginal density of the data computed in the full, unidentified parameter space:

$$p(y) = \iint p(y|\delta_1, \delta_2) p_\delta(\delta_1, \delta_2)\, d\delta_1\, d\delta_2$$

$$= \int \left[\int p(y|\delta_2) p_{\delta_2}(\delta_2)\, d\delta_2 \right] p(\delta_1|\delta_2)\, d\delta_1$$

$$= \int p'(y) p(\delta_1|\delta_2)\, d\delta_1 = p'(y).$$

Here we are using the fact that $p(y|\delta_1, \delta_2) = p(y|\delta_2)$ since δ_1 represents the unidentified parameters.

Thus, we are theoretically justified in using MCMC draws from the unidentified parameter space in the methods considered in Section 6.6. However, in practice we recognize that the draws of the unidentified parameters may exhibit a great deal of variation, especially in situations with vague or very diffuse priors. Again, there is a pay-off to assessing realistic priors.

6.8 CHIB'S METHOD

Chib (1995) proposes a method which uses MCMC output to estimate the marginal density of the data. This method is particularly appropriate for models that have a conjugate structure conditional on the value of augmented latent variables. Chib starts with the basic identity relating the normalized and unnormalized posteriors:

$$p(y) = \frac{\tilde{p}(\theta^*|y)}{p(\theta^*|y)} = \frac{p(y|\theta^*)p(\theta^*)}{p(\theta^*|y)}. \tag{6.8.1}$$

This identity holds for any value of θ, indicated by θ^* above. The key insight is that, for certain conditionally conjugate models, the denominator of (6.8.1) can be expressed as an average of densities which are known up to and including normalizing constants.

For example, consider the archetypal data augmentation model:

$$\begin{aligned} &y|z, \\ &z|\theta, \\ &\theta. \end{aligned} \tag{6.8.2}$$

For this model we can write the ordinate of the posterior at θ^* as the average of the posterior conditional on the latent z over the marginal posterior distribution of z:

$$\begin{aligned} p(\theta|y) &= \int \frac{p(y, z, \theta)}{p(y)}\, dz = \int \frac{p(\theta|y, z)p(y, z)}{p(y)}\, dz \\ &= \int \frac{p(\theta|y, z)p(z|y)p(y)}{p(y)}\, dz = \int p(\theta|y, z)p(z|y)\, dz. \end{aligned} \tag{6.8.3}$$

This suggests that we can estimate the marginal density of the data as follows:

$$\hat{p}(y) = \frac{p(y|\theta^*)p(\theta^*)}{\hat{p}(\theta^*|y)} = \frac{p(y|\theta^*)p(\theta^*)}{R^{-1}\sum_r p(\theta^*|y, z^r)} \tag{6.8.4}$$

The $\{z^r\}$ are draws from the marginal posterior of the latent variables. θ^* is usually taken to be the posterior mean or mode, computed from the MCMC draws. Equation (6.8.4) requires that we simply save the latent draws from our MCMC run and that we be able to evaluate the data density (likelihood) and the prior densities with all normalizing constants. We should note that (6.8.4) requires evaluation of the marginal likelihood (without the latents). For some models, this can be computationally challenging. However, we should point out that the Chib method only requires one likelihood evaluation, whereas Gelfand–Dey style methods would require R evaluations.

The Chib method in (6.8.4) requires that we be able to evaluate $p(\theta^*|y, z)$, including all normalizing constants. For some models, we can use the fact that $p(\theta|y, z) = p(\theta|z)$ and that we have a conjugate set-up conditional on z. However, in other applications, this will not be possible. In some cases, $\theta|z$ is not fully conjugate but can be broken into two conjugate blocks. Consider the case where θ is partitioned into (θ_1, θ_2). We can then estimate $p(\theta^*|y, z)$ using the identity $p(\theta_1, \theta_2|y) = p(\theta_1|y)p(\theta_2|\theta_1, y)$. We can compute the marginal posterior density of θ_1 by averaging the conditional density:

$$p(\theta_1|y) = \int p(\theta_1|\theta_2, z, y)p(\theta_2, z|y)\, dz\, d\theta_2. \tag{6.8.5}$$

This can be estimated by averaging the conditional density over the MCMC draws of θ_2 and z:

$$\hat{p}(\theta_1^*|y) = \frac{1}{R}\sum_r p(\theta_1^*|\theta_2^r, z^r). \tag{6.8.6}$$

However, the conditional posterior density of $\theta_2|\theta_1$ cannot be estimated by averaging the conditional with respect to the marginal posterior as in (6.8.5):

$$p(\theta_2|\theta_1, y) = \int p(\theta_2|\theta_1, z, y)p(z|\theta_1, y)\, dz. \tag{6.8.7}$$

To estimate this density at the point (θ_1^*, θ_2^*) requires a modified MCMC sampler for (z, θ_2) given $\theta_1 = \theta_1^*$. As Chib points out, this is simple to achieve by shutting down the draws for θ_1. If $z_{\theta_1^*}^r$ are draws from the marginal posterior of z given $\theta_1 = \theta_1^*$, then we can estimate (6.8.7) by

$$\hat{p}(\theta_2^*|\theta_1^*, y) = \frac{1}{R}\sum_r p(\theta_2^*|\theta_1^*, z_{\theta_1^*}^r, y) \tag{6.8.8}$$

To estimate the marginal density of the data, we put together (6.8.6) and (6.8.8):

$$\begin{aligned}\hat{p}(y) &= \frac{p(y|\theta^*)p(\theta^*)}{\hat{p}(\theta_1^*|y)\hat{p}(\theta_2^*|\theta_1^*, y)} \\ &= \frac{p(y|\theta^*)p(\theta^*)}{\left(R^{-1}\sum_r p(\theta_1^*|\theta_2^r, z^r)\right)\left(R^{-1}\sum_r p(\theta_2^*|\theta_1^*, z_{\theta_1^*}^r, y)\right)}.\end{aligned} \tag{6.8.9}$$

For cases in which there is no convenient decomposition into conjugate blocks, the method of Chib and Jeliazhov (2001) can be used.

6.9 AN EXAMPLE OF BAYES FACTOR COMPUTATION: DIAGONAL MULTINOMIAL PROBIT MODELS

We will illustrate the importance sampling, Newton–Raftery, and Chib methods using an example of comparison of different MNP models. The estimation of the off-diagonal covariance (correlation) elements in the MNP is often difficult. We typically find that the diagonal elements (relative variances) are much more precisely estimated. In addition, we often find large differences in relative variance between choice alternatives. This is consistent with a view that X variables explain different portions of the utility of each choice. For example, some choice alternatives may have utility explained well by price while others may have utility that depends on attributes not measured in our data. For this reason, the MNP model with a non-scalar but diagonal covariance matrix could be considered a central model:

$$
\begin{aligned}
y_i &= \sum_{j=1}^{p} I(z_{ij} = \max(z_i)),\\
z_i &= X_i\beta + \varepsilon_i,\\
\varepsilon_i &\sim N(0, \Lambda).
\end{aligned}
$$

We will develop Bayes factors to compare the diagonal MNP, where

$$
\Lambda = \begin{bmatrix} 1 & & & \\ & \sigma_{22} & & \\ & & \ddots & \\ & & & \sigma_{pp} \end{bmatrix},
$$

with an 'identity' MNP or a model with an identity covariance matrix, $\varepsilon_i \sim N(0, I_p)$. This would closely approximate the independence of irrelevant alternatives properties of the multinomial logit model.

We use normal priors on β and independent scaled inverted chi-squared priors on the diagonal elements $2, \ldots, p$. The first diagonal element is fixed at 1:

$$
\begin{aligned}
\beta &\sim N(\bar{\beta}, A^{-1}),\\
\sigma_{jj} &\sim \text{ind}\ \nu_0 s_0^2/\chi^2_{\nu_0}.
\end{aligned}
\tag{6.9.1}
$$

To apply the Newton–Raftery method, we must implement MCMC samplers for each of the two MNP models and compute the likelihood at each draw as in (6.5.4). The diagonal MNP sampler and identity MNP samplers are special cases of the algorithms given in section 4.2. To compute the likelihood, we must compute the choice probabilities for each observation. The general MNP likelihood requires evaluation of the integral of a correlated normal random variable over a rectangular region as discussed in Section 4.2.2. However, for the diagonal MNP model, the choice probabilities can be simplified to the average of normal cumulative distribution functions with respect to a univariate normal distribution. Both importance sampling and NR methods will be computationally intensive due to the evaluation of the likelihood over the set of MCMC draws or over draws from the importance density. This argues in favor of the Chib method, which only requires one evaluation of the likelihood.

To implement an importance function approach, we first transform the variance parameters to an unrestricted space. This only applies to the diagonal MNP model. If θ is the stacked vector of the parameters of the diagonal MNP model, $\theta' = (\beta, \text{diag}(\Lambda)')$, then we define the transformed vector η by

$$\eta = \begin{bmatrix} \eta_1 \\ \eta_2 \end{bmatrix} = \begin{bmatrix} \beta \\ \ln(\text{diag}(\Lambda)) \end{bmatrix}, \qquad \theta = \begin{bmatrix} \eta_1 \\ e^{\eta_2} \end{bmatrix}, \tag{6.9.2}$$

with Jacobian

$$J_{\theta\to\eta} = \left\| \begin{matrix} I_{\dim(\beta)} & & 0 & \\ & e^{\eta_{22}} & & \\ 0 & & \ddots & \\ & & & e^{\eta_{2p}} \end{matrix} \right\| = \prod_{i=2}^{p} e^{\eta_{2i}}. \tag{6.9.3}$$

To implement the importance sampling method, we use a normal importance function with location and scale chosen using the MCMC draws in the transformed parameter η:

$$\begin{aligned} q(\eta) &= \phi(\overline{\eta}, s^2\hat{V}), \\ \overline{\eta} &= \frac{1}{R}\sum_r \eta^r_{\text{mcmc}} \\ \hat{V} &= \frac{1}{R}\sum_r (\eta^r_{\text{mcmc}} - \overline{\eta})(\eta^r_{\text{mcmc}} - \overline{\eta})', \end{aligned} \tag{6.9.4}$$

where ϕ denotes a normal density. We approximate the marginal density of the data using draws from q as follows:

$$\hat{p}(y) = \frac{1}{N}\sum_i \frac{p(y|\theta(\eta^i))p(\theta(\eta^i))J_{\theta\to\eta}(\eta^i)}{q(\eta^i)} = \frac{1}{N}\sum_i w_i. \tag{6.9.5}$$

We note that the q density in the denominator of (6.9.5) is the full normalized density. The importance sampling method can be tuned by choice of the scaling parameter s.

The Chib method can be implemented directly using (6.8.4) for the identity MNP model. However, for the diagonal MNP model, we use the variant of the Chib method which breaks the parameter vector into two parts, $\theta_1 = \beta$, $\theta_2 = \text{diag}(\Lambda)$, in (6.8.9).

As an illustration, we simulate data from a diagonal MNP model with $p = 3$ and $\text{diag}(\Lambda) = (1, 2, 3)$, $N = 500$. X contains $p - 1$ intercepts and two regressors which are simulated from Unif(-1,1). We compute the Bayes factors for the diagonal MNP versus the identity MNP. We use 150,000 draws from the MCMC samplers and the importance sampling density. We assess modestly informative priors with $\overline{\beta} = 0$, $A = 0.25I$, $\nu_0 = 10$, and $s_0^2 = 1$.

Figure 6.1 shows the log-likelihood values for the diagonal MNP (dark) and the identity MNP (light) plotted for every 75th draw. The figure clearly shows that the

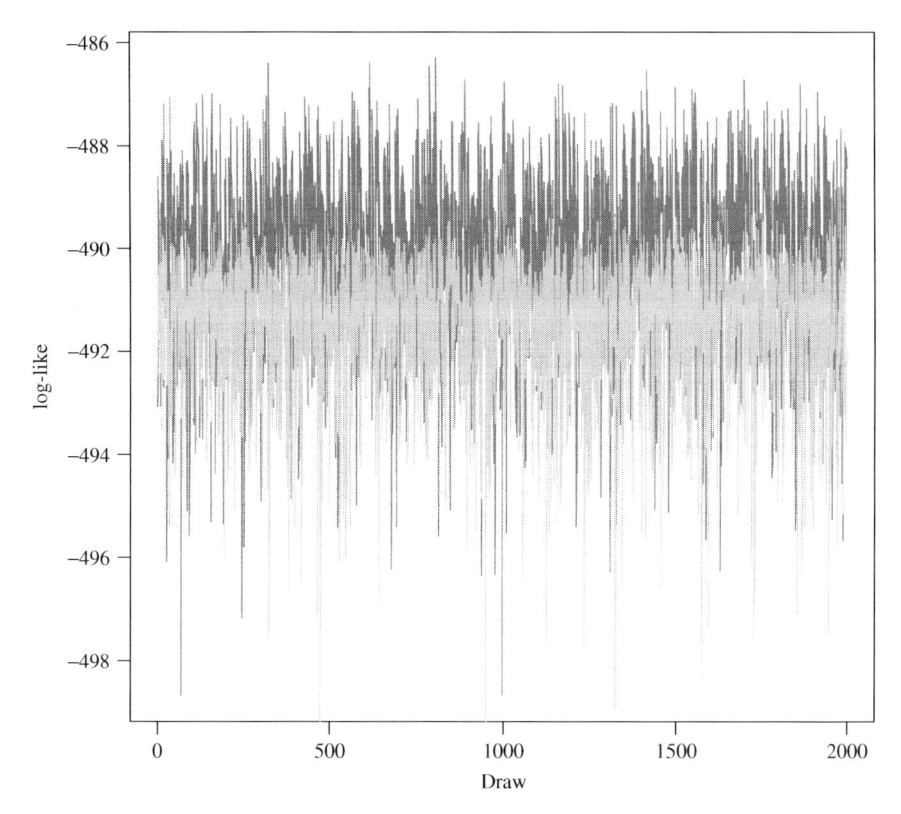

Figure 6.1 Log-likelihood values: diagonal versus identity MNP

diagonal MNP fits the data better than the identity MNP. Even though inspection of this plot is somewhat 'informal', we highly recommend this as a rough assessment of the models as well as the convergence of the MCMC algorithm. In fact, in our experience this sort of plot can be more informative than formal Bayes factor computations.

Below we show the results of Bayes factor computations using all three methods:

Method	BF
NR	3.4
IS ($s = 1.5$)	2.9
IS ($s = 2.0$)	2.9
Chib	3.4

These numbers do not convey the sampling error in each of the estimates. As an informal assessment, we made more than 10 different runs and observed very little variation in the importance sampling or Chib numbers and a wide range of NR Bayes factors from 2.0 to 3.5. We note that our prior is more informative than most and we expect that with more traditional 'vague' prior settings we would see even more variation in the NR numbers.

The NR approach is frequently criticized in the literature due to the fact that the estimator has an infinite variance under some conditions. However, the convenience of the NR approach accounts for its widespread popularity. Figure 6.2 illustrates the problems with the NR approach more dramatically. The NR estimate is driven by a small portion of the values of the reciprocal of the likelihood, as shown by the histogram of the reciprocal of the likelihood for every 10th draw (right-hand panel). While the distribution of log-likelihood values is not particularly abnormal, the harmonic mean estimator is particularly vulnerable to outlying observations.

The importance sampling approach does not have as severe a problem with outlying observations as the NR approach. However, even with careful choice of scaling constant,[4] the weights have outliers as shown in Figure 6.3. Figure 6.3 shows the distribution of the importance sampling weights normalized by the median of the weight. This figure does not instill confidence in the importance sampling approach either. The parameter space is only of dimension 7 in this example, and we might expect the problems with the importance sampling approach to magnify in higher-dimensional parameter spaces.

The Chib approach to calculating Bayes factors relies on various estimates of the ordinate of posterior or conditional posterior densities. For the diagonal MNP, the

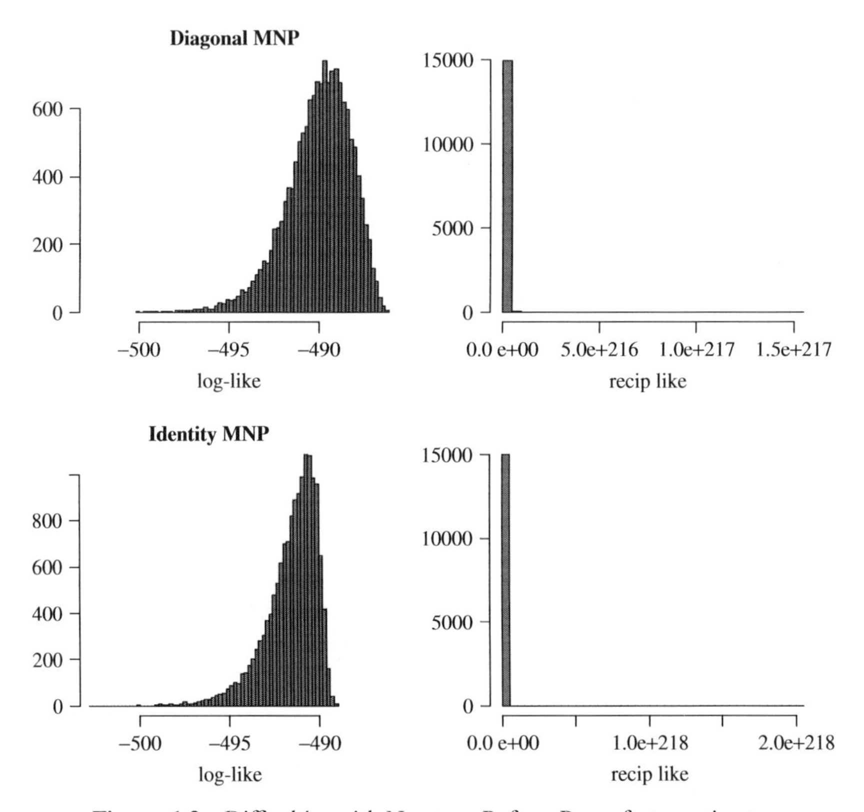

Figure 6.2 Difficulties with Newton–Raftery Bayes factor estimator

[4] In this application, choice of the scaling constant is critical. We experimented with Student t importance densities and found little benefit.

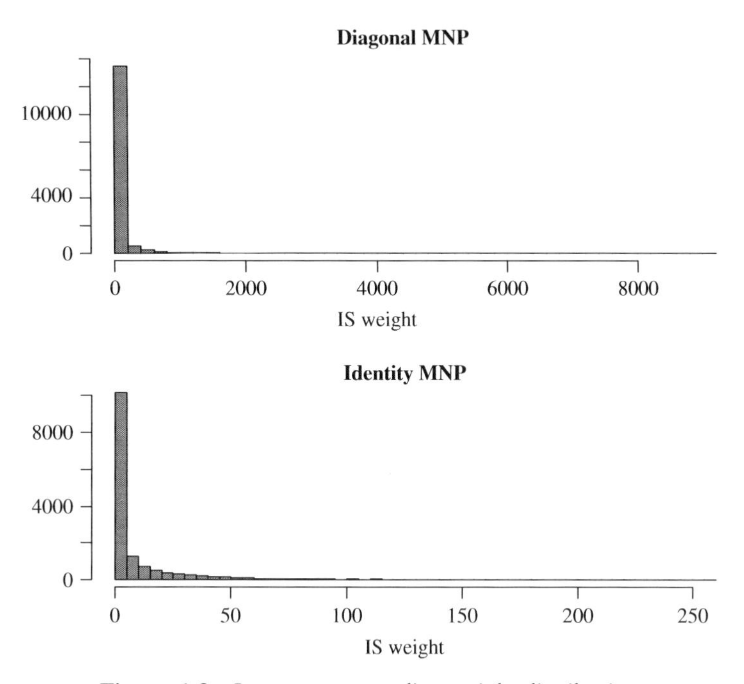

Figure 6.3 Importance sampling weight distribution

Chib approach averages $p(\beta^*|\Lambda, z, y)$ over MCMC draws of Λ and z. $p(\Lambda^*|\beta^*, z, y)$ is averaged over draws of $z|\beta^*, y$. For the identity MNP, only the density of β is averaged. Figure 6.4 shows the distribution of these densities (right-hand side) and the log of these densities. Again, there are very large outliers which influence the averages. These outliers contribute to instability of this approach.

In this example, the importance sampling approach provides the most reliable results. However, it is important to note that all three approaches are sensitive to outliers. The 'numerical standard error' formulas used to measure the sampling errors in the importance sampling and Chib approaches are unlikely to be reliable in the presence of such large outliers.

6.10 MARKETING DECISIONS AND BAYESIAN DECISION THEORY

Computation of posterior model probabilities is motivated as a special case of Bayesian decision theory with a zero–one loss function. In marketing problems, the profit function of the firm provides a more natural choice of loss. In addition, there is often considerable parameter or modeling uncertainty. Bayesian decision theory is ideally suited for application to many marketing problems in which a decision must be made given substantial parameter or modeling uncertainty. In these situations, the uncertainty must factor into the decision itself. The marketing decision-maker takes an action by setting the value of various variables designed to quantify the marketing environment facing the consumer (such as price or advertising levels). These decisions should be affected by the level of uncertainty facing the marketer. To make this concrete, begin

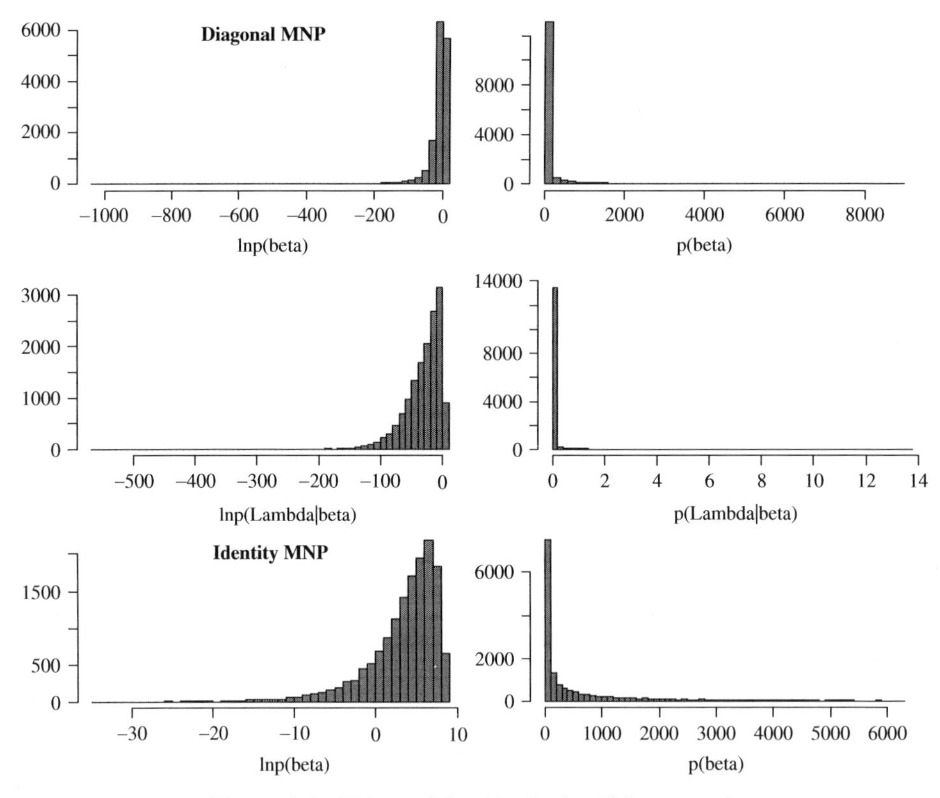

Figure 6.4 Values of densities in the Chib approach

with a probability model that specifies how the outcome variable (y) is driven by the explanatory variables (x) and parameters θ:

$$p(y|x, \theta).$$

The decision maker has control over a subset of the x vector, $x' = [x'_d, x'_{cov}]$. x_d represents the variables under the decision-maker's control and x_{cov} are the covariates. The decision-maker chooses x_d so as to maximize the expected value of profits where the expectation is taken over the distribution of the outcome variable. In a fully Bayesian decision-theoretic treatment, this expectation is taken with respect to the posterior distribution of θ as well as the predictive conditional distribution $p(y|x_d, x_{cov})$:

$$\begin{aligned}\pi^*(x_d|x_{cov}) &= \mathrm{E}_\theta[\mathrm{E}_{y|\theta}[\pi(y|x_d)]] \\ &= \mathrm{E}_\theta\left[\int \pi(y|x_d)p(y|x_d, x_{cov}, \theta)\,\mathrm{d}y\right] \\ &= \mathrm{E}_\theta[\overline{\pi}(x_d|x_{cov}, \theta)]. \qquad (6.10.1)\end{aligned}$$

The decision-maker chooses x_d to maximize profits π^*. In general, the decision-maker can be viewed as minimizing expected loss which is frequently taken as profits but need not be in all cases (see, for example, Steenburgh *et al.* 2003).

6.10.1 Plug-in versus Full Bayes Approaches

The use of the posterior distribution of the model parameters to compute expected profits is an important aspect of the Bayesian approach. In an approximate or conditional Bayes approach, the integration of the profit function with respect to the posterior distribution of θ is replaced by an evaluation of the function at the posterior mean or mode of the parameters:

$$\pi^*(x_{\mathrm{d}}) = \mathrm{E}_{\theta|y}[\overline{\pi}(x_{\mathrm{d}}|\theta)] \neq \overline{\pi}(x_{\mathrm{d}}|\hat{\theta} = \mathrm{E}_{\theta|y}[\theta]). \qquad (6.10.2)$$

This approximate approach is often called the 'plug-in' approach, or according to Morris (1983), 'Bayes empirical Bayes'. When the uncertainty in θ is large and the profit function is nonlinear, errors from the use of the plug-in method can be large. In general, failure to account for parameter uncertainty will overstate the potential profit opportunities and lead to 'overconfidence' that results in an overstatement of the value of information (see Montgomery and Bradlow 1999).

6.10.2 Use of Alternative Information Sets

One of the most appealing aspects of the Bayesian approach is the ability to incorporate a variety of different sources of information. All adaptive shrinkage methods utilize the similarity between cross-sectional units to improve inference at the unit level. A high level of similarity among units leads to a high level of information shared. Since the level of similarity is determined by the data via the first-stage prior, the shrinkage aspects of the Bayesian approach adapt to the data. For example, Neelamegham and Chintagunta (1999) show that similarities between countries can be used to predict the sales patterns following the introduction of new products.

The value of a given information set can be assessed using a profit metric and the posteriors of θ corresponding to the two information sets. For example, consider two information sets A and B along with corresponding posteriors $p_A(\theta)$, $p_B(\theta)$. We solve the decision problem using these two posterior distributions:

$$\Pi_\ell = \max_{x_{\mathrm{d}}} \pi_\ell^*(x_{\mathrm{d}}|x_{\mathrm{cov}}) = \max_{x_{\mathrm{d}}} \int \overline{\pi}(x_{\mathrm{d}}|x_{\mathrm{cov}}, \theta) p_\ell(\theta)\, \mathrm{d}\theta, \qquad \ell = A, B. \qquad (6.10.3)$$

We now turn to the problem of valuing disaggregate information.

6.10.3 Valuation of Disaggregate Information

Once a fully decision-theoretic approach has been specified, we can use the profit metric to value the information in disaggregate data. We compare profits that can be obtained via our disaggregate inferences about $\{\theta_i\}$ with profits that could be obtained using only aggregate information. The profit opportunities afforded by disaggregate data will depend both on the amount of heterogeneity across the units in the panel data and on the level of information at the disaggregate level.

To make these notions explicit, we will lay out the disaggregate and aggregate decision problems. As emphasized in Chapter 5, Bayesian methods are ideally suited for inference about the individual or disaggregate parameters as well as the common parameters. Recall the profit function for the disaggregate decision problem:

$$\pi_i^*(x_{\mathrm{d},i}|x_{\mathrm{cov},i}) = \int \overline{\pi}(x_{\mathrm{d},i}|x_{\mathrm{cov},i}, \theta_i) p(\theta_i|\mathrm{Data})\, \mathrm{d}\theta_i. \qquad (6.10.4)$$

Here we take the expectation with respect to the posterior distribution of the parameters for unit i. Total profits from the disaggregate data are simply the sum of the maximized values of the profit function above, $\Pi_{\mathrm{disagg}} = \sum \pi_i^*(\tilde{x}_{\mathrm{d},i}|x_{\mathrm{cov},i})$, where $\tilde{x}_{\mathrm{d},i}$ is the optimal choice of $x_{\mathrm{d},i}$.

Aggregate profits can be computed by maximizing the expectation of the sum of the disaggregate profit functions with respect to the predictive distribution of θ_i

$$\pi_{\mathrm{agg}}(x_{\mathrm{d}}) = \mathrm{E}_\theta \left[\sum \overline{\pi}(x_{\mathrm{d}}|x_{\mathrm{cov},i}, \theta) \right] = \int \sum \overline{\pi}(x_{\mathrm{d}}|x_{\mathrm{cov},i}, \theta) \overline{p}(\theta)\, \mathrm{d}\theta,$$

$$\Pi_{\mathrm{agg}} = \pi_{\mathrm{agg}}(\tilde{x}_{\mathrm{d}}). \qquad (6.10.5)$$

The appropriate predictive distribution of θ, $\overline{p}(\theta)$, is formed from the marginal of the first-stage prior with respect to the posterior distribution of the model parameters:

$$\overline{p}(\theta) = \int p(\theta|\tau) p(\tau|\mathrm{Data})\, \mathrm{d}\tau.$$

Comparison of Π_{agg} with Π_{disagg} provides a metric for the achievable value of the disaggregate information.

6.11 AN EXAMPLE OF BAYESIAN DECISION THEORY: VALUING HOUSEHOLD PURCHASE INFORMATION

As emphasized in Section 6.10, the valuation of information must be made within a decision context. Rossi *et al.* (1996) consider the problem of valuing household purchase information using a targeted couponing problem. In traditional couponing exercises, neither retailers nor manufacturers have access to information about individual consumers. Instead, a 'blanket' coupon is distributed (via mass mailings or inserts in newspapers). This blanket coupon is, in principle, available to all consumers (note that the large-denomination 'rebates' available on consumer durable goods are the same idea with a different label). In order to be profitable, the issuer of the coupon relies on an indirect method of price discrimination where the consumers with lower willingness to pay use the coupon with higher probability than those with higher willingness to pay. In the late 1980s, the technology for issuing customized coupons became available. For example, Catalina Marketing Inc. started a highly successful business by installing coupon printers in grocery stores that were connected to the point of sale terminal. This technology opened the possibility of issuing coupons based on purchase history information. At the most elementary level, it was now possible to issue coupons to consumers who exhibited interest in a product category by purchasing some product in this category. For example, an ice cream manufacturer could pay Catalina to issue coupons for its ice cream products to anyone observed to be purchasing ice cream.

Frequent shopper programs adopted by many retailers also provide a source of purchase and demographic information. By linking purchase records via the frequent shopper ID, it is possible to assemble a panel of information about a large fraction of the retailers' customers. This allows for much more elaborate coupon trigger strategies which involve estimated willingness to pay. That is, we might be able to directly estimate a consumer's willingness to pay for a product by estimating a demand model using purchase history data. In addition, frequent shopper programs sometimes collect limited demographic information either directly on enrollment applications or via indirect inference from the member's address (so-called geo-demographic information).

These developments mean that the issuer of the coupon now has access to a rich information set about individual consumers. How much should the issuer be willing to pay for this information? Or, what is the value of targeted couponing relative to the traditional blanket approach? To answer these questions, a model of purchase behavior is required as well as a loss function. The natural loss function for this problem is expected incremental profits from the issue of a coupon. Rossi *et al.* postulate a hierarchical probit model similar to the hierarchical logit model of Chapter 5 in which demographic information enters the hierarchy. The hierarchical probit model allows the issue of the coupon to make inferences about consumer-level purchase probabilities given various information sets. Clearly, only a very limited set of information is available for each consumer. These inferences will be imprecise and the degree of 'targeting' or customization of the face value of the coupon will depend not only on how different consumers are in terms of willingness to pay but also on how precisely we can measure these differences. This full accounting for parameter uncertainty is a key feature of a Bayesian decision-theoretic approach. This will guard against the overconfidence that can arise from 'plug-in' approaches that simply estimate parameters without accounting for uncertainty.

Rossi *et al.* start with a multinomial probit model at the household or consumer level. Households are confronted with a choice from among p brands, with covariates taken to be measures of the marketing mix facing the consumer (price and advertising variables):

$$y_{ht} = j \qquad \text{if } \max(z_{ht} = X_{ht}\beta_h + \varepsilon_{ht}) = z_{jht}, \tag{6.11.1}$$

$$\varepsilon_{ht} \sim N(0, \Lambda). \tag{6.11.2}$$

In Rossi *et al.* (in contrast to the standard MNP in Section 4.2), the covariance matrix of the latent errors (Λ) is taken to be diagonal. This means that we do not have to difference the system and that identification is achieved by setting $\lambda_{11} = 1$. Rossi *et al.* make this simplification for practical and data-based reasons. Given β_h, Λ, choice probabilities for the diagonal MNP model are simple to compute, requiring only univariate integration. Since these choice probabilities figure in the loss/profit function and will be evaluated for many thousands of draws, it is important that these probabilities can be computed at low cost. This is not as much of a consideration now as it was in the early 1990s when this research was conducted. However, this is not the only reason for using a diagonal covariance. Much of the observed correlation in nonhierarchical probit models can be ascribed to heterogeneity. Once heterogeneity is taken into account via the hierarchy, the errors in the unit-level probit are much less correlated. However, they can often be very heteroskedastic, as pointed out in Section 6.9.

This 'unit-level' model is coupled with the by now standard normal model of heterogeneity:

$$\beta_h = \Delta' z_h + \nu_h, \qquad \nu_h \sim N(0, V_\beta). \tag{6.11.3}$$

z_h is a vector of demographic information. The key task is to compute the predictive distribution of the household parameters and choice probabilities for various information sets. That is we must compute, $p(\beta_h, \Lambda|\Omega^*)$, where Ω^* denotes a particular information set. This predictive distribution will be used in the decision problem to determine what is the optimal face-value coupon to issue to household h.

Given the proper choice of priors on the common parameters, Λ, Δ, V_β, we can define an MCMC method to draw from the joint posterior of $\{\beta_1, \ldots, \beta_h, \ldots, \beta_H\}$, Λ, Δ, V_β. Rossi *et al.* observed that one can develop a Gibbs sampler for this problem by using the McCulloch and Rossi (1994) sampler coupled with the standard normal hierarchical set-up. That is, given $\{\beta_h\}$, we have standard normal, inverted Wishart draws for Δ, V_β as in Section 3.7 or 5.3. A DAG for the model is given below:

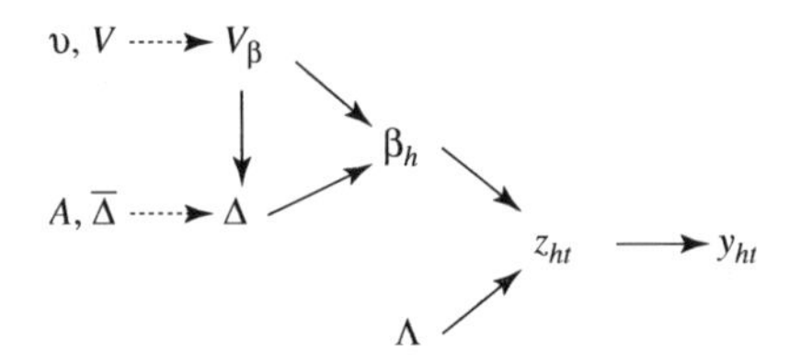

Thus, we can obtain draws from the marginal posterior distributions of all household and common parameters.

Consider three information sets:

1. 'Full' – purchase history information on household h and demographics.
2. 'Demographics only' – just knowledge of the demographic vector for household h.
3. 'Blanket' – no information about household h but only information about the 'population' distribution of households obtained from a sample of households.

Rossi *et al.* consider a 'choices-only' information set which we will not discuss here. We must compute the 'predictive' or posterior distribution of β_h for each of the three information sets. The first and richest information set is a natural by-product of the Gibbs sampler defined above. That is, when we compute the posterior distribution of each β_h, we will take into account each household's demographics and purchase history. In addition, information from other households will influence our views about a particular β_h via inferences about the common parameters. We note that by looking at the marginal distribution of the posterior draws, we are integrating out or averaging over draws of the common parameters, accounting for uncertainty in these quantities. That is,

$$\begin{aligned} &p(\beta_h|\{y_1, \ldots, y_h, \ldots, y_H, X_1, \ldots, X_h, \ldots, X_H\}, Z) \\ &\quad = \int p(\beta_h|y_h, X_h, \Lambda, \Delta, V_\beta) p(\Lambda, \Delta, V_\beta|\{y_1, \ldots, y_h, \ldots, y_H, \\ &\quad X_1, \ldots, X_h, \ldots, X_H\}, Z)\, \mathrm{d}\Lambda\, \mathrm{d}\Delta\, \mathrm{d}V_\beta. \end{aligned}$$

Z is the matrix of all household demographics. We cannot simply 'plug in' estimates of the common parameters:

$$p(\beta_h|\{y_1, \ldots, y_h, \ldots, y_H, X_1, \ldots, X_h, \ldots, X_H\}, Z) \neq p(\beta_h|y_h, X_h, \hat{\Lambda}, \hat{\Delta}, \hat{V}_\beta).$$

The second and third information sets require some thought. If we only have demographic information but no purchase information on household h, we then must compute an appropriate predictive distribution from the model in (6.11.3):

$$\int p(\beta_h|z_h, \Delta, V_\beta)p(\Delta, V_\beta|\text{Info})\,\mathrm{d}\Delta\,\mathrm{d}V_\beta. \tag{6.11.4}$$

In order to undertake the computation in (6.11.4), we must specify an information set on which we base our inferences on the parameters of the heterogeneity distribution. Our idea is that we might have a 'pet' panel of households, not including household h, on which we observed purchases and demographics so that we can gauge the distribution of probit parameters. That is, we have a sample of data which enables us to gauge the extent of differences among households in this population. This distribution could also simply reflect prior beliefs on the part of managers regarding the distribution. To implement this, we use the posterior distribution from our sample of households. We can define a simulator for (6.11.4) by using all R draws of Δ, V_β and drawing from the appropriate normal draw:

$$\beta_h^r|\Delta^r, V_\beta^r \sim N(\Delta^{r'} z_h, V_\beta^r). \tag{6.11.5}$$

The third and coarsest information set is the same information set used in setting a blanket coupon face value. No information is available about household h. The only information available is information regarding the distribution of the β parameters and demographics in the population. In this situation, we simply integrate over the distribution of demographics as well as the posterior distribution of Δ, V_β:

$$\int p(\beta_h|z_h, \Delta, V_\beta)p(z_h)p(\Delta, V_\beta|\text{Info})\,\mathrm{d}z_h\,\mathrm{d}\Delta\,\mathrm{d}V_\beta. \tag{6.11.6}$$

Rossi *et al.* use the empirical distribution of z to perform this integral.

Thus, we now have the ability to compute the predictive distribution of each household parameter vector for each of the three information sets. We must now pose the decision problem. The problem is to choose the face value of a coupon so as to maximize total profits. We model the effect of a coupon as simply a reduction in price for the brand for which the coupon is issued. If the coupon is issued for brand i, the decision problem can be written as follows:

$$\max_F \pi(F) = \int \Pr[i|\beta_h, \Lambda, X(F)](M - F)p(\beta_h, \Lambda|\Omega^*)\,\mathrm{d}\beta_h\,\mathrm{d}\Lambda. \tag{6.11.7}$$

$p(\beta_h, \Lambda|\Omega^*)$ is the predictive or posterior distribution for information set Ω^*. M is the margin on the sale of brand i without consideration of the coupon cost. $X(F)$ denotes the value of the marketing mix variables with a coupon of face value F. We assume that the effect of the coupon is to reduce the price of alternative i by F. In many cases, the

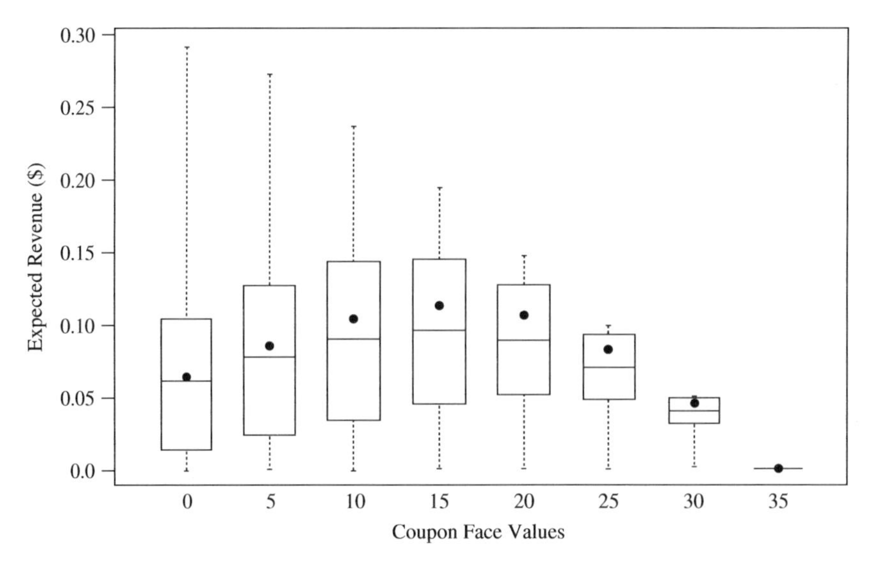

Figure 6.5 Posterior distribution of expected revenues for various coupon values (Reprinted by permission, Rossi, McCulloch and Allenby, The value of Purchase History Data in Target Marketing, *Marketing Science*, **15**, 321–340, 1996. Copyright 1996, the Institute for Operations Research and the Management Sciences, 7240 Parkway Drive, Suite 310, Hanover, MD 21076 USA.)

amount of information available about β_h is very small so that the predictive distribution will be very diffuse. It is therefore, extremely important that the integration in (6.11.7) be performed. If 'plug-in' estimates of β_h are used, profits can be dramatically overstated due to overconfidence. At the plug-in estimate, the probabilities can be more extreme, suggesting that the coupon will have a greater effect on purchase behavior. In the case of the most popular Catalina product, coupons are issued on the basis of only *one* purchase observation so the extent of the effect can be huge.

Figure 6.5 shows the distribution of expected revenue for a specific household and various coupon face values in an example drawn from Rossi *et al.* The product is canned tuna fish, and the coupon face values are given in cents and restricted to be multiples of 5 cents. The boxplots show the considerable uncertainty regarding expected revenue based on very imprecise inferences about β_h. The solid dots in the figure correspond to predicted revenue as the 'plug-in' estimate of β_h equal to the posterior mean. The figure shows the 'overconfidence' aspect of plug-in estimates.

Rossi *et al.* demonstrate that, relative to no household-specific information, various information sets regarding households have the potential for large value. That is, even with a small amount of information, the ability to customize the face value of the coupon is high. Revenues from even one observation are over 50 % higher than in the blanket coupon condition. Revenues from longer purchase histories can be even greater, exceeding 100 % larger.

7
Simultaneity

Using this Chapter

This chapter discusses the problem of Bayesian inference for models in which both the response variable and some of the marketing mix variables are jointly determined. We can no longer focus only on the model of the distribution of the response variable conditional on the marketing mix variables. We must build a model which (at least implicitly) specifies the joint distribution of these variables conditional on a set of driving or 'exogenous' variables. Section 7.1 provides a Bayesian treatment of the linear 'instrumental' variables problem, a problem for which standard classical asymptotic methods have proved inadequate. Section 7.2 considers a system of supply and demand where the demand system is built up by aggregating consumer-level choice models. Section 7.3 considers the situation in which simultaneity is present in a hierarchical model.

At the base of all the models considered so far is the distribution of a dependent variable (or vector) conditional on a set of independent factors. The classic marketing example is a sales response model in which quantity demanded is modeled conditional on marketing mix variables which typically include price and advertising measures. However, we should recognize that firms may set these marketing mix variables in a strategic fashion. For example, firms may consider the response of other competitors in setting price. Firms may also set the levels of marketing mix variables by optimizing a profit function that involves the sales response parameters. These considerations lead toward a joint or 'simultaneous' model of the entire vector of sales responses and marketing mix variables. In this chapter, we will consider three variants of this problem. We will first consider a Bayesian version of the instrumental variables or 'limited information' approach. We will then consider a joint or simultaneous approach. Finally, we will consider the implications of optimization in the selection of the marketing mix on the estimation of conditional sales response models.

7.1 A BAYESIAN APPROACH TO INSTRUMENTAL VARIABLES

Instrumental variables techniques are widely used in economics to avoid the "endogeneity" bias of including an independent variable that is correlated with the error term. It is

Bayesian Statistics and Marketing P. E. Rossi, G. M. Allenby and R. McCulloch

probably best to start with a simple example:[1]

$$x = \delta z + \varepsilon_1, \tag{7.1.1}$$

$$y = \beta x + \varepsilon_2. \tag{7.1.2}$$

If ε_1, ε_2 are independent, then both (7.1.2) and (7.1.1) are valid regression equations and form what is usually termed a recursive system. From our perspective, we can analyze each equation separately using the standard Bayesian treatment of regression. However, if ε_1, ε_2 are dependent, then (7.1.2) is no longer a regression equation in the sense that the conditional distribution of ε_2 given x depends on x. In this case, x is often referred to as an 'endogenous' variable. If (7.1.1) is still a valid regression equation, econometricians call z an 'instrument'. z is a variable related to x but independent of ε_2.

The system in (7.1.1)–(7.1.2) can be motivated by the example of a sales response model. If y is sales volume in units and x is price, then there are situations in which x can depend on the value of ε_2. One situation (see Villas-Boas and Winer 1999) has a shock which affects the demand of all consumers and which is known to the firm setting prices. For example, a manufacturer coupon will be dropped in a market. The drop of a blanket coupon will presumably increase retail demand for the product for all consumers. If retailers know about the coupon drop, they may adjust retail price (x) in order to take advantage of this knowledge. This would create a dependence or correlation between x and ε_2. Another classic example, in this same vein, is the 'unobserved' characteristics argument of Berry *et al.* (1995). In this example, there are characteristics of a product which drive demand but are not observed by the econometrician. These unobserved characteristics also influence price. If we are looking across markets, then these characteristics could be market-specific demand differences. If the data is a time series then these characteristics would have to vary across time. In both cases, this motivates a dependence between x and ε_2 in the same manner as the common demand shock argument.

These examples provide an 'omitted' variables interpretation of the 'endogeneity' problem. If we were able to observe the omitted variable, then we would have a standard multivariate regression system:

$$x = \delta z + \alpha_x w + u_1, \tag{7.1.3}$$

$$y = \beta x + \alpha_y w + u_2. \tag{7.1.4}$$

If w is unobserved, this induces a dependence or correlation between x and $\varepsilon_2 = \alpha_y w + u_2$. w would be interpreted as the common demand shock or unobserved characteristic(s) in the examples mentioned above. The consequence of ignoring the dependence between x and ε_2 and analyzing (7.1.2) using standard conditional models is a so-called 'endogeneity' bias. Only part of the movement in x helps us to identify β. The part of the movement in x which is correlated with ε_2 should not be used in inferring about β. This means that if the correlation between ε_1 and ε_2 is positive, there will be a positive 'endogeneity' bias.

[1] See also Lancaster (2004, Chapter 8) for an excellent introduction to instrumental variables as well as a lucid discussion of the classic returns to education endogeneity problem.

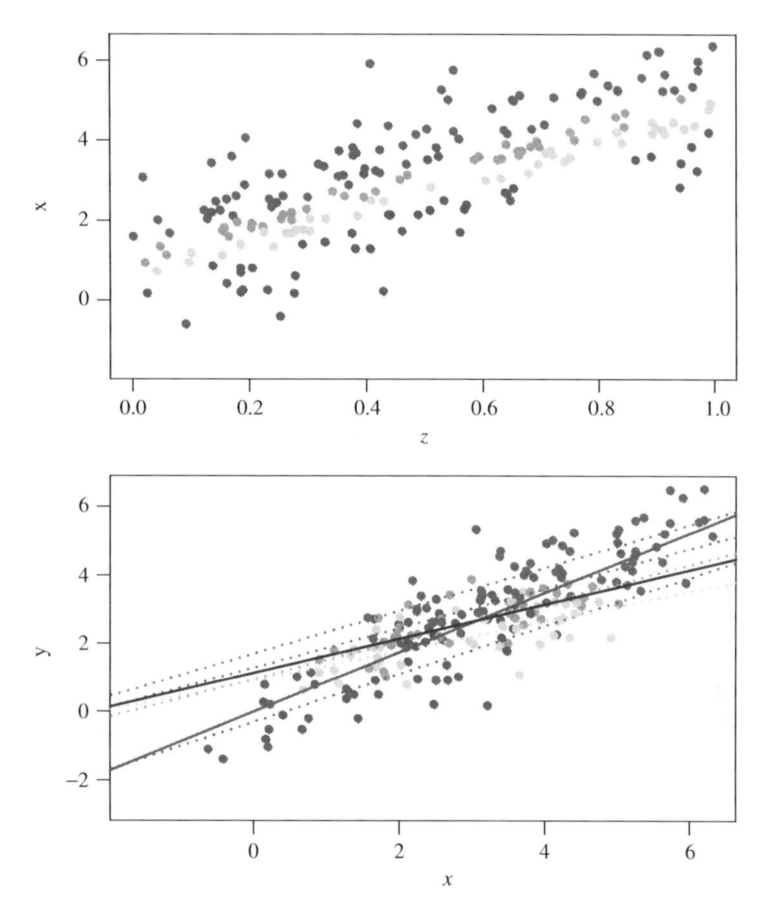

Figure 7.1 Illustration of instrumental variables method

To illustrate this situation, Figure 7.1 plots x against z and y against x using a 'brushing' technique. The data are simulated from a model in which errors are bivariate normal with correlation 0.8. Points of the same shade are linked between the two scatterplots. In the graph of x versus z, we hold the variation in the residuals from a regression of x on z relatively constant. Thus, within each color group we see variation in x that has been created by movements in the instrument z. The dotted lines in the y versus x graph show a least squares estimate of β for each brushing group. The solid gray line shows the biased least squares estimate of β from a simple regression of y on x while the solid black line shows the true value of β. The role of the instrument is to inject variation into x which is independent of ε_2.

It is easy to see that if the instrument is 'weak' in the sense that it induces little variation in x, then there will be little information in the sample regarding β. As instruments become weaker and weaker, we approach an unidentified case. To see this, we need to write down the likelihood for this model. The likelihood is the joint distribution of (x, y) given z. This is often called the 'reduced' form. In order to proceed further, we will need to make assumptions regarding the joint distribution of $\varepsilon_1, \varepsilon_2$. Given the regression specification, a natural starting point would be to assume that $\varepsilon_1, \varepsilon_2$ are bivariate normal.

The joint distribution of (x, y) given z can be written down by substituting (7.1.1) into (7.1.2):

$$\begin{aligned} x &= \delta z + \varepsilon_1, \\ y &= \beta\delta z + (\beta\varepsilon_1 + \varepsilon_2) \end{aligned} \tag{7.1.5}$$

or

$$\begin{aligned} x &= \pi_x z + \nu_1, \\ y &= \pi_y z + \nu_2. \end{aligned} \tag{7.1.6}$$

Here $\beta = \pi_y/\pi_x$ and we can think of the model as a multivariate regression with a restriction on the coefficient matrix. This is the approach taken by Lancaster (2004). However, we must recognize that the covariance of ν_1, ν_2 depends on β and elements of the covariance matrix of the 'structural' equation errors, ε_1 and ε_2. This is also the source of potential identification problems in this model. Consider the case with $\delta = 0$. Here

$$\begin{aligned} x &= \varepsilon_1, \\ y &= \beta\varepsilon_1 + \varepsilon_2. \end{aligned} \tag{7.1.7}$$

Variances and covariances are identified via observation of x and y. Equations (7.1.7) imply that

$$\frac{\operatorname{cov}(x, y)}{\operatorname{var}(x)} = \beta + \frac{\sigma_{12}}{\sigma_{11}}, \qquad \text{with } \sigma_{12} = \operatorname{cov}(\varepsilon_1, \varepsilon_2). \tag{7.1.8}$$

The identified quantity on the left-hand side of (7.1.8) can be achieved by many different combinations of β and σ_{12}/σ_{11}. Thus, the model, in the limiting case of weak instruments, is not identified. In the case where variation in δz is a small fraction of the total variation in x, there will be a 'ridge' in the likelihood, reflecting the trade-off between β and σ_{12}/σ_{11}.

Consider a more general version of (7.1.1)–(7.1.2),

$$x = z'\delta + \varepsilon_1, \tag{7.1.9}$$

$$y = \beta x + w'\gamma + \varepsilon_2, \tag{7.1.10}$$

$$\begin{pmatrix} \varepsilon_1 \\ \varepsilon_2 \end{pmatrix} \sim N(0, \Sigma). \tag{7.1.11}$$

This system shows the case of a structural equation (7.1.10) with one 'endogenous' variable, multiple instruments, and an arbitrary number of other regressors. We will put standard conditionally conjugate priors on these parameters:

$$\begin{aligned} \Sigma &\sim \text{IW}(\nu, V), \\ \delta &\sim N(\bar{\delta}, A_\delta^{-1}), \end{aligned} \tag{7.1.12}$$

$$\begin{pmatrix} \beta \\ \gamma \end{pmatrix} \sim N\left(\begin{pmatrix} \bar{\beta} \\ \bar{\gamma} \end{pmatrix}, A_{\beta\gamma}^{-1}\right).$$

We believe that it is appropriate to put a prior on the covariance matrix of the structural errors. Lancaster (2004) puts an inverted Wishart prior on the covariance matrix of the reduced-form errors and makes this prior independent of the prior on β. The reduced-form errors are related to the structural errors by the transformation

$$\begin{pmatrix} \nu_1 \\ \nu_2 \end{pmatrix} = \begin{bmatrix} 1 & 0 \\ \beta & 1 \end{bmatrix} \begin{pmatrix} \varepsilon_1 \\ \varepsilon_2 \end{pmatrix}. \tag{7.1.13}$$

Given the relationship in (7.1.13), we do not feel that a prior on the covariance of the reduced-form errors should be independent of the prior on β. In our prior, we induce a dependence via (7.1.13).

We can easily develop a Gibbs sampler for the model in (7.1.9)–(7.1.12).[2] The basic three sets of conditionals are given by:

$$\beta, \gamma | \delta, \Sigma, x, y, w, z \tag{7.1.14}$$

$$\delta | \beta, \gamma, \Sigma, x, y, w, z \tag{7.1.15}$$

$$\Sigma | \beta, \gamma, \delta, x, y, w, z \tag{7.1.16}$$

The first conditional, in (7.1.14), can easily be accomplished by a standard Bayesian regression analysis. The key insight is to recognize that given δ, we can 'observe' ε_1. We can then condition our analysis of (7.1.10) on ε_1:

$$\begin{aligned} y &= \beta x + w'\gamma + \varepsilon_2 | \varepsilon_1 \\ &= \beta x + w'\gamma + \frac{\sigma_{12}}{\sigma_{11}} \varepsilon_1 + \nu_{2|1}. \end{aligned} \tag{7.1.17}$$

$\varepsilon_2 | \varepsilon_1$ denotes the conditional distribution of ε_2 given ε_1. Since $\text{var}(\nu_{2|1}) \equiv \sigma_{2|1}^2 = \sigma_{22} - \sigma_{12}^2/\sigma_{11}$, we can rewrite (7.1.17) so that we can use standard Bayes regression with a unit variance error term:

$$\frac{(y - (\sigma_{12}/\sigma_{11})\varepsilon_1)}{\sigma_{2|1}} = \beta x/\sigma_{2|1} + (w/\sigma_{2|1})'\gamma + \zeta, \qquad \zeta \sim N(0, 1). \tag{7.1.18}$$

The second conditional, in (7.1.15), can be handled by transforming to the reduced form which can be written as a regression model with 'double' the number of observations:

$$\begin{aligned} x &= z'\delta + \varepsilon_1 \\ \tilde{y} &= \left(\frac{y - w'\gamma}{\beta}\right) = z'\delta + \left(\varepsilon_1 + \frac{\varepsilon_2}{\beta}\right). \end{aligned} \tag{7.1.19}$$

[2] Geweke (1996) considers a model similar to ours but with a 'shrinkage' prior that specifies that the regression coefficients are each independent with zero mean and the same variance.

We can transform the system above to an uncorrelated set of regressions by computing the covariance matrix of the vector of errors in (7.1.19):

$$\text{var}\begin{pmatrix} \varepsilon_1 \\ \varepsilon_1 + \frac{1}{\beta}\varepsilon_2 \end{pmatrix} = A\Sigma A' = \Omega = U'U, \qquad A = \begin{bmatrix} 1 & 0 \\ 1 & \frac{1}{\beta} \end{bmatrix}. \tag{7.1.20}$$

Pre-multiplying (7.1.19) with $(U^{-1})'$ reduces the system to a bivariate system with unit covariance matrix and we can simply stack it up and perform a Bayes regression analysis with unit variance. That is,

$$(U^{-1})'\begin{pmatrix} x \\ \tilde{y} \end{pmatrix} = (U^{-1})'\begin{bmatrix} z' \\ z' \end{bmatrix}\delta + u, \qquad \text{var}(u) = I_2. \tag{7.1.21}$$

The draw of Σ given the other parameters can be accomplished by computing the matrix of residuals and doing a standard inverted Wishart draw:

$$\begin{aligned} &\Sigma|\delta, \beta, \gamma, x, y, w \sim \text{IW}(\nu + n, S + V), \\ &S = \sum_{i=1}^{n} \varepsilon_i \varepsilon_i', \\ &\varepsilon_i = \begin{pmatrix} \varepsilon_{1i} \\ \varepsilon_{2i} \end{pmatrix}. \end{aligned} \tag{7.1.22}$$

The Gibbs sampler defined by (7.1.14)–(7.1.16) is implemented in the function **R** `rivGibbs` in *bayesm*. In the case of weak instruments and high correlation between the two structural errors (high 'endogeneity'), we might expect that the Gibbs sampler will exhibit the highest autocorrelation due to the 'ridge' in the likelihood between β and σ_{12}/σ_{11}. The 'hem-stitching' behavior of the Gibbs sampler might induce slower navigation and, hence, autocorrelation in this case.

To illustrate the functioning of the sampler and to gain insight into this model, we consider a simulated example:

$$\begin{aligned} x &= \lambda_x + \delta z + \varepsilon_1, \\ y &= \lambda_y + \beta x + \varepsilon_2, \\ \begin{pmatrix} \varepsilon_1 \\ \varepsilon_2 \end{pmatrix} &\sim N\left(0, \begin{bmatrix} 1 & \rho \\ \rho & 1 \end{bmatrix}\right). \end{aligned} \tag{7.1.23}$$

Z is a vector of 200 Unif(0,1) draws and $\lambda_x = \lambda_y = 1$. We use relatively diffuse priors for coefficients, $A = 0.04I$, and for the covariance matrix, $\Sigma \sim \text{IW}(3, 3I_3)$. We first consider the case of 'strong instruments' and a high degree of 'endogeneity'. This is achieved by simulating data with $\delta = 4$ and $\rho = 0.8$ In this situation, much of the variation in x is 'exogenous' due to variation in δz and we would expect to see good performance of the sampler. Figure 7.2 shows plots of the posterior draws for each pair of the three key parameters, $(\beta, \delta, \sigma_{12}/\sigma_{11})$. In addition, we display the sequence of the beta draws on the bottom right. Five thousand draws were used, with every fifth draw plotted. The dotted lines in each of the scatterplots represent the true values used to simulate

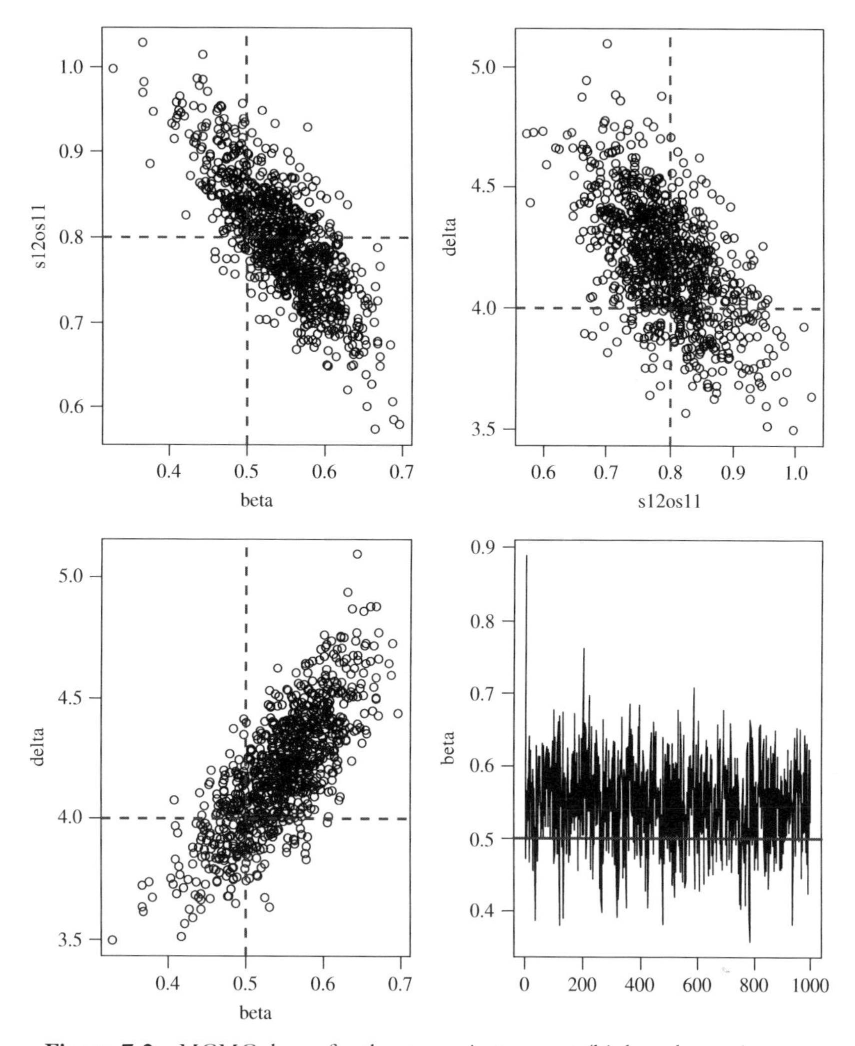

Figure 7.2 MCMC draws for the strong instrument/high endogeneity case

the data. The sampler performs very well with a very short 'burn-in' period and high numerical efficiency relative to an iid sampler ($\sqrt{f} = 1.6$). As expected, there is a negative correlation between β and σ_{12}/σ_{11}, but it is not too high.

It is well known in the classical econometrics literature that 'weak' instruments present problems for the standard asymptotic approximations to the distribution of instrumental variables estimators. This problem is caused by the fact that the instrumental variables estimator is a ratio of sample moments. In a likelihood-based procedure such as ours, the weak instruments case creates a situation of near nonidentification, as indicated in (7.1.8). This means there will be a ridge in the likelihood function which will create problems for asymptotic approximations. Since our methods do not rely on asymptotic approximations of any kind, we should obtain more accurate inferences. However, we should point out that since there is a near nonidentification problem, the role of the prior will be critical in determining the posterior. In addition, our intuition suggests that the MCMC draws may become more autocorrelated due to the ridge in the likelihood function.

In the simulated example, the weak instruments case corresponds to situations in which the variation in δz is small relative to the total variation in x. If we set $\delta = 0.1$, we create a situation with extremely weak instruments (the population R^2 of the instrument regression is 0.01). Figure 7.3 shows the distribution of the posterior draws for $\delta = 0.1$ and $\rho = 0.8$ which we dub the case of weak instruments and high endogeneity. As expected, there is a very high negative correlation between the draws of β and σ_{12}/σ_{11}. This creates a sampler with higher autocorrelation ($\sqrt{f} = 10$), or a relative numerical efficiency of one-tenth of an iid sampler. However, more than 1 million draws of the

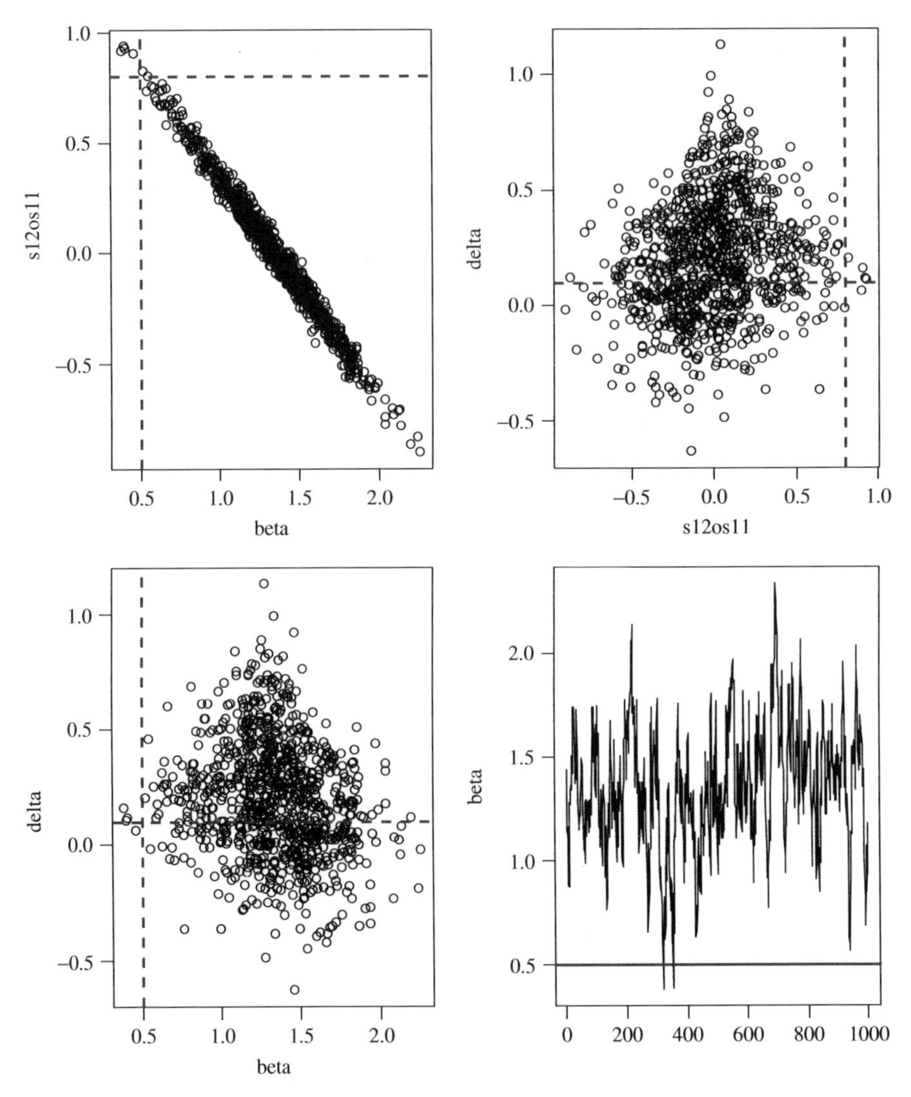

Figure 7.3 MCMC draws for the weak instrument/high endogeneity case: diffuse but proper prior

sampler can be achieved in less than 1 hour of computing time, so we do not think this is an important issue (Figure 7.3 is based on 10,000 draws with every tenth draw plotted).

It is important to note the influence of the prior for the weak instrument case. We note that the posterior distribution of β and σ_{12}/σ_{11} is centered away from the true values. This is due to the prior on Σ. Even though it is set to barely proper values, the prior is centered on the identity matrix. This means that σ_{12}/σ_{11} will be 'shrunk' toward zero. There is so little information in the likelihood that this prior 'shows' through. However, it is important to realize that the posterior of both quantities is very diffuse, revealing the near lack of identification. Figure 7.4 shows the same situation with an improper prior on Σ. Now the posterior is centered closer to the true values but with huge variability.

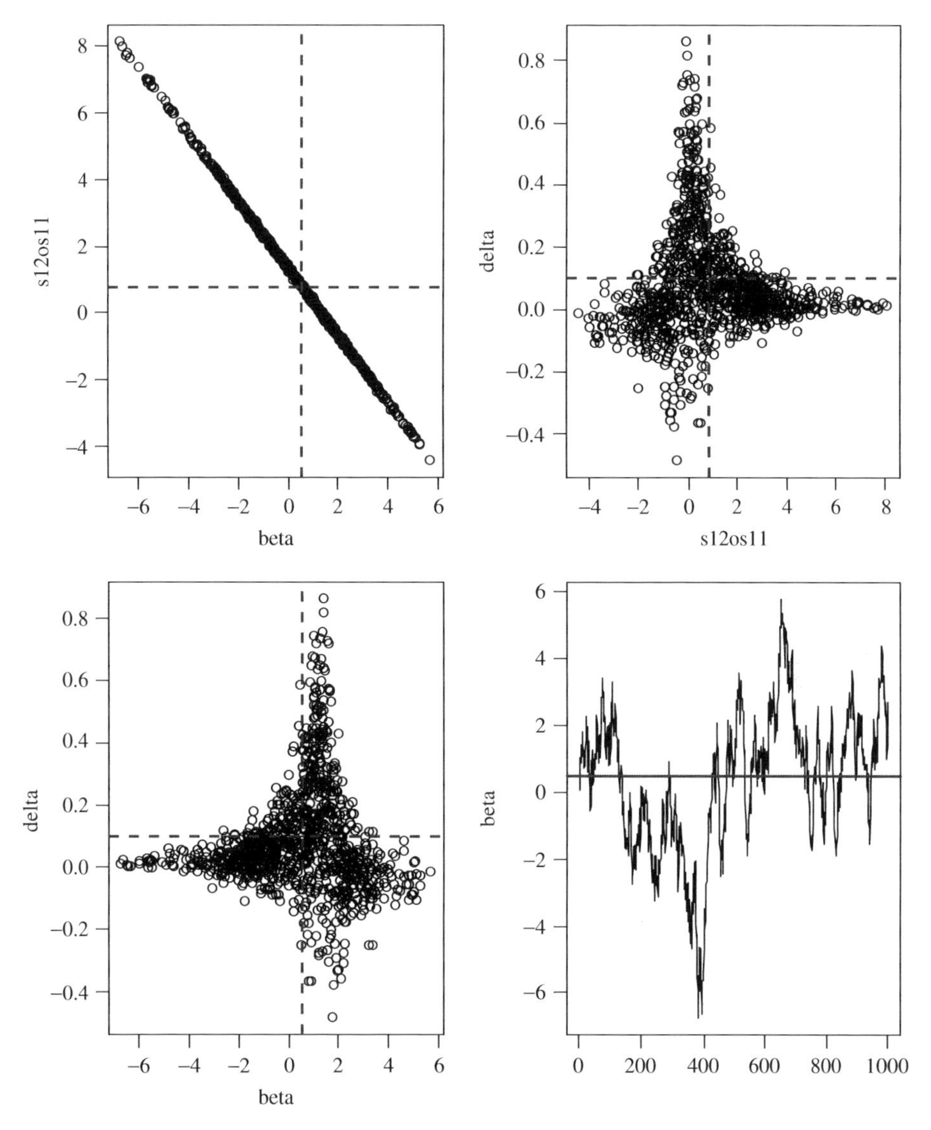

Figure 7.4 MCMC draws for the weak instrument/high endogeneity case: improper prior

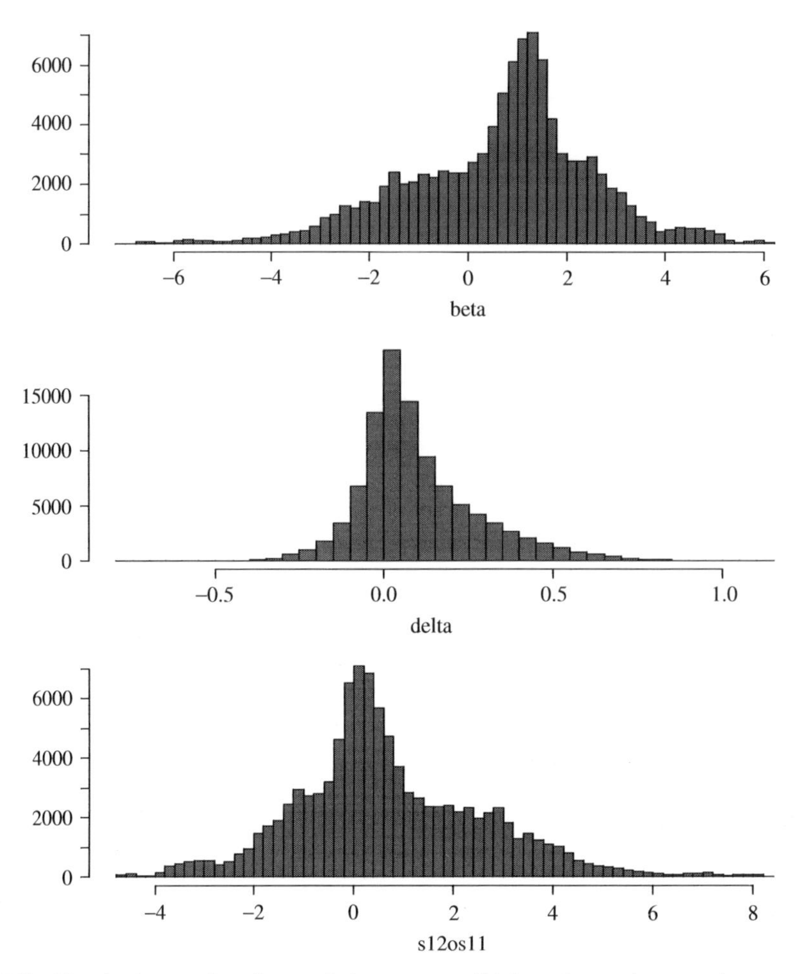

Figure 7.5 Marginal posteriors for weak instruments/high endogeneity case: improper prior, 100000 draws

Figure 7.5 shows the marginal posterior distributions for a longer run of 100,000 draws. We can see that the marginal posteriors for β and σ_{12}/σ_{11} have a mode at the true value and a huge 'shoulder' representing high uncertainty.

One might argue that the case of weak instruments and a modest or low degree of endogeneity is more representative of the applications of instrumental variable models. For this reason, we considered the case $\delta = 0.1$ and $\rho = 0.1$. Figure 7.6 shows a plot of the posterior draws for these parameter settings, using the proper but diffuse prior on Σ. There is still a high degree of dependence between β and σ_{12}/σ_{11}, reflecting the low amount of information in the weak instrument case. The sampler is slightly less autocorrelated than the weak instruments/high endogeneity case with a relative numerical efficiency of one-eighth of an iid sample. The posterior distributions are very spread out, properly reflecting the small amount of sample information. In addition, the bivariate posteriors have a decidedly non-elliptically symmetric shape, reminiscent of the shape of a double exponential distribution.

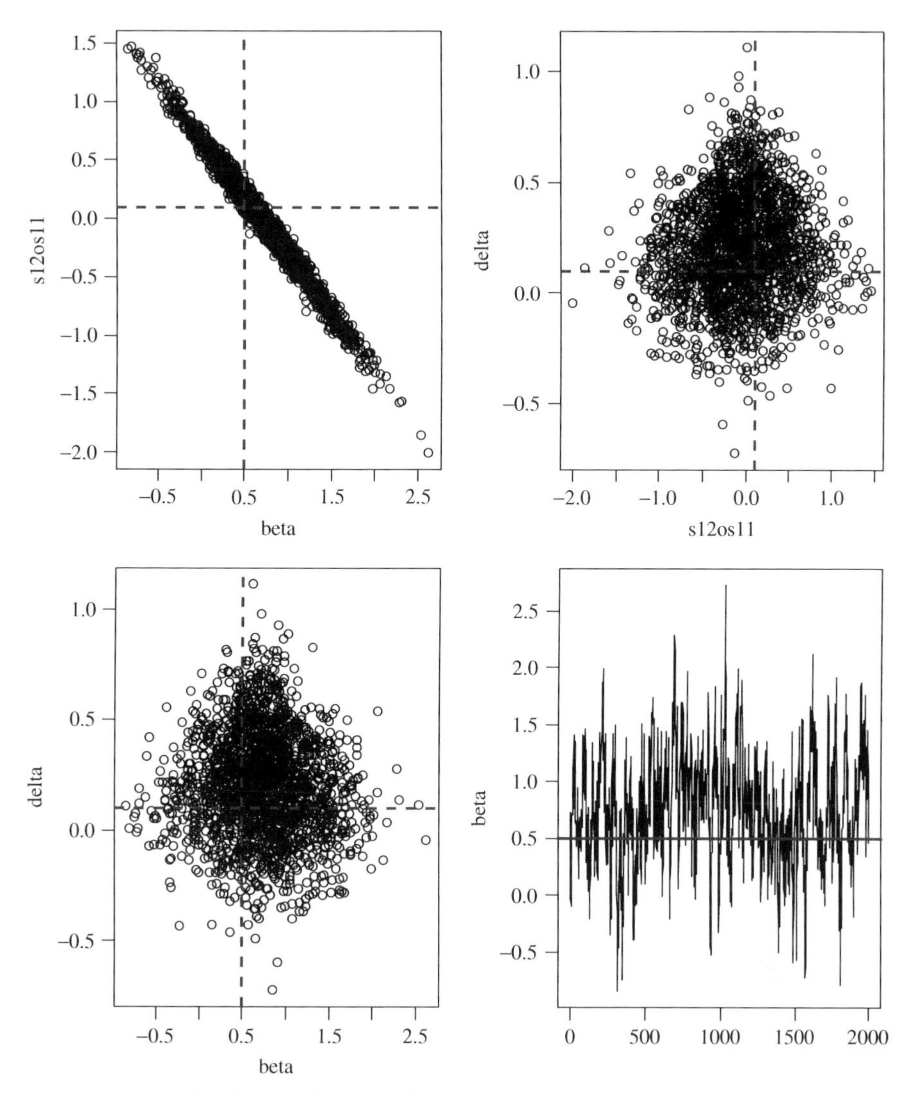

Figure 7.6 MCMC draws for the weak instrument/low endogeneity case

7.2 STRUCTURAL MODELS AND ENDOGENEITY/SIMULTANEITY

The basic 'instrumental variable' model can also be interpreted as resulting from a structural model with attention 'limited' to one structural equation. The classic example is the example of supply and demand. If we observe market clearing quantities and prices along with a 'demand' shifter or 'exogenous' variable, then we can write down a model that is identical to the simple example with which we started Section 7.1:

$$\begin{aligned} q_d &= \alpha_0 + \alpha_1 p + \alpha_2 z + \varepsilon_d, \\ q_s &= \beta_0 + \beta_1 p + \varepsilon_s. \end{aligned} \tag{7.2.1}$$

Here z is a variable such as advertising or promotion which shifts demand but does not alter supply conditions. If we impose the market clearing condition that $q_d = q_s$, then we can rewrite the first equation as a regression and obtain the 'limited' information or instrumental variables model:

$$p = \frac{\alpha_0 - \beta_0}{\beta_1 - \alpha_1} + \frac{\alpha_2}{\beta_1 - \alpha_1} z + \frac{\varepsilon_d - \varepsilon_s}{\beta_1 - \alpha_1},$$
$$q = \beta_0 + \beta_1 p + \varepsilon_s. \tag{7.2.2}$$

Thus, one possible justification of the 'instrumental' variables approach is that the model arises from some joint or simultaneous model. The joint model imposes the condition of market equilibrium and makes an assumption about the joint distribution of $(\varepsilon_s, \varepsilon_d)$. This allows us to formulate the likelihood which is the joint distribution of (q, p) given z via the Jacobian of the transformation from $(\varepsilon_s, \varepsilon_d)$ to (q, p). This idea can be extended to more 'realistic' models of demand in which the demand equation at the aggregate level is the sum of the demands over heterogeneous consumers. Yang *et al.* (2003) tackle this problem with a demand model aggregated from heterogenous logits.

The estimation of simultaneous demand and supply with household heterogeneity is challenging. In a full information analysis, where specific assumptions are made about the nature of competition among firms, the supply-side equation is a complex function of household-level parameters and common error terms or shocks. To date, likelihood-based approaches have not been developed for models with unobserved heterogeneity specified as random effects. The reason is that a frequentist (non-Bayesian) approach to analysis does not view the shocks and random effects as parameters, and analysis proceeds by first integrating them out of the likelihood function. This integration is computationally demanding when the shocks and random effects are jointly present, and researchers have instead controlled for consumer heterogeneity by incorporating past purchases, or other exogenous (for example, demographic) variables, into the model specification. In the context of demand analysis using aggregate level data, it has been shown that it is important to control for both price endogeneity and heterogeneity to avoid potential biases in demand-side parameter estimates.

We first set notation by developing the demand and supply models used in our study of consumer brand choice and retailer pricing. Consumers are assumed to make brand choice decisions according to a standard discrete choice model. On the supply side, we develop specifications derived from profit-maximizing assumptions made about manufacturer and retailer behavior.

7.2.1 Demand Model

The disaggregate demand function is specified as a logistic normal regression model. Suppose we observe purchase incidences and choices (y) for a group of individuals ($i = 1, \ldots, I$) for J brands ($j = 0, \ldots, J$) in a product category over T time periods ($t = 1, \ldots, T$). The utility of consumer i for brand j at time t is specified as

$$u_{ijt} = \beta_i' x_{jt} + \alpha_i p_{jt} + \xi_{jt} + \varepsilon_{ijt}, \tag{7.2.3}$$

and in the case of no purchase from the J available brands, we denote $j = 0$ and the associated utility function as

$$u_{i0t} = \varepsilon_{i0t}, \tag{7.2.4}$$

where x_{jt} is a vector with observed product characteristics including brand intercepts, feature and display variables, p_{jt} is the unit price for brand j at time t, β_i and α_i are individual-level response coefficients, ξ_{jt} is an unobserved demand shock for brand j at time period t, and ε_{ijt} is the unobserved error term that is assumed to be uncorrelated with price. We make assumptions on the error terms and response coefficients as follows:

$$\begin{aligned} \varepsilon_{ijt} &\sim \mathrm{EV}(0, 1), \\ \xi_t &\sim N(0, \Sigma_{\mathrm{d}}), \\ \theta_i = (\alpha_i, \beta_i')' &\sim N(\bar{\theta}, \Sigma_\theta). \end{aligned} \tag{7.2.5}$$

The type I extreme value specification of ε_{ijt} leads to a standard logit choice probability for person i choosing brand j at time t:

$$\Pr(y_{ijt} = 1) = s_{ijt} = \frac{\exp(V_{ijt})}{1 + \sum_k \exp(V_{ikt})} \tag{7.2.6}$$

where $V_{ijt} = \beta_i' x_{jt} + \alpha_i p_{jt} + \xi_{jt}$. Assuming the sample is representative of the market and households do not make multiple purchases, we obtain market share for brand j at time t as

$$s_{jt} = \frac{\sum_i s_{ijt}}{I}. \tag{7.2.7}$$

If firms use expected demand (s_{jt}) to set price, then price is not exogenously determined. Price and the demand-side error, ξ_t in equation (7.2.3), will be correlated because expected demand, used to set prices, is a function of both. That is, it is not possible to write the joint distribution of prices and demand as a conditional demand distribution and a marginal price distribution. Demand and prices are both functions of consumer price sensitivity and other model parameters. Not accounting for the endogenous nature of price will produce biased estimates of model parameters, including household price sensitivity.

7.2.2 Supply Model – Profit-Maximizing Prices

We illustrate the use of Bayesian methods to estimate simultaneous demand and supply models using a simple supply-side model. More complicated models of manufacturer and retailer behavior can be found in Yang *et al.* (2003). We assume that each manufacturer produces only one product and maximizes the following objective function:

$$\max_{w_i} \ \pi_i = M s_i (w_i - c_{\mathrm{m}i}) \tag{7.2.8}$$

where M is the potential market size, w is the wholesale price, and $c_{\mathrm{m}i}$ is manufacturer i's marginal cost. The first-order condition for the manufacturers implies

$$w - c_{\mathrm{m}} = (HQ)^{-1}(-s), \tag{7.2.9}$$

where

$$H_{ik} = \frac{\partial s_i}{\partial p_k}, \quad Q_{ik} = \frac{\partial p_i}{\partial w_k} \qquad i = 1, \ldots, J, \text{ and } k = 1, \ldots, J.$$

Next, we turn to the retailer's pricing strategy. For the purpose of illustration, we only model a single retailer's pricing behavior even though competition among multiple retailers is possible. One simple rule the retailer can use is to simply charge a fixed mark-up over wholesale price for each brand, resulting in the specification

$$p_i = w_i + m_i, \tag{7.2.10}$$

where m stands for the fixed mark-up. This pricing strategy implies that $\partial p_i/\partial w_i = 1$ and $\partial p_i/\partial w_j = 0$. Substituting those two conditions into (7.2.9), we obtain the pricing equation

$$p = c_{\mathrm{m}} + m - (H)^{-1}s. \tag{7.2.11}$$

Finally, we can specify the manufacturer and retailer cost c_{m} as a brand-specific linear function of cost shifters Z, that is,

$$c_{\mathrm{m}t} = Z_t'\delta_j + \eta_t, \tag{7.2.12}$$

where η_t is the supply-side error for which we assume a multivariate normal distribution, that is, $\eta_t \sim N(0, \Sigma_{\mathrm{s}})$, or

$$p_t = m + Z_t'\delta_j - H_t^{-1}s_t + \eta_t. \tag{7.2.13}$$

The distribution of observed prices is obtained from the distribution of the supply-side error by using change-of-variable calculus. This distribution is of nonstandard form because (7.2.13) is implicit in price – that is, price appears on both the left and right side of the equal sign in the terms H and s. The distribution of observed prices is obtained by defining a new variable $r = p + H^{-1}s - m - Z\delta$ that is distributed normally with mean 0 and covariance Σ_{s}. The likelihood for price is obtained in the standard way as the likelihood for r multiplied by the determinant of the Jacobian ($J = \{\partial r_i/\partial p_j\}$).

7.2.3 Bayesian Estimation

Data augmentation is used to facilitate estimation of the model. We introduce household-specific coefficients $\{\theta_i\}$ and supply shock realizations $\{\xi_t\}$ as augmented, latent variables, and use them as conditioning arguments in the model hierarchy. The dependent variables are choice (y_{it}) and prices (p_t), and the model can be written in hierarchical form:

$y_{it}\vert p_t, \theta_i, \xi_t, \varepsilon_{it}$	observed demand,
$p_t\vert\{\theta_i\}, \{\xi_t\}, \delta, \eta_t$	observed prices,
$\theta_i\vert\bar{\theta}, \Sigma_\theta$	heterogeneity,
$\xi_t\vert\Sigma_{\mathrm{d}}$	demand shock,
$\eta_t\vert\Sigma_{\mathrm{s}}$	supply shock,
ε_{it}	extreme value (logit) error,

where $\theta_i = (\alpha_i', \beta_i')'$. Observed demand for the ith household is dependent on the household coefficients (θ_i), the demand shock (ξ_t), the unobserved error (ε_{it}) and the explanatory variables, including prices (equation (7.2.3)). Observed prices are determined by the set of household coefficients $\{\theta_i\}$, cost shifter coefficients (δ) and the supply shock (η_t), and are set in response to the expected demand across the heterogeneous households (7.2.13). Household coefficients are specified as random effects, and the demand and supply shocks specified as normally distributed.

Given $\{\theta_i\}$ and $\{\xi_t\}$, the joint distribution of demand and prices is obtained by multiplying the conditional (on prices) demand density by the marginal price density. The marginal price density, given $\{\theta_i\}$ and $\{\xi_t\}$, is derived from the supply-side error term, η_t. The conditional demand density, given $\{\theta_i\}$, $\{\xi_t\}$ and prices, is multinomial with logit probabilities (7.2.6). The joint density of all model parameters is then:

$$\begin{aligned} & f(\{\theta_i\}, \{\xi_t\}, \delta, \bar{\theta}, \Sigma_\text{d}, \Sigma_\text{s}, \Sigma_\theta | \{y_{it}\}, \{p_t\}) \\ & \quad \propto \prod_{t=1}^{T} \prod_{i=1}^{I} \Pr(y_{it}|\theta_i, p_t, \xi_t) \pi_1(\xi_t|\Sigma_\text{d}) \pi_2(p_t|\{\theta_i\}, \{\xi_t\}, \delta, \Sigma_\text{s}) \\ & \quad \times \pi_3(\theta_i|\bar{\theta}, \Sigma_\theta) \pi_4(\delta, \bar{\theta}, \Sigma_\text{d}, \Sigma_\text{s}, \Sigma_\theta), \end{aligned} \tag{7.2.14}$$

where $\Pr(y_{it}|\theta_i, p_t, \xi_t)$ is the logit choice probability for household i at time t, π_1 is the density contribution of the demand error ξ_t, π_2 is the density contribution of the observed prices at time t that depend on consumer preferences and price sensitivities $\{\theta_i\}$, demand errors $\{\xi_t\}$, cost variables and coefficients (Z, δ) and the supply-side error (η_t), π_3 is the distribution of heterogeneity and π_4 is the prior distribution on the hyperparameters.

Given these augmented variables, estimation proceeds using standard distributions for heterogeneity ($\bar{\theta}$ and Σ_θ) and error covariance matrices (Σ_d and Σ_s). Draws of the augmented variables are obtained from the full conditional distribution for ξ_t and θ_t:

$$[\xi_t|^*] \propto \prod_i \Pr(\text{choice}_{it}) \times |J_t| \times (r_t \sim N(0, \Sigma_\text{s})) \times (\xi_t \sim N(0, \Sigma_\text{d})),$$

$$[\theta_i|^*] \propto \prod_{t=1}^{T} \Pr(\text{choice}_{it}) \times \prod_{t=1}^{T} |J_t| \times \prod_{t=1}^{T} (r_t \sim N(0, \Sigma_\text{s})) \times (\theta_i \sim N(\bar{\theta}, \Sigma_\theta)),$$

where

$$\Pr(\text{choice}_{it} \neq 0) = \prod_{j=1}^{J} (s_{ijt})^{I(\text{choice}_{ijt}=1)} \qquad \text{(choose one of the brands)},$$

$$\Pr(\text{choice}_{it} = 0) = \frac{1}{1 + \exp(V_{it})} \qquad \text{(choose outside good)},$$

$$V_{ijt} = \beta_i' x_{jt} + \alpha_i p_{jt} + \xi_{jt},$$

$$s_{ijt} = \frac{\exp(V_{ijt})}{1 + \sum_{k=1}^{J} \exp(V_{ikt})},$$

$$|J_t| \text{ is the Jacobian} = \begin{vmatrix} \dfrac{\partial r_{1t}}{\partial p_{1t}} & \cdots & \dfrac{\partial r_{1t}}{\partial p_{Jt}} \\ \vdots & \ddots & \vdots \\ \dfrac{\partial r_{Jt}}{\partial p_{1t}} & \cdots & \dfrac{\partial r_{Jt}}{\partial p_{Jt}} \end{vmatrix},$$

$$r_t = p_t + H_t^{-1} s_t - m - Z_t \delta$$

and

$$H_t^{-1} s_t = \frac{\sum_{i=1}^{I} s_{ijt}}{\sum_{i=1}^{I} \alpha_i s_{ijt}(1 - s_{ijt})}.$$

Bayesian analysis does not require the integration of the random effects (θ_i) and demand shocks (ξ_t) to obtain the marginalized or unconditional likelihood. Such marginalization is difficult to evaluate because the integral is of high dimension and involves highly nonlinear functions of the model parameters, including the Jacobian needed to obtain the distribution of observed prices. An advantage of using a Bayesian MCMC estimator is that these variables can be used as conditioning arguments for generating other model parameters – resulting in significant simplification in model estimation.

7.3 NONRANDOM MARKETING MIX VARIABLES

Thus far, we have focused on examples where supply-side behavior has resulted in an 'endogeneity' problem in which the marketing mix variables such as price can be functions of unobserved variables. Typically, these ideas are applied to a problem in which the marketing mix variable is being set on a uniform basis for a given market or retailer. However, if firms have access to information regarding the response parameters of a given customer, then the marketing mix variables can be customized to that specific account. This creates a related situation in which the levels of the marketing mix variables are set as a function of the response parameters. Manchanda *et al.* (2004) explore this problem and offer a general approach as well as a specific application to the problem of allocating salesforce effort in the pharmaceutical industry.

7.3.1 A General Framework

Consider the general setting in which the marketing mix or X variables are chosen strategically by managers. The basic contribution is to provide a framework for situations in which the mix variables are chosen with some knowledge of the response parameters of the sales response equation.

Sales response models can be thought of as particular specifications of the conditional distribution of sales (y) given the marketing mix x,

$$y_{it} | x_{it}, \beta_{it}, \tag{7.3.1}$$

where i represents the individual customer/account and t represents the time index. For example, a standard model would be to use the log of sales or the logit of market share and specify a linear regression model,

$$\ln(y_{it}) = x'_{it}\beta_i + \varepsilon_{it}, \qquad \varepsilon_{it} \sim N(\cdot, \cdot).$$

Here the transform of y is specified as conditionally normal with sales response parameters, β_i. Analysis of (7.3.1) is usually conducted under the assumption that the marginal distribution of x is independent of the conditional distribution in (7.3.1). In this case the marginal distribution of x provides no information regarding β_i and the likelihood factors. If $x_{it}|\theta$ is the marginal distribution of x, then the likelihood factors as follows:

$$\ell(\{\beta_i\}, \theta) = \prod_{i,t} p(y_{it}|x_{it}, \beta_i)p(x_{it}|\theta) = \prod_{i,t} p(y_{it}|x_{it}, \beta_i) \prod_{i,t} p(x_{it}|\theta). \tag{7.3.2}$$

This likelihood factorization does not occur once the model is changed to build dependence between the marginal distribution of x and the conditional distribution. There are many forms of dependence possible, but in the context of sales response modeling with marketing mix variables a particularly useful form is to make the marginal distribution of x dependent on the response parameters in the conditional model. Thus, our general approach can be summarized as follows:

$$\begin{aligned} &y_{it}|x_{it}, \beta_i, \\ &x_{it}|\beta_i, \tau. \end{aligned} \tag{7.3.3}$$

This approach is a generalization of the models developed by Chamberlain (1980, 1984) and applied in a marketing context by Bronnenberg and Mahajan (2001). Chamberlain considers situations in which the X variables are correlated to random intercepts in a variety of standard linear and logit/probit models. Our random effects apply to all of the response model parameters and we can handle nonstandard and nonlinear models. However, the basic results of Chamberlain regarding consistency of the conditional modeling approach apply. Unless T grows, any likelihood-based estimator for the *conditional* model will be inconsistent. The severity of this asymptotic bias will depend on the model, the data and on T. For small T, these biases have been documented to be very large.

What is less well appreciated is that the model for the marginal distribution of x in (7.3.3) provides additional information regarding β_i. The levels of x now are used to infer about slope coefficients.

The general data augmentation and Metropolis–Hasting MCMC approach is ideally suited to exploit the conditional structure of (7.3.3). That is, we can alternate between draws of $\beta_i|\tau$ (here we recognize that the $\{\beta_i\}$ are independent conditional on τ) and $\tau|\{\beta_i\}$. With some care in the choice of the proposal density, this MCMC approach can handle a very wide range of specific distributional models for both the conditional and the marginal distributions in (7.3.3).

To further specify the model in (7.3.3), it is useful to think about the interpretation of the parameters in the β vector. We might postulate that in the marketing mix application, the important quantities are the level of sales given some 'normal' settings of x (for example, the baseline sales) and the derivative of sales with respect to various marketing mix variables. In many situations, decision-makers are setting marketing mix variables proportional to the baseline level of sales. More sophisticated decision-makers might recognize that the effectiveness of the marketing mix is also important in allocation of marketing resources. This means that the specification of the marginal distribution of x should make the level of x a function of the baseline level of sales and the derivatives of sales with respect to the elements of x.

7.3.2 An Application to Detailing Allocation

Manchanda *et al.* (2004) consider an application to the problem of allocation of salesforce effort ('detailing') across 'customers' (physicians). The data is based on sales calls (termed 'details') made to physicians for the purpose of inducing them to prescribe a specific drug. In theory, sales managers should allocate detailing efforts across the many thousands of regularly prescribing physicians so as to equalize the marginal impact of a detail across doctors (assuming equal marginal cost, which is a reasonable assumption according to industry sources).

The barrier to implementing optimal allocation of detailing effort is the availability of reliable estimates of the marginal impact of a detail. While individual physician-level data is available on the writing of prescriptions from syndicated suppliers such as IMS and Scott-Levin, practitioners do not fit individual physician models due to the paucity of detailing data and extremely noisy coefficient estimates obtained from this data. Instead, practitioners pool data across physicians in various groups, usually on the basis of total drug category volume. Detailing targets are announced for each group. Generally speaking, higher-volume physicians receive greater detailing attention. Even if detailing had no effect on prescription behavior, volume-based setting of the detailing independent variable would create a spurious detailing effect in pooled data. In addition to general rules which specify that detailing levels are related to volume, it is clear that individual salesforce managers adjust the level of detailing given informal sources of knowledge regarding the physician. This has the net effect of making the levels of detailing a function of baseline volume and, possibly, detailing responsiveness. Thus, the independent variable in our analysis has a level which is related to parameters of the sales response function.

Given the need for physician-specific detailing effects, it might seem natural to apply Bayesian hierarchical models to this problem. Bayesian hierarchical models 'solve' the problem of unreliable estimates from individual physician models by a form of 'shrinkage' or partial pooling in which information is shared across models. A Bayesian hierarchical model can be viewed as a particular implementation of a random coefficients model.

If detailing levels are functions of sales response parameters, then standard Bayesian hierarchical models will be both biased and inefficient. The inefficiency, which can be substantial, comes from the fact that the *level* of the independent variable has information about the response coefficients. This information is simply not used by the standard approach. The sales response function is supplemented by an explicit model for the distribution of detailing which has a mean related to response coefficients. In our application, given that sales (prescriptions) and detailing are count data, we use a negative binomial regression as the sales response function and a Poisson distribution for detailing. This joint model provides much more precise estimates of the effects of detailing and improved predictive performance. Rather than imposing optimality conditions on our model, we estimate the detailing policy function used by the sales managers.

7.3.3 Conditional Modeling Approach

A conditional model for the distribution of prescriptions written given detailing and sampling is the starting point for our analysis. The data we have is count data, with most observations at less than 10 prescriptions in a given month. Manchanda *et al.* provide some evidence that the distribution of the dependent variable is overdispersed relative to the Poisson distribution. For this reason, we will adopt the negative binomial as the base model for the conditional distribution and couple this model with a model of the distribution of coefficients over physicians. The negative binomial model is flexible in the sense that it can exhibit a wide range of degrees of overdispersion, allowing the data to resolve this issue. A negative binomial distribution with mean λ_{it} and overdispersion parameter α is given by

$$\Pr(y_{it} = k|\lambda_{it}) = \frac{\Gamma(\alpha + k)}{\Gamma(\alpha)\Gamma(k+1)} \left(\frac{\alpha}{\alpha + \lambda_{it}}\right)^{\alpha} \left(\frac{\lambda_{it}}{\alpha + \lambda_{it}}\right)^{k}, \tag{7.3.4}$$

where y_{it} is the number of new prescriptions written by physician i in month t. As α goes to infinity, the negative binomial distribution approaches the popular Poisson distribution.

The standard log link function specifies that the log of the mean of the conditional distribution is linear in the parameters:

$$\lambda_{it} = \mathrm{E}[y_{it}|x_{it}] = \exp(x_{it}'\beta_i), \tag{7.3.5}$$

$$\ln(\lambda_{it}) = \beta_{0i} + \beta_{1i}\mathrm{Det}_{it} + \beta_{2i}\ln(y_{i,t-1} + d). \tag{7.3.6}$$

The lagged log-prescriptions term, $\ln(y_{i,t-1} + d)$, in (7.3.6) allows the effect of detailing to be felt not only in the current period, but also in subsequent periods. We add d to the lagged level of prescriptions to remove problems with zeros in the prescription data. The smaller the number added, the more accurate the Koyck solution is as an approximation. The problem here is that the log of small numbers can be a very large in magnitude, which would have the effect of giving the zeros in the data undue influence on the carry-over coefficients. We choose $d = 1$ as the smallest number which will not create large outliers in the distribution of $\ln(y_{it} + d)$.

To complete the conditional model, a distribution of coefficients across physicians is specified. This follows a standard hierarchical formulation:

$$\beta_i = \Delta z_i + \nu_i, \qquad \nu_i \sim N(0, V_\beta).$$

The z vector includes information on the nature of the physicians' practice and level of sampling (note that we use primary care physicians as the base physician type):

$$z' = (1, \mathrm{SPE}, \mathrm{OTH}, \mathrm{SAMP}).$$

SAMP is the mean number of monthly samples (per physician) divided by 10. This specification of z and the model in (7.3.4) allows there to be a main effect and an interaction for both physician specialty type and sampling. We might expect that physicians with a specialty directly relevant to the therapeutic class of drug X will have a different level of prescription writing. In addition, detailing may be more or less effective, depending on the physician's specialty. Sales calls may include the provision of free drug X samples. The effect of sampling is widely debated in the pharmaceutical industry, with some arguing that it enhances sales and others arguing for cannibalization as a major effect. Most believe that sampling is of secondary importance to detailing. Sampling is conditional on detailing in the sense that sampling cannot occur without a detail visit. For this reason, we include the average sampling variable in the mean of the hierarchy which creates an interaction term between detailing and sampling.

7.3.4 Beyond the Conditional Model

The company producing drug X does not set detailing levels randomly. The company contracts with a consultant to help optimize the allocation of their national salesforce. The consultant recognizes that detailing targets for the salesforce should be set at the physician level and not at some higher level such as the sales territory. According to the consultant, detailing is set primarily on the basis of the physician decile computed for the quarter prior to the annual planning period. Each physician is assigned to a decile based on the physician's total prescription writing for all drugs in the therapeutic class. These annual targets are then adjusted quarterly based on the previous quarter's deciles. These quarterly adjustments tend to be minor.

Conditional modeling approaches rely on the assumption that the marginal distribution of the independent variables does not depend on the parameters of the conditional distribution specified in (7.3.6). If total category volume is correlated with the parameters of the conditional response model, then this assumption will be violated. We think it is highly likely that physicians who write a large volume of drug X prescriptions regardless of detailing levels (for example, have high value of the intercept in (7.3.6)) will also have higher than average category volume. This means that the marginal distribution of detailing will depend at the minimum on the intercept parameter in (7.3.6). This dependence is the origin of the spurious correlation that can occur if higher-volume physicians are detailed more.

It is also clear that, although detailing targets are set on an annual basis (and revised quarterly), there is much month-to-month variation in detailing due to factors outside the control of the salesforce managers. In addition, even though detailing targets are set at a high level in the firm, salesforce district or territory managers may change the actual level of detailing on the basis of their own specialized knowledge regarding specific physicians. If salesforce managers had full knowledge of the functional form and parameters of the detailing response function, then detailing would be allocated so as to equalize the marginal effects across physicians. Given that the current industry practice is not to compute individual physician estimates, it is unreasonable to assume that firms are using a full information optimal allocation approach.

A specification of the detailing distribution that allows for some partial knowledge of detailing response parameters is required. A simple but flexible approach would be to assume that detailing is iid with mean set as a function of the long-run response parameters from (7.3.6). Note that the average first-order autocorrelation for detailing is less than 0.3. Monthly detailing is a count variable, with rarely more than 5 details per month. Detailing is modeled as an iid draw from a Poisson distribution with a mean that is a function of baseline sales and the long-run response to detailing:

$$\Pr(\text{Det}_{it} = m|\eta_i) = \frac{\eta_i^m \exp(-\eta_i)}{m!}. \tag{7.3.7}$$

The mean of this Poisson distribution is a function of the (approximate long-run) coefficients:

$$\ln(\eta_i) = \gamma_0 + \gamma_1 \left(\frac{\beta_{0i}}{(1-\beta_{2i})}\right) + \gamma_2 \left(\frac{\beta_{1i}}{(1-\beta_{2i})}\right). \tag{7.3.8}$$

The specification in (7.3.8) allows for a variety of different possibilities. If detailing is set with no knowledge of responsiveness to detailing, then we should expect γ_2 to be zero. On the other hand, if detailing is set with some knowledge of responsiveness to detailing, then γ_1 and γ_2 should have posteriors massed away from zero. There are a variety of different functional forms for the relationship between the mean level of detailing and the response parameters. This specification should be regarded as exploratory and as a linear approximation to some general function of long-run effects.

To summarize, our approach is to enlarge the conditional model by specifying a model for the marginal distribution of detailing. The marginal distribution of detailing depends on conditional response parameters. Using the standard notation for conditional distributions in hierarchical models, the new model can be expressed as follows:

$$y_{it}|\text{Det}_{it}, y_{i,t-1}, \beta_i, \alpha \qquad \text{(negative binomial regression)} \tag{7.3.9}$$

$$\text{Det}_{it}|\beta_i, \gamma \qquad \text{(Poisson marginal).} \tag{7.3.10}$$

This dependence of the marginal distribution on the response parameters alters the standard conditional inference structure of hierarchical models. In the standard conditional model given only by (7.3.9), inference about the response parameters, β_i, is based on time series variation in detailing for the same physician and via similarities between physicians as expressed by the random effects or first-stage prior. However, when (7.3.10) is added to the model, inferences about β_i will change as new information is available from the level of detailing. The marginal model in (7.3.10) implies that the level of detailing is informative about responsiveness and this information is incorporated into the final posterior on β_i. For example, suppose $\gamma_2 < 0$ in (7.3.8). Then detailing is set so that less responsive physicians are detailed at higher levels, and this provides an additional source of information which will be reflected in the β_i estimates. Thus, the full model consisting of (7.3.9) and (7.3.10) can deliver improved estimates of physician-level parameters by exploiting information in the levels of detailing. The model specified is conditional on β_i. We add the standard heterogeneity distribution on β_i.

Another way of appreciating this modeling approach is to observe that likelihood for β_i has two components – the negative binomial regression and the Poisson marginal model:

$$\ell(\{\beta_i\}) = \prod_i \prod_t p_{\text{NB}}(y_{it}|\text{Det}_{it}, \beta_i, \alpha) p_{\text{Poisson}}(\text{Det}_{it}|\beta_i, \gamma). \tag{7.3.11}$$

β_i is identified from both the negative binomial and the Poisson portions of the model. Examination of the form of the likelihood in (7.3.11) and the mean function in (7.3.8) indicates some potential problems for certain data configurations. In the

Poisson portion of the model, elements of the γ vector and the collection of β_i values enter multiplicatively. In terms of the Poisson likelihood, $[\gamma_1, \{\beta_{1i}\}]$ and $[-\gamma_1, -\{\beta_{1i}\}]$, for example, are observationally equivalent. What identifies the signs of these parameters is the negative binomial regression. In other words, if the signs of the detailing coefficients are flipped, then the negative binomial regression fits will suffer, lowering the posterior at that mode. This suggests that in data sets where there is only weak evidence for the effects of detailing (or, in general, any independent variable), there may exist two modes in the posterior of comparable height. Navigating between these modes can be difficult for Metropolis–Hastings algorithms. To gauge the magnitude of this problem, we simulated a number of different data sets with varying degrees of information regarding the effect of the independent variable. For moderate amounts of information, similar to that encountered in the data, the multimodality problem was not pronounced. However, for situations with little information regarding the $\{\beta_i\}$, there could potentially be two modes.

Manchanda *et al.* (2004) provide extensive analysis of this model applied to detailing and sales (prescription) data for a specific drug category. They find ample evidence that detailing is set with at least partial knowledge of the sales response parameters. They confirm that detailing is set on the basis of expected volume of prescriptions but also on the responsiveness of sales to detailing activities. Finally, the information in the detailing levels improves the physician-level parameter estimates substantially.

Case Study 1
A Choice Model for Packaged Goods: Dealing with Discrete Quantities and Quantity Discounts

Modeling demand for packaged goods presents several challenges. First, quantities are discrete and, at least for some products, this discreteness can be important. Second, different package sizes are used to implement quantity discounts which result in a non-convex budget set. Finally, in many product categories, form and size combinations create a large set of products from which consumers choose. Allenby *et al.* (2004) propose a demand model that allows for quantity discounts created by different package size, discrete demand, and a large number of package size and form combinations.

BACKGROUND

Economic models of brand choice must deal with two complexities when applied to the study of packaged goods. In many product categories, there are many different size and format combinations for the same brand. For example, in the application studied below, demand for light beer is modeled. Beer is available in individual bottles or cans, six-packs, 12-packs and cases. Demand for brands is defined on a grid of available brand–pack combinations rather than a continuum. Second, the unit price of a brand often depends on the quantity purchased. Quantity discounts are typically available where the per-unit price declines with the size of the package. Thus, first-order conditions cannot be used to identify utility-maximizing solutions. First-order conditions may not hold at the available package sizes. Furthermore, constraints imposed by the nonlinear budgetary allotment can lead to first-order conditions that identify the minimum rather than the maximum utility. Allenby *et al.* (2004) propose a random utility choice model capable of dealing with discrete quantities and quantity discounts.

Quantity discounts are a common marketing practice. They occur with packaged goods if the unit price declines with larger size packages, and with services if the price declines as usage increases. Quantity discounts afford manufacturers the ability to price-discriminate between high-volume and low-volume users. An important issue in managing discounts is in determining the expected increase in sales due to increased product usage versus substitution from competitive brands. This requires a model which is capable of handling nonlinear price schedules as well as a comprehensive model of demand at the stock-keeping unit (SKU) rather than brand level.

Consider a brand choice model arising from a linear utility structure ($u(x) = \psi' x$) where the price of the brand depends on quantity. In this case, the vector of marginal utility, ψ, is constant, and utility, u, is maximized subject to the budget constraint

$$\sum_{k=1}^{k} p_k(x_k) < y,$$

where $p_k(x_k)$ is the price of x_k units of brand k. Assume that the price function $p_k(x_k)$ reflects quantity discounts. For example, the price schedule could be a marginally decreasing function of quantity:

$$p\left(x_k = x_k^{1/a}\right), \qquad a > 1, x \geq 0. \tag{CS1.1}$$

The identification of the utility-maximizing solution using first-order conditions typically proceeds by differentiating the Lagrangian function,

$$L = \psi' x - \lambda \left(\sum_{k=1}^{K} p_k(x_k) - y \right), \tag{CS1.2}$$

setting first derivatives to zero, and solving for the values of x_k at which the vector of marginal utility is tangent to the budget curve. We note immediately that if demand is restricted to the available pack sizes, first-order conditions will not necessarily apply at the observed quantity demanded. However, even when quantity is not restricted to a fixed number of values, the use of first-order conditions can lead to a solution identifying a point of utility minimization, not utility maximization. As illustrated in Figure CS1.1 for the case of linear utility and quantity discounts, the point of tangency is not associated with the utility-maximizing solution. Utility is greater at the intersection of the budget curve and either axis.

It can be verified that the first-order conditions lead to a minimum not a maximum by inspecting the concavity of the Lagrangian. The Lagrangian function in (CS1.2) is convex. The second derivative of (CS1.2) when prices follow (CS1.1) can be shown to be equal to

$$\frac{\partial^2 L}{\partial x_i^2} = -\lambda \left(\frac{1}{a} \right) \left(\frac{1-a}{a} \right) x_i^{(1-2a)/a}, \tag{CS1.3}$$

which is positive for $a > 1$ and $x > 0$. More generally, while the concavity of the Lagrangian function is guaranteed when the utility function is concave and the budget

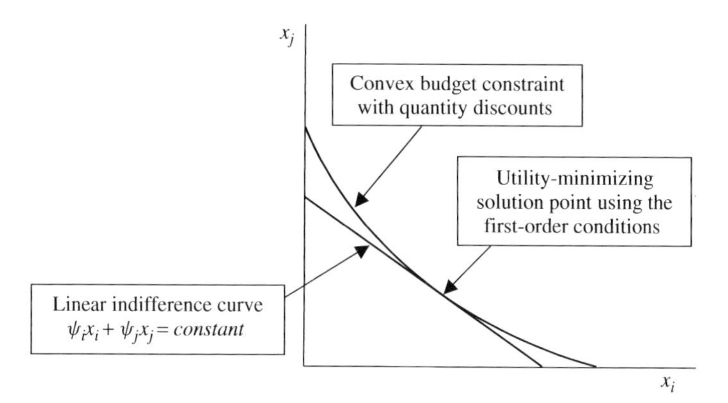

Figure CS1.1 Illustration of first-order conditions and utility minimization

constraint is linear, this property is not necessarily present when prices are nonlinear. It is certainly not true when the utility function is linear and the budget constraint is convex.

In packaged good applications, quantity discounts are frequently implemented via different package sizes. Larger sizes frequently are priced at a discount per consumption unit relative to smaller sizes. This creates a piecewise linear pricing schedule with discontinuities at each package size:

$$p_k(x_k) = \begin{cases} p_{k,M_k+1}x_k = p_{k,\mathrm{Low}}x_k, & \text{for } x_k \geq x_{k,\mathrm{Cut}_{M_k}}, \\ p_{kM_k}x_k & \text{for } x_{k,\mathrm{Cut}_{M_k}-1} \leq x_k < x_{k,\mathrm{Cut}_{M_k}}, \\ \vdots & \\ p_{k1}x_k = p_{\mathrm{k,Hi}}x_k & \text{for } x_k < x_{k,\mathrm{Cut}_1}; \end{cases} \tag{CS1.4}$$

then the utility-maximizing point in the (x_k, z) plane may be at the point of discontinuity of the budget line and the identification becomes more complicated. z is the quantity purchased of the outside good, which is assumed to be nonzero. Figure CS1.2 illustrates that the utility-maximizing value need not be at a point of tangency, but can be located at a point of discontinuity in the price schedule. When using a grid search procedure to identify the point of utility maximization, it is therefore important to include the points of discontinuity in the grid.

MODEL

A Cobb–Douglas utility function is used to model the trade-off in expenditure between the product class and other relevant products:

$$\ln u(x, z) = \alpha_0 + \alpha_x \ln u(x) + \alpha_z \ln(z), \tag{CS1.5}$$

where $x = (x_1, \ldots, x_K)$ is the vector of the amount of each brand purchased, K represents the number of brands available in the product class, z represents the amount of the outside good purchased, and $u(x)$ denotes a subutility function. The Cobb–Douglas

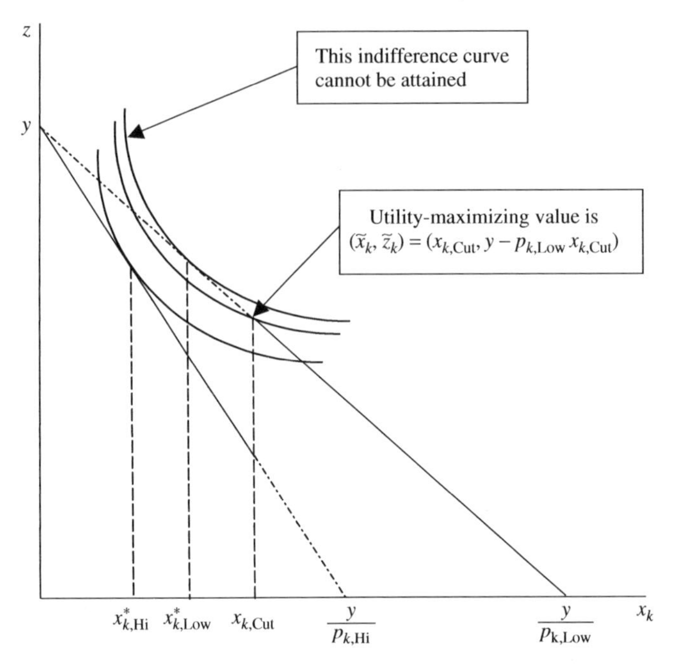

Figure CS1.2 Utility maximization in the (x, z) plane for a price schedule

utility function ensures that there will be an interior solution in which both the outside good and at least one good in the product class will be consumed.

A linear subutility function is used to represent the near-perfect substitute nature of brands within a product class:

$$u(x) = \psi' x, \tag{CS1.6}$$

where ψ_k is the marginal utility for brand k. This structure results in parallel linear indifference curves. Alternatively, one could adopt the nonhomothetic structure of Allenby and Rossi (1991) (see also Chapter 4):

$$u(x) = \sum_{k=1}^{K} \psi_k(\bar{u}) x_k, \qquad \ln \psi_k(\bar{u}) = \delta_k - \kappa_k \bar{u}(x, z), \tag{CS1.7}$$

in which the deterministic part of marginal utility, $\bar{u}$, is implicitly defined. We note that (CS1.7) differs from the nonhomothetic structure of Allenby and Rossi (1991) by including the outside good z. If the coefficient vector κ is strictly positive, this implicitly defined utility function has linear indifference curves that fan out but do not intersect in the positive orthant. The ratio of the marginal utility of two brands, $\psi_i(\bar{u})/\psi_j(\bar{u}) = \exp[(\delta_i - \delta_j) - \bar{u}(x, z)(\kappa_i - \kappa_j)]$, indicates that if $\kappa_i < \kappa_j$ then brand i is relatively superior to brand j. As utility increases, the ratio of marginal utility for brand i increases relative to brand j. The κ parameters govern the relative rates of rotation of the indifference curves. In a standard choice model, both the linear and the nonhomothetic utility specifications result in corner solutions where only one of the available brands is selected.

The consumer determines expenditure on the K brands and the outside good by maximizing the utility function in (CS1.5) subject to the budget constraint:

$$\sum_{k=1}^{K} p_k(x_k) + z = y, \qquad \text{(CS1.8)}$$

where $p_k(x_k)$ is the price of x_k units of brand k, z is the expenditure on the outside good, and the price of the outside good is 1. The price function $p_k(x_k)$ allows for the possibility of price discounts. Note that if $p_k(x_k) = q_k x_k$ for each k, where q_k is a constant, then the budget constraint reduces to the standard linear budget constraint.

In contrast to approaches that rely on first-order conditions to relate model parameters to the data, the approach taken here is to directly evaluate the above expression at all feasible solution points. This could be a formidable task for large numbers of products. However, if the utility-maximizing solution is restricted to corner solutions where only one element of x is nonzero, we obtain a solution that is easy to evaluate. The solution strategy and resulting likelihood is outlined first, followed by a proof of corner solutions with either linear subutility or nonhomothetic subutility.

Solution Procedure and Likelihood Specification

The likelihood function is derived from assumptions about stochastic elements in the utility function. In the standard random utility model, the vector of log marginal utility ($\ln \psi$) is assumed stochastic with an additive error, $\ln \psi_k = \nu_k + \varepsilon_k$. Assumptions about the random utility error are used to derive the likelihood of observed demand. In the case of a linear utility model, $u(x) = \psi' x$, discrete choice probabilities are derived from the utility-maximizing solution (choose x_k if ψ_k / p_k is maximum) that links observed demand to the latent utility parameters.

In this random utility formulation, the stochastic element has a distribution over the K brands even though there exist many more than K product offerings. For each brand, there are $a = 1, \ldots, A_k$ product offerings. To derive the demand for each product offering, we recognize that the stochastic elements only affect brand choice and that given a brand we simply search over all offerings of that brand to find the offering which has maximum utility.

The likelihood for the observed demand for the product category and the outside good is derived from distributional assumptions of random utility. We note that the budget restriction in (CS1.8) imposes an "adding-up" constraint that induces a singularity in the distribution of observed, utility maximizing demand (x,z) (see Kim *et al.* 2002). Therefore, only K error terms – one for each brand – are needed to derive the likelihood of observed demand, and we substitute $z = y - \Sigma_k p_k(x_k)$ for the outside good. We follow standard convention by assuming that the vector of log marginal utility is stochastic with an additive error, $\ln(\psi_k) = \nu_k + \varepsilon_k$.

To derive demand, we recognize that our utility formulation results in only one good in the product category being consumed. The solution strategy when only one element of x is nonzero involves two steps. In the first step the optimal product quantity is determined for each brand separately. That is, we search over all product offerings of brand k for the offering that maximizes utility:

$$\begin{aligned} a^* &= \arg\max_a \{\alpha_0 + \alpha_x \ln \psi_k x_{ka} + \alpha_z \ln(y - p_k(x_{ka}))\} \\ &= \arg\max_a \{\alpha_0 + \alpha_x(\nu_k + \varepsilon_k) + \alpha_x \ln(x_{ka}) + \alpha_z \ln(y - p_k(x_{ka}))\} \\ &= \arg\max_a \{\alpha_x x_{ka} + \alpha_z \ln(y - p_k(x_{ka}))\}, \end{aligned} \tag{CS1.9}$$

where y is the budgetary allotment for the product and outside good, and $p_k(x_{ka})$ denotes the price of brand k with pack size x_{ka}. We note that if a retailer offers only a limited selection of package sizes, the consumer can consider alternative combinations of the available offerings. For example if a retailer only offers six-packs of a beverage, the consumer can consider six-, 12-, 18-, 24-, ... package bundles. The stochastic element, ε_k, is the same for each of the package bundles in (CS1.9) and cancels from the expression. That is, since the same good is contained in each of the possible package bundles, the determination of the utility-maximizing quantity is deterministic. Furthermore, by substituting the expression $y - p_k(x_{ka})$ for z, we ensure that the evaluated solution points correspond to the budget restriction.

A limitation of using a linear utility structure $u(x) = \psi' x$ is that the optimal quantity for each brand depends only on the prices for the brand, not on the quality of the brand (ν_k). The terms α_0 and $\ln \psi$ are constant over the alternative bundles of brand k in (CS1.9), shifting the intercept of utility but not affecting the optimal allocation of the budget between alternatives in the product class (x) and the outside good (z). This limitation can be overcome by using the nonhomothetic function in (CS1.7). For the nonhomothetic function the optimal quantity for each brand proceeds by solving for the package size that maximizes the expression

$$\begin{aligned} &\arg\max_a \{\alpha_0 + \alpha_x \ln \psi_k(\overline{u}) x_{ka} + \alpha_z \ln(y - p_k(x_{ka}))\} \\ &= \arg\max_a \{\alpha_x \ln \psi_k(\overline{u}) x_{ka} + \alpha_z \ln(y - p_k(x_{ka}))\}, \end{aligned} \tag{CS1.10}$$

where $\ln \psi_k(\overline{u}) = \nu_k + \varepsilon_k = \delta_k - \kappa_k \overline{u}(x, z) + \varepsilon_k$. As with the linear utility structure, the determination of the optimal quantity for brand k is deterministic because ε_k is the same for each package size. The quantity depends on the brand because κ_k does not cancel from the evaluation. The optimal quantity of the product demanded therefore depends on the brand under consideration.

The first step of the solution procedure, outlined above, identifies the optimal quantity given that brand k is purchased. The second step of our method involves determining the probability that brand k is purchased. In this second step, the stochastic element of marginal utility takes on a traditional role of generating a choice probability. We have

$$\begin{aligned} \max_{ka}\{\ln u(x_{ka}, y - p_k(x_{ka}))\} &= \max_k [\max_{a|k}\{\ln u(x_{ka}, y - p_k(x_{ka}))\}] \\ &= \max_k [\alpha_0 + \alpha_x \ln u(x_k^*) + \alpha_z \ln(y - p_k(x_k^*))], \end{aligned} \tag{CS1.11}$$

where x_k^* indicates the optimal quantity for brand k that is identified in the first step. Substituting the linear subutility expressions in (CS1.9) results in the expression

$$\max_k [\alpha_0 + \alpha_x(\nu_k + \varepsilon_k) + \alpha_x \ln(x_k^*) + \alpha_z \ln(y - p_k(x_k^*))], \tag{CS1.12}$$

and assuming that $\varepsilon_k \sim \mathrm{EV}(0,1)$ results in the choice probability,

$$\Pr(x_i) = \frac{\exp[\nu_i + \ln(x_i) + (\alpha_z/\alpha_x)\ln(y - p_i(x_i))]}{\sum_{k=1}^{K}\exp[\nu_k + \ln(x_k^*) + (\alpha_z/\alpha_x)\ln(y - p_k(x_k^*))]}, \tag{CS1.13}$$

where x_i is the observed demand. In the estimation procedure, the observed x_i is used for the selected brand, while x_k^* is used for the brands not selected. Alternatively, assuming nonhomothetic subutility in (CS1.10) results in the expression

$$\Pr(x_i) = \frac{\exp[\delta_i + \kappa_i \overline{u}^i + \ln(x_i) + (\alpha_z/\alpha_x)\ln(y - p_i(x_i))]}{\sum_{k=1}^{K}\exp[\delta_k + \kappa_k \overline{u}^k + \ln(x_k^*) + (\alpha_z/\alpha_x)\ln(y - p_k(x_k^*))]} \tag{CS1.14}$$

where $\overline{u}^i = \overline{u}^i(x_i^*, y - p_i(x_i^*))$ is the deterministic part of utility derived from the consumption of x_i^* units of brand i, with the remainder of the budget allocation, $y - p_i(x_i^*)$, devoted to the outside good. From (CS1.7), we see that $\overline{u}^i$ is the implicit solution to the equation

$$\ln \overline{u}^i = \delta_i - \kappa_i \overline{u}^i + \ln(x_i^*) + \frac{\alpha_z}{\alpha_x}\ln(y - p_i(x_i^*)). \tag{CS1.15}$$

Details of the properties of the nonhomothetic function and its estimation can be found in Allenby and Rossi (1991).

We note that not all parameters of the Cobb–Douglas specification are identified. Since an arbitrary rescaling of utility by a positive constant can represent the same preference ordering, we set α_0 to zero, α_x to one, and estimate $\alpha_z^* = \alpha_z/\alpha_x$ subject to the constraint that it takes on a positive value.

The above strategy is dependent on conducting an initial search along each of the axes of demand. We next provide a proof that corner solutions are consistent with utility maximization when indifference curves are linear and prices are either constant or monotonically decreasing with quantity.

Proof of Corner Solution

Assume that the amount of expenditure for the product is fixed at y^*. The proof holds for any value of y^*, and this assumption therefore does not restrict the generality of the proof. For any continuous convex budget set defined as $\sum_k p(x_k) < y^*$, there exists a less restrictive linear set $\sum_k \hat{p}_k x_k < y^*$ that contains the convex set, where $\hat{p}_k = y^*/x_k^*$ is the per-unit price of allocating all the expenditure to brand k, that is, x_k^* such that $p(x_k^*) = y^*$. As shown in Figure CS1.3, the dominating linear budget set is equal to the convex budget set at each axis, and is greater than the convex set at all interior points. For the price schedule with discounts at specific quantities (CS1.4), a dominating linear budget set also exists that is equal to the actual set at each axis, and is greater than or equal to the actual set in the interior. Figure CS1.4 illustrates such a budget set when one price cut exists. Since the dominating linear budget set is less restrictive than the convex set, the maximum utility in the linear set is greater than or equal to the maximum utility in the convex set. We note that the utility-maximizing solution for the linear budget set is a corner solution when the indifference curves are linear. Therefore the utility-maximizing

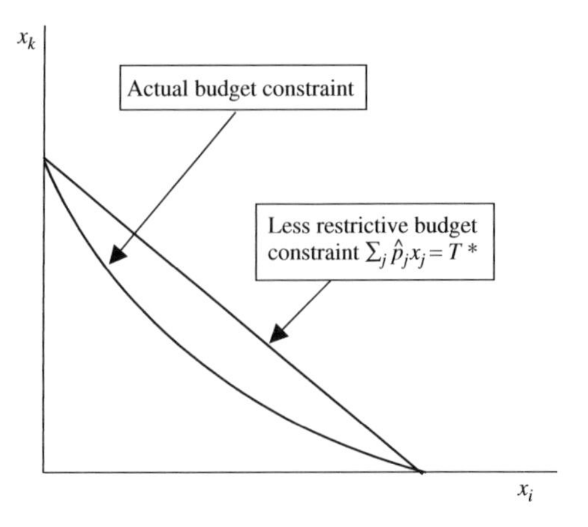

Figure CS1.3 Continuous convex budget set and dominating linear Set

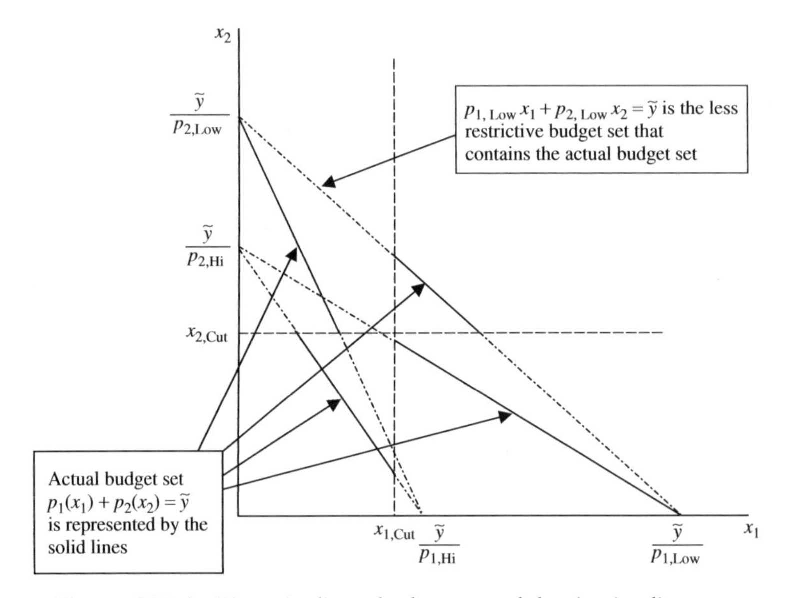

Figure CS1.4 Piecewise linear budget set and dominating linear set

solution for any convex budget set contained by the linear set must also be at the same corner.

DATA

The model is illustrated with a scanner panel data set of light beer purchases. Three dominant brands in the domestic product category are studied – Miller Lite, Bud Light and Coors Light. The data set has 20,914 purchase occasions for 2282 households making beer purchases at grocery stores. Table CS1.1 reports the packages sizes and

Table CS1.1 Description of the data

Item	Brand	Oz/unit	Units/pack	Pack form	Volume/pack	Price/oz ($)	Display frequency	Feature frequency	Choice share	Average unit price ($) by brand	Average choice share by brand
1	1	7	1	2	7	0.0967	0.0000	0.0000	0.0000	0.0536	0.3504
2	1	12	1	2	12	0.0649	0.0000	0.0000	0.0014		
3	1	16	1	2	16	0.0564	0.0000	0.0000	0.0002		
4	1	22	1	2	22	0.0483	0.0000	0.0000	0.0019		
5	1	24	1	1	24	0.0482	0.0004	0.0007	0.0003		
6	1	32	1	2	32	0.0495	0.0001	0.0000	0.0041		
7	1	32	1	1	32	0.0393	0.0000	0.0004	0.0031		
8	1	40	1	2	40	0.0511	0.0000	0.0000	0.0002		
9	1	7	6	2	42	0.0706	0.0000	0.0000	0.0009		
10	1	8	6	1	48	0.0550	0.0000	0.0000	0.0000		
11	1	10	6	1	60	0.0641	0.0000	0.0000	0.0001		
12	1	16	4	2	64	0.0544	0.0000	0.0000	0.0008		
13	1	16	4	1	64	0.0526	0.0000	0.0000	0.0000		
14	1	12	6	1	72	0.0587	0.0011	0.0011	0.0166		
15	1	12	6	2	72	0.0588	0.0028	0.0007	0.0328		
16	1	16	6	1	96	0.0525	0.0014	0.0002	0.0036		
17	1	10	12	1	120	0.0603	0.0016	0.0000	0.0000		
18	1	12	12	1	144	0.0538	0.1355	0.1016	0.0951		
19	1	12	12	2	144	0.0541	0.0897	0.0743	0.0553		
20	1	12	15	1	180	0.0420	0.0242	0.0070	0.0085		
21	1	16	12	1	192	0.0428	0.0015	0.0003	0.0002		
22	1	12	18	1	216	0.0497	0.0735	0.0281	0.0272		
23	1	12	18	2	216	0.0475	0.0005	0.0000	0.0000		
24	1	10	24	1	240	0.0559	0.0039	0.0006	0.0000		
25	1	12	24	1	288	0.0493	0.1318	0.0784	0.0788		

(*continued overleaf*)

Table CS1.1 *(continued)*

Item	Brand	Oz/unit	Units/pack	Pack form	Volume/pack	Price/oz ($)	Display frequency	Feature frequency	Choice share	Average unit price ($) by brand	Average choice share by brand
26	1	12	24	2	288	0.0485	0.0001	0.0000	0.0002		
27	1	12	30	1	360	0.0396	0.0922	0.0305	0.0191		
28	1	16	24	2	384	0.0427	0.0000	0.0000	0.0000		
29	1	16	24	1	384	0.0474	0.0000	0.0000	0.0000		
30	2	12	1	2	12	0.0591	0.0000	0.0000	0.0006	0.0548	0.4327
31	2	16	1	2	16	0.0610	0.0000	0.0000	0.0000		
32	2	22	1	2	22	0.0521	0.0000	0.0000	0.0044		
33	2	24	1	1	24	0.0539	0.0008	0.0002	0.0052		
34	2	32	1	2	32	0.0565	0.0000	0.0000	0.0040		
35	2	40	1	2	40	0.0507	0.0009	0.0000	0.0020		
36	2	7	6	2	42	0.0724	0.0001	0.0000	0.0051		
37	2	8	6	2	48	0.0632	0.0000	0.0000	0.0005		
38	2	7	8	2	56	0.0744	0.0000	0.0000	0.0000		
39	2	10	6	1	60	0.0642	0.0002	0.0000	0.0000		
40	2	16	4	2	64	0.0561	0.0029	0.0000	0.0011		
41	2	8	8	1	64	0.0541	0.0000	0.0000	0.0011		
42	2	12	6	1	72	0.0591	0.0015	0.0015	0.0302		
43	2	12	6	2	72	0.0594	0.0031	0.0004	0.0328		
44	2	16	6	1	96	0.0536	0.0121	0.0000	0.0048		
45	2	16	6	2	96	0.0549	0.0000	0.0000	0.0000		
46	2	10	12	1	120	0.0633	0.0009	0.0001	0.0002		
47	2	12	12	1	144	0.0541	0.1984	0.1236	0.1589		
48	2	12	12	2	144	0.0547	0.1392	0.0696	0.0599		
49	2	7	24	2	168	0.0597	0.0000	0.0000	0.0000		
50	2	12	15	1	180	0.0431	0.0083	0.0018	0.0065		
51	2	12	18	1	216	0.0512	0.0989	0.0310	0.0273		

52	2	10	24	1	240	0.0567	0.0032	0.0006	0.0011		
53	2	12	20	2	240	0.0441	0.0138	0.0018	0.0024		
54	2	12	24	1	288	0.0502	0.2091	0.0948	0.0740		
55	2	12	24	2	288	0.0475	0.0044	0.0008	0.0005		
56	2	24	12	1	288	0.0439	0.0005	0.0000	0.0000		
57	2	22	15	2	330	0.0438	0.0000	0.0000	0.0000		
58	2	12	30	1	360	0.0405	0.0390	0.0164	0.0100		
59	2	16	24	1	384	0.0468	0.0000	0.0000	0.0000		
60	3	7	1	2	7	0.0732	0.0000	0.0000	0.0000	0.0593	0.2170
61	3	12	1	2	12	0.0604	0.0000	0.0000	0.0007		
62	3	16	1	2	16	0.0609	0.0000	0.0000	0.0003		
63	3	18	1	2	18	0.0732	0.0018	0.0000	0.0001		
64	3	22	1	2	22	0.0496	0.0000	0.0000	0.0006		
65	3	24	1	1	24	0.0501	0.0002	0.0000	0.0007		
66	3	32	1	2	32	0.0557	0.0000	0.0000	0.0019		
67	3	40	1	2	40	0.0511	0.0000	0.0000	0.0011		
68	3	7	6	2	42	0.0712	0.0000	0.0000	0.0019		
69	3	8	6	1	48	0.0904	0.0000	0.0000	0.0000		
70	3	10	6	1	60	0.0646	0.0000	0.0000	0.0000		
71	3	16	4	2	64	0.0597	0.0003	0.0000	0.0009		
72	3	12	6	1	72	0.0589	0.0043	0.0026	0.0294		
73	3	12	6	2	72	0.0593	0.0008	0.0002	0.0227		
74	3	18	4	2	72	0.0748	0.0125	0.0000	0.0006		
75	3	16	6	1	96	0.0533	0.0001	0.0001	0.0007		
76	3	10	12	1	120	0.0633	0.0006	0.0000	0.0004		
77	3	12	12	1	144	0.0540	0.1255	0.1041	0.0966		
78	3	12	12	2	144	0.0540	0.0575	0.0460	0.0314		
79	3	7	24	2	168	0.0773	0.0000	0.0000	0.0000		
80	3	12	18	1	216	0.0499	0.0695	0.0243	0.0183		
81	3	18	12	2	216	0.0358	0.0000	0.0000	0.0000		
82	3	10	24	1	240	0.0560	0.0018	0.0005	0.0001		

(*continued overleaf*)

Table CS1.1 *(continued)*

Item	Brand	Oz/unit	Units/pack	Pack form	Volume/pack	Price/oz ($)	Display frequency	Feature frequency	Choice share	Average unit price ($) by brand	Average choice share by brand
83	3	12	24	2	288	0.0477	0.0003	0.0000	0.0000		
84	3	12	30	1	360	0.0392	0.0395	0.0165	0.0085		

Note:

Variable list	Description
Brand ID	1 = Miller Lite, 2 = Bud Light, 3 = Coors Light
Oz/Unit	Ounces per bottle or can
Units/pack	Number of units of cans or bottles included in the pack
Pack form	1 = Cans, 2 = Nonreturnable bottles
Volume/Pack	Total volume contained in the pack

package types under study. Our analysis includes beer packaged in nonreturnable bottles and cans, for which there are 84 different varieties for the three brands. Table CS1.1 also reports the choice shares, average prices and the frequency of merchandising activity for each of the choice alternatives.

Inspection of the table reveals the following data characteristics. Bud Light has the highest overall choice share at 0.43, followed by Miller Lite at 0.35 and Coors Light at 0.22. The choice shares for the smaller package sizes are approximately equal, while the choice share for larger packages sizes (such as 12-packs) favor Bud Light. For example, the choice share of Miller Lite six-packs of 12 oz bottles and cans (items 14, 15) is 0.0494, Bud Light (items 42, 43) is 0.0630, and Coors Light (items 72, 73) is 0.0521, while the choice shares of 12-packs are 0.1504 for Miller Lite (items 18, 19), 0.2188 for Bud Light (items 47, 48), and 0.1280 for Coors Light (items 77, 78). The data also indicates that bottles are somewhat more preferred to cans for six-packs (for example, items 14 versus 15, 42 versus 43, and 72 versus 73), but that cans are preferred to bottles when consumers purchase larger quantities of beer (for example, items 18 versus 19, 47 versus 48, 77 versus 78). These results indicate that the nonhomothetic specification may fit the data better because the marginal utility for the brands depends on the level of expenditure. However, the shift in brand and package preferences could also be attributed to differences in consumer tastes (for example, households who consume large quantities of beer may simply prefer Bud Light). We therefore fit both the linear subutility model and the nonhomothetic subutility model.

We note that Budweiser engages in significantly more display and feature activity than either Miller or Coors. Some form of Bud Light is on display in 74 % of the purchases compared to 56 % for Miller Lite and 31 % for Coors Light. Feature activity is approximately equal for Bud Light and Miller Lite (33 % of the time) but lower for Coors Light (19 %). An interesting issue to investigate is whether the superior performance of Bud Light in large package sizes is due to it being a superior product offering, as measured by the nonhomothetic coefficients in (CS1.7), versus engaging in more frequent merchandising activity.

There exists large intra-household variation in grocery expenditures, with many shopping trips totaling more than $200 while others are much less. Across these purchase occasions, the non-beer expenditures take on a varied meaning. For small shopping trips, the outside good includes snacks and a few miscellaneous items, while for the larger shopping trips the outside good is comprised of a much broader array of items. Rather than equating the observed grocery expenditure to the budgetary allotment (y) in (CS1.13) and (CS1.14), we treat the budgetary allotment as an unknown parameter and estimate it from the data. In our analysis, we specify a prior distribution for y and derive posterior estimates. The prior specification reflects managerial judgment about the amount of money consumers allocate to beer and substitutable goods.

RESULTS

The linear and nonhomothetic subutility structures were embedded in the Cobb–Douglas function and estimated as a hierarchical Bayes model. As discussed above, some parameters must be restricted to ensure algebraic signs that conform to economic theory. In the Cobb–Douglas model, we set $\alpha_x = 1$ and $\alpha_z = \exp(\alpha_z^*)$ and estimate α_z^* unrestricted. In the linear subutility function, we follow standard convention and set one of the brand intercepts (Miller Lite) to zero. In the nonhomothetic subutility function, we restrict κ to be positive by estimating $\kappa^* = \ln \kappa$ with κ^* unrestricted, and set κ^* for Miller Lite to zero.

The deterministic portion of log marginal utility for the linear model is specified as

$$\nu_i = \beta_{0i} + \beta_b \text{bottle}_i + \beta_d \text{display}_i + \beta_f \text{feature}_i, \quad \text{(CS1.16)}$$

where β_{0i} is the preference coefficient for brand i. The three covariates – 'bottle', 'display', and 'feature' – are coded as dummy variables for each brand. This specification assumes that the marginal utility of a brand can be influenced by the package type and merchandising activity of the retailer. The nonhomothetic model specification is:

$$\begin{aligned} \nu_i &= \delta_i - \kappa_i \bar{u}(x, z), \\ \delta_i &= \beta_{0i} + \beta_d \text{display}_i + \beta_f \text{feature}_i, \quad \text{(CS1.17)} \\ \kappa_i &= \exp[\kappa_{0i} + \beta_b \text{bottle}_i]. \end{aligned}$$

As reported in Allenby and Rossi (1991), the β_{0i} intercept parameters are redundant if the κ_i parameters take on nonzero values, and are therefore set to zero. Household heterogeneity was allowed for all parameters $\theta = (\kappa', \alpha, \beta')'$ and specified as a multivariate normal distribution:

$$\theta \sim N(\bar{\theta}, V). \quad \text{(CS1.18)}$$

A total of 50,000 iterations of the Markov chain were executed, with a burn-in of 25,000 iterations. Convergence was checked by starting the chain from multiple start points and noting common convergence, and through inspection of time series plots.

The log marginal likelihood for the linear homothetic model is estimated to be −4864.99 and for the nonhomothetic specification −4859.79, indicating a Bayes factor in favor of the nonhomothetic model. The marginal densities are high and translate to an in-sample hit probability of approximately 0.75 across the 20,914 observations. A reason for the exceptional model fit, in the presence of over 80 choice alternatives, is that there are only three error terms in the model – one for each brand.

The model fit statistics indicate that the nonhomothetic subutility model fits the data better than the linear subutility model. This implies that the shift in brand preference to Bud Light, and package preference to bottles, reported in Table CS1.1 cannot be solely attributed to differences in household tastes. Bottles weigh more than cans, which may partially explain the shift in preference to cans during grocery trips with larger beer expenditures. Bud Light is the market share leader and is rated by consumers as less bitter than Miller Lite. Having a less bitter taste decreases the likelihood of satiation when larger quantities are consumed.

Table CS1.2 reports aggregate parameter estimates for the nonhomothetic subutility model. Bud Light has the smallest aggregate estimate of κ, indicating that at higher levels of utility ($u(x, z)$) it is preferred to Miller Lite and Coors Light. Coors Light has the largest aggregate estimate of κ, indicating that it is the least preferred brand when total utility is high. The aggregate estimate of $\alpha_z^* = \ln \alpha_z = -0.548$ translates to an estimate of α_z equal to 0.578. Recall that we have restricted our analysis to household shopping trips in which beer is purchased, which accounts for the large amount of utility

Table CS1.2 Aggregate coefficient estimates (posterior standard deviations)

Parameter	Estimate
Bud Light κ^*	−1.230
	(0.228)
Coors Light κ^*	3.006
	(0.322)
$\alpha^* = \ln(\alpha_z/\alpha_x)$	−0.548
	(0.068)
Bottle	0.233
	(0.032)
Feature	1.026
	(0.131)
Display	0.766
	(0.090)

derived from the beer category ($\alpha_x = 1.00$) relative to the outside good ($\alpha_z = 0.578$). We discuss the implications of the Cobb–Douglas estimates below. On average, we find that households prefer cans when the grocery trip involves large expenditures, and bottles when the total expenditure is smaller. Finally, the feature and display coefficients have positive algebraic sign, as expected, with the effect of feature advertisement larger than the effect of a display.

The covariance matrix of the distribution of heterogeneity is reported in Table CS1.3. Variances and covariances are reported in the lower triangle, and the associated correlations are reported in the upper triangle. The diagonal entries for the brands are large, indicating that there is a wide dispersion of preferences for the brands. In addition, we find that the off-diagonal entries are small in magnitude, indicating that brand preferences are not strongly associated with feature and display sensitivity, preference for bottle versus can, nor preference for the outside good, z.

Table CS1.3 Covariance (lower triangle) and correlation (upper triangle) matrix, with posterior standard deviations in parentheses

Bud Light κ^*	72.52	−0.03	−0.02	−0.03	−0.14	0.04
	(6.74)					
Coors Light κ^*	−2.34	84.38	0.03	0.02	0.29	0.10
	(2.60)	(4.01)				
α^*	−0.11	0.18	0.49	0.00	0.03	0.07
	(0.70)	(0.60)	(0.05)			
Bottle	−0.11	0.08	0.00	0.24	0.02	−0.03
	(0.32)	(0.32)	(0.03)	(0.02)		
Feature	−1.13	2.46	0.02	0.01	0.85	0.01
	(1.72)	(1.57)	(0.05)	(0.03)	(0.18)	
Display	0.25	0.72	0.04	−0.01	0.01	0.61
	(0.89)	(0.80)	(0.04)	(0.02)	(0.06)	(0.07)

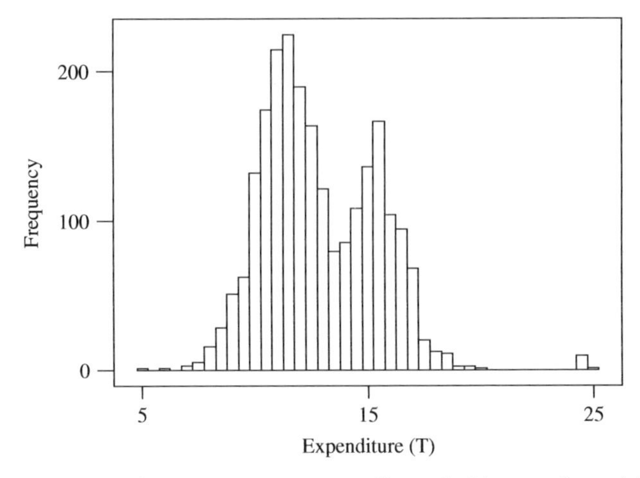

Figure CS1.5 Posterior estimates of household expenditure (y)

Posterior estimates of household expenditures (y) are displayed in Figure CS1.5. The mean of the distribution is equal to \$12.86, the interquartile range is (\$10.97, \$14.98) and the distribution is bimodal. The correlation between the posterior estimates and the Cobb–Douglas parameter α_z^* is approximately zero, implying that there is little evidence in the data of an association between the amount of budget allocation and price sensitivity. Hence retailers and manufacturers have little ability to price-discriminate between high-volume and low-volume users. Additional implications of the parameter estimates are explored below.

DISCUSSION

Issues related to pricing and product assortment are investigated. The first issue relates to the expected demand for a particular package size. Manufacturers often speculate about the existence of price points, or threshold values at which individual households become willing to consider a given package size. Our economic model is capable of identifying the price at which a particular package size becomes the most preferred quantity for a household. Finally, our model provides a valid measure of utility that can be used to assess the change in consumer welfare and compensating value associated with changes in the assortment of package sizes. We use the model to measure the utility consumers derive from the various package sizes by considering the impact of removing alternative package sizes from the current mix of offerings.

The derivation of model implications is a function of the model parameters, the available offerings, and the current pricing policies of the retailers. In the analysis presented below, we do not attempt to explore supply-side implications of our model by identifying the optimal number and size of a brand's offerings. In addition, because our analysis conditions on light beer purchases, the pricing implications do not investigate issues such as purchase timing, stockpiling, and substitution from the regular beer category. In order to address these issues, our basic model must be extended, and we leave this for future work.

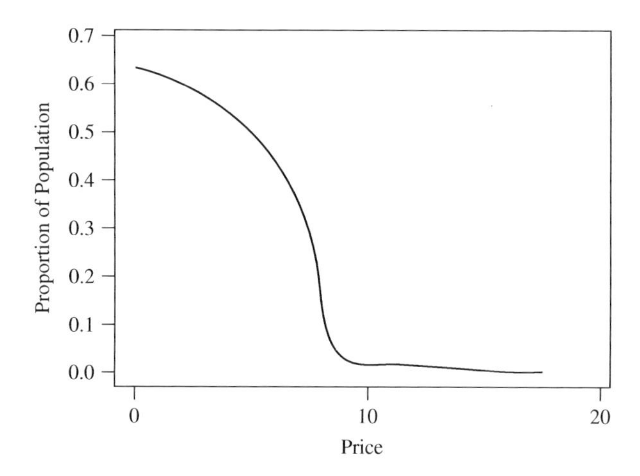

Figure CS1.6 Cumulative distribution of 12-pack price points for Miller Lite

Price Points

Changes in price result in a redistribution of the quantity demanded. The demand for a specific quantity (package size) is a function of the prices of the other alternatives and household parameters. Our model permits calculation of the price at which a particular package size is most preferred by a household by identifying the highest price at which a particular quantity (x^*) yields maximum utility in (CS1.9) or (CS1.10). Figure CS1.6 displays the cumulative distribution of prices at which the 12-pack of Miller Lite is the most preferred Miller offering. When priced at \$10, the 12-pack is the most preferred package size for 2 % of the population, and at \$5 it is most preferred package size for 49 % of the population. Even at very low prices, not everyone is willing to purchase a 12-pack. The reason is that many households are estimated to have high utility for the other brands. When coupled with low utility for the outside good (α_z), it is not possible to reduce the price of Miller Lite low enough to induce these individuals to purchase a 12-pack.

Utility of an Assortment

A common problem in retailing is determining the variety provided by alternative packages and variants of an offering. The depth and breadth of an assortment is not well reflected by the total number of SKUs in a product category because many of the offerings are nearly identical to each other. Since our model is derived from an economic model, we can use utility as a scalar measure in assessing the value of specific offerings. That is, we can compute the average consumer utility for the current set of offerings across the purchases, and investigate the decrease in utility as specific offerings are removed. These differences in utility can be translated into a dollar metric by computing the compensating value, defined as the increase in the budget (y) needed to return to the original level of utility.

Table CS1.4 The utility of alternative assortments

Assortment	Utility per purchase	Compensating value ($)
Current offerings	3.3516	0.000
Remove 6-packs	3.3434	0.005
Remove 12-packs	3.0477	0.300
Remove 18-packs	3.3161	0.026
Remove 24-packs	3.3094	0.033

Table CS1.4 reports on the change in utility and associated compensating value of removing alternative package sizes across all brands. The current utility (per purchase) of 3.35, for example, declines to 3.05 when 12-packs are no longer made available to consumers. This translates to a compensating value of 30 cents per purchase, indicating a substantial decrease in consumer utility. Our analysis indicates that consumers would be least affected by the removal of six-packs from the shelves.

R IMPLEMENTATION

R We have implemented the homothetic pack-size model in R. The file `packsize.R` contains an R function to estimate the model. `Run packsize.R` contains R code to read in a subset of the original data and run the MCMC sampler. Both files and the data are available on the book website. We have extracted a subset of data for the 67 households who are observed to purchase one of the three brands – Miller Lite, Bud Lite and Coors Lite. These 67 households make 1272 purchases. We also narrowed the list of SKUs to 14, spread among these three brands. The full data set used in the article is extremely large (over 20,000 observations and some 90 SKUs). Analysis of data sets of this size is not feasible with the R implementation we provide. Customized programming in a lower-level language would be required. The full data set also has problems with a large number of brand loyal households. Most households are observed to purchased only SKUs of one brand. This means that attempts to fit hierarchical models to these data can result in unreasonable values of some of the demand parameters at the household level. As pointed out in Chapter 5, the household-level likelihood will force brand intercept parameters to very large positive or negative numbers in order to drive purchase probabilities as close to zero as possible. The hierarchical prior structure reins in this tendency. However, if the vast majority of households are completely brand loyal, the variance term in the hierarchical prior may be driven off to a large value as well. We advise that researchers do not attempt to fit our packsize models to data of this sort.

Case Study 2
Modeling Interdependent Consumer Preferences

An individual's preference for an offering can be influenced by the preferences of others in many ways, ranging from the psychological benefits of social identification and inclusion to the benefits of network externalities. Yet models of unobserved heterogeneity using random effects specifications typically assume that draws from the mixing distribution are independently distributed. If the preferences of an individual are dependent on the preferences of others, then it will be useful to incorporate these dependencies in our standard models. The challenge in modeling dependencies between consumers is to choose a flexible yet parsimonious model of dependence. That is, some structure must be imposed on the covariance across consumers in order to make the model tractable.

Yang and Allenby (2003) use a Bayesian spatial autoregressive discrete choice model to study the preference interdependence among individual consumers. The autoregressive specification can reflect patterns of heterogeneity where influence propagates within and across networks. These patterns can be difficult to capture with covariates in a standard random effects model. Interdependent preferences are illustrated with data on automobile purchases, where preferences for Japanese-made cars are shown to be related to geographic and demographically defined networks.

BACKGROUND

Quantitative models of consumer purchase behavior often do not recognize that preferences and choices are interdependent. Economic models of choice typically assume that an individual's latent utility is a function of brand and attribute preferences, not the preferences of others. Preferences are assumed to vary across consumers in a manner described either by exogenous covariates, such as demographics (for example, household income), or by independent draws from a mixing distribution in random effects models. However, if preferences in a market are interdependently determined, their pattern will

Bayesian Statistics and Marketing P. E. Rossi, G. M. Allenby and R. McCulloch

not be well represented by a simple linear model with exogenous covariates. Failure to include high-order interaction terms in the model to reflect interdependent influences will result in a correlated structure of unobserved heterogeneity, where the draws from a random effects mixing distribution are dependent, not independent.

Preferences and choice behavior are influenced by a consumer's own tastes and also the tastes of others. People who identify with a particular group often adopt the preferences of the group, resulting in choices that are interdependent. Examples include the preference for particular brands (Abercrombie and Fitch) and even entire product categories (minivans). Interdependence may be driven by social concerns, by endorsements from respected individuals that increase a brand's credibility, or by learning the preference of others who may have information not available to the decision-maker. Moreover, since people engage in multiple activities with their families, co-workers, neighbors and friends, interdependent preferences can propagate across and through multiple networks. In this case study, a Bayesian model of interdependent preferences is developed within a consumer choice context.

MODEL

We observe choice information for a set of individuals ($i = 1, \ldots, m$). Assume that the individual is observed to make a selection between two choice alternatives ($y_i = 1$ or 0) that is driven by the difference in latent utilities, U_{ik}, for the two alternatives ($k = 1, 2$). The probability of selecting the second alternative over the first is

$$\Pr(y_i = 1) = \Pr(U_{i2} > U_{i1}) = \Pr(z_i > 0), \qquad z_i = x_i'\beta + \varepsilon_i,\ \varepsilon_i \sim N(0, 1), \qquad \text{(CS2.1)}$$

where z_i is the latent preference for the second alterative over the first alternative, x_i is a vector of covariates that captures the differences of the characteristics between the two choice alternatives and characteristics of the individual, β is the vector of coefficients associated with x_i, and ε_i reflects unobservable factors modeled as error. Preference for a durable offering, for example, may be dependent on the existence of local retailers who can provide repair service when needed. This type of exogenous preference dependence can easily be incorporated into the model via covariates.

The presence of interdependent networks creates preferences that are mutually dependent, resulting in a covariance matrix of the ε across consumers with nonzero off-diagonal elements. Recognizing that the scale of each ε_i should be set to 1 to achieve identification, various dependence structures specify the correlation matrix of the stacked vector of errors,

$$\varepsilon = \begin{bmatrix} \varepsilon_1 \\ \varepsilon_2 \\ \vdots \\ \varepsilon_m \end{bmatrix}, \qquad \operatorname{cov}(\varepsilon) = R. \qquad \text{(CS2.2)}$$

The presence of off-diagonal elements in the covariance matrix leads to conditional and unconditional expectations of preferences that differ. The unconditional expectation of latent preference, z, in (CS2.1) is equal to $X\beta$ regardless of whether preferences of other individuals are known. However, if the off-diagonal elements of the covariance matrix are nonzero, then the conditional expectation of the latent preference of one individual is correlated with the revealed preference of another individual:

$$E[z_2|z_1] = X_2'\beta + \Sigma_{21}\Sigma_{11}^{-1}(z_1 - X_1\beta). \quad \text{(CS2.3)}$$

Positive covariance leads to a greater expectation of preference z_2 if it is known that z_1 is greater than its mean, $X_1\beta$. In the probit model, we do not directly observe the latent utilities. The choices will be dependent via the dependence in latent utility. For example, consider the probability that $y_1 = 1|y_2 = 1$:

$$\begin{aligned} \Pr[y_1 = 1|y_2 = 1] &= \int_0^\infty \left[\int_0^\infty p(z_1|z_2)p(z_2|y_2 = 1)\mathrm{d}z_2\right]\mathrm{d}z_1 \\ &\neq \Pr[y_1 = 1] = \int_0^\infty p(z_1)\mathrm{d}z_1. \end{aligned} \quad \text{(CS2.4)}$$

An approach to modeling the structure of correlation in the matrix R is to add in a second error term from an autoregressive process (LeSage 2000) to the usual independent vector of errors:

$$z_i = x_i'\beta + \varepsilon + \theta, \quad \text{(CS2.5)}$$

$$\theta = \rho W\theta + u, \quad \text{(CS2.6)}$$

$$\varepsilon \sim N(0, I), \quad \text{(CS2.7)}$$

$$u \sim N(0, \sigma^2 I). \quad \text{(CS2.8)}$$

θ is a vector of autoregressive parameters, and the matrix ρW reflects the interdependence of preferences across individuals. Co-dependence is captured by nonzero entries appearing in both the upper and lower triangular submatrices of W; contrast this with time series models in which nonzero elements occur only in the lower triangle of W. It is assumed that the diagonal elements w_{ii} are equal to zero and each row sums to one. There can be multiple ways to specify the weight matrix W, each according to specific definitions of a network. In the empirical section below, we will explore various specifications of W.

The coefficient ρ measures the degree of overall association among the units of analysis beyond that captured by the covariates X. A positive (negative) value of ρ indicates positive (negative) correlation among people. It should be noted that, unlike in the time series context, ρ is not a unitless quantity. The network matrix reflects the structure of the network (zero or nonzero entries) as well as the strength of relationships within a network. Therefore, ρ is in units of the network distance metric.

The augmented error model results in latent preferences with nonzero covariance:

$$z \sim N\left(X\beta, I + \sigma^2(I - \rho W)^{-1}(I - \rho W')^{-1}\right). \quad \text{(CS2.9)}$$

This specification is different from that encountered in standard spatial data models (see Cressie 1991, p. 441) where the error term ε is not present and the covariance term is equal to $\sigma^2(I - \rho W)^{-1}(I - \rho W')^{-1}$. The advantage of specifying the error in two parts is that it facilitates a data augmentation strategy as discussed in Chapter 3. The parameter θ is responsible for the nonzero covariances in the latent preferences, z, but is not present in the marginal likelihood specification (CS2.9). By augmenting the parameters with θ, we avoid direct evaluation of (CS2.9).

The elements of the autoregressive matrix, $W = \{w_{ij}\}$, reflect the potential dependence between units of analysis. A critical part of the autoregressive specification concerns the construction of W. Spatial models, for example, have employed a coding scheme where the unnormalized elements of the autoregressive matrix equal one if i and j are neighbors and zero otherwise. An alternative specification for the model could involve other metrics, such as Euclidean distance.

The model described by (CS2.5)–(CS2.8) is statistically identified. This can be seen by considering the choice probability $\Pr(y_i = 1) = \Pr(z_i > 0) = \Pr(x_i'\beta + \varepsilon_i + \theta_i > 0)$. The fact that the variance of ε_i is one identifies the probit model. Multiplying by an arbitrary positive scalar quantity would alter the variance of ε_i. However, we note that care must be exercised in comparing estimates of β from the independent probit model with that from the autoregressive model because of the differences in the magnitude of the covariance matrix. We return to the issue of statistical identification below when demonstrating properties of the model in a simulation study.

Prior Specification and MCMC Inference

We introduce prior distributions over the model parameters in standard conjugate form where possible:

$$\beta \sim N(\bar{\beta}, A^{-1}), \qquad \bar{\beta} = 0, A = 0.01I, \tag{CS2.10}$$

$$\rho \sim \text{Unif}\left[\frac{1}{\lambda_{\text{max}}}, \frac{1}{\lambda_{\text{min}}}\right], \tag{CS2.11}$$

$$\sigma^2 \sim \frac{\nu_0 s_0^2}{\chi^2_{\nu_0}}, \qquad \nu_0 = 5, s_0^2 = 0.1. \tag{CS2.12}$$

The restricted support of the uniform prior on ρ is necessary to ensure that the matrix $I - \rho W$ is invertible, where λ denotes the eigenvalue of this matrix.

We define a hybrid MCMC sampler suggested by the DAG for this model.

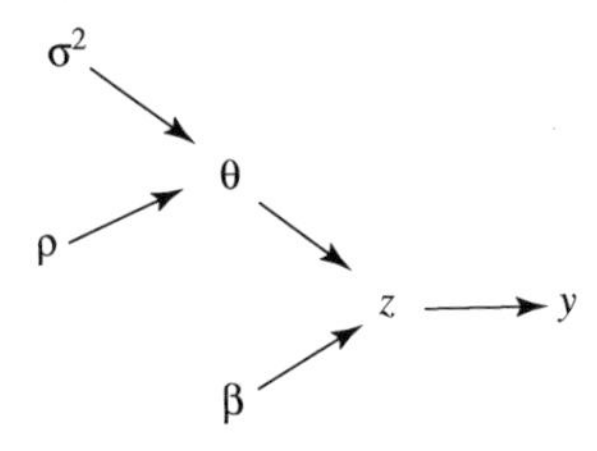

The hybrid MCMC sampler is based on the following set of conditional distributions:

$$z|y, \theta, \beta$$

$$\beta | z, \theta$$

$$\theta | \rho, \sigma^2, z, \beta$$

$$\sigma^2 | \rho, \theta$$

$$\rho | \sigma^2, \theta$$

DATA

Data were collected by a marketing research company on purchases of mid-sized cars in the United States. We obscure the identity of the cars for the purpose of confidentiality. The cars are functionally substitutable, priced in a similar range, and are distinguished primarily by their national origin: Japanese and non-Japanese. Japanese cars have a reputation for reliability and quality in the last twenty years, and we seek to understand the extent to which preferences are interdependent among consumers.

We investigate two sources of dependence – geographic and demographic neighbors. Geographic neighbors are created by physical proximity and measured in terms of geographic distance among individuals' places of residence. Demographic neighbors are defined by membership in the same cluster defined by demographic variables. Young people, for example, are more likely to associate with other young people, obtain information from them, and may want to conform to the beliefs to their reference group to gain group acceptance and social identity. We empirically test these referencing structures and analyze their importance in driving the preferences.

The different referencing schemes are based on different distance metrics. We consider three such distance metrics. The data include the longitude and latitude information of each person's residence, and we can calculate the Euclidean distance between person i and j, denoted $d(i, j)$. We assume that geographic influence is an inverse function of the geographic distance:

$$w_{ij}^{g1} = \frac{1}{\exp(\mathrm{d}(i, j))}. \tag{CS2.14}$$

An alternative geographic specification of W that leads to a symmetric matrix is to identify neighbors by the zip code of their home mailing address:

$$w_{ij}^{g2} = \begin{cases} 1, & \text{if person } i \text{ and person } j \text{ have the same zip code,} \\ 0, & \text{otherwise.} \end{cases} \tag{CS2.15}$$

We can also define a notion of distance in terms of demographic characteristics such as education, age, income, etc. Individuals in the data set are divided into groups defined by age of the head of the household (3 categories), annual household income (3 categories), ethnic affiliation (2 categories) and education (2 categories). This leads us to a maximum of 36 groups, 31 of which are present in our sample. The demographic specification of W becomes:

$$w_{ij}^{d} = \begin{cases} 1, & \text{if person } i \text{ and person } j \text{ belong to the same demographic group,} \\ 0, & \text{otherwise.} \end{cases} \tag{CS2.16}$$

Table CS2.1 Sample statistics

Variable	Mean	Standard deviation
Car choice (1 = Japanese, 0 = Non-Japanese)	0.856	0.351
Difference in price (× \$1000)	−2.422	2.998
Difference in options (× \$1000)	0.038	0.342
Age of buyer (years)	48.762	13.856
Annual Income of Buyer (× \$1000)	66.906	25.928
Ethnic origin (1 = Asian, 0 = Non-Asian)	0.117	0.321
Education (1 = College, 0 = Below college)	0.349	0.477
Latitude (relative to 30 = original −30)	3.968	0.484
Longitude (relative to −110 = original + 110)	−8.071	0.503

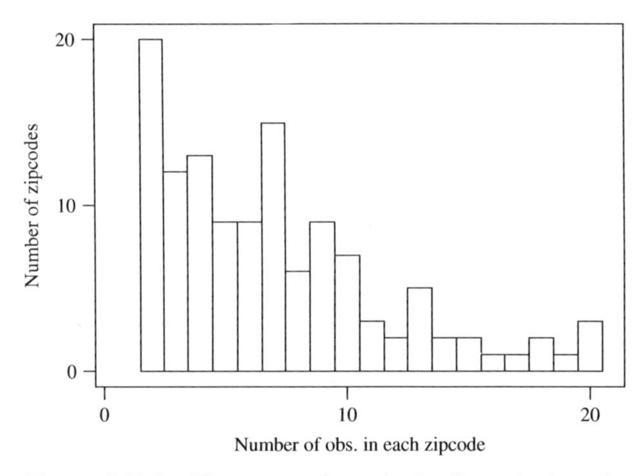

Figure CS2.1 Histogram of sample size for each zip code

The data consist of 857 consumers who live in 122 different zip codes. Table CS2.1 provides summary statistics. Approximately 85 % of people purchased a Japanese car. On average, the price of a Japanese car is \$2400 cheaper than a non-Japanese car and there is little difference between the optional accessories purchased with the cars. We note, however, that the sample standard deviations are nonzero, indicating intra-group variation. The average age of the consumer is 49 years, and average annual household income is approximately \$67,000. Approximately 12 % of the consumers are of Asian origin, and 35 % have earned a college degree. The choices of 666 individuals from 100 zip codes are used to calibrate the model, and 191 observations form a hold-out sample. Figure CS2.1 is a histogram showing the number of zip codes containing a specified number of respondents.

RESULTS

In-sample and out-of-sample fit are assessed in the following way. When possible, we report fit statistics conditional on the augmented parameter θ_i. For the independent probit model in which $\theta_i = 0$, the latent preferences are independent conditional on the covariates which drive the mean of the latent distribution. Knowledge that an individual actually purchased a Japanese car provides no help in predicting the preferences and choices of other individuals. When preferences are interdependent, information about others' choices is useful in predicting choices, and this information is provided through θ_i as described in (CS2.6).

In-sample fit is assessed using the method of Newton and Raftery (see Chapter 6). Conditional on θ, this evaluation involves the product of independent probit probabilities and is easy to compute. Computing the out-of-sample fit is more complicated. Our analysis proceeds by constructing the autocorrelation matrices W for the entire data set (857 observations) and then estimating the model parameters using the first 666 observations. We obtain the augmented parameters for the hold-out sample, θ^p, by noting that

$$\begin{pmatrix} \theta \\ \theta^p \end{pmatrix} \approx N(0, \hat{\sigma}^2 (I_{857} - \hat{\rho}\hat{W})^{-1}(I_{857} - \hat{\rho}\hat{W}')^{-1}) = N\left(0, \begin{matrix} \Sigma_{11} & \Sigma_{12} \\ \Sigma_{21} & \Sigma_{22} \end{matrix}\right). \quad \text{(CS2.17)}$$

Σ_{12} is the covariance matrix between θ and θ^p, and we can obtain the conditional distribution of θ^p given θ using the properties of the multivariate normal distribution:

$$(\theta^p|\theta) \sim N(\mu, \Omega), \quad \text{(CS2.18)}$$

where $\mu = \Sigma_{21}\Sigma_{11}^{-1}\theta$ and $\Omega = \Sigma_{22} - \Sigma_{21}\Sigma_{11}^{-1}\Sigma_{12}$

Table CS2.2 reports the in-sample and out-of-sample fit statistics for six different models. Model 1 is an independent binary probit model where the probability of purchasing a Japanese car is associated with the feature differences between cars, demographics information for the individual, geographic information (longitude, latitude), and dummy variables for the demographic groups. The dummy variables can be viewed as an attempt to capture high-order interactions of the covariates. Dummy variables for only 19 of the 31 groups are included because the proportion of buyers of Japanese cars in the remaining groups is at or near 100 %. Thus, the first model attempts to approximate the structure of heterogeneity using a flexible exogenous specification.

Model 2 is a random effects model that assumes people living in the same zip code have identical price and option coefficients. Geographic and demographic variables are incorporated into the model specification to adjust the model intercept and the mean of the random effects distribution. The model represents a standard approach to modeling preferences, incorporating observed and unobserved heterogeneity.

Models 3–6 specify four variations of models in which consumer preferences are interdependent. Models 3 and 4 are alternative specifications for geographic neighbors (CS2.14) and (CS2.15), whereas model 5 specifies the autoregressive matrix in terms of the 31 demographic groups (CS2.16). Model 6 incorporates both geographic and demographic structures using the finite-mixture model of both geographic and demographic specifications of the weight matrix (for details, see Yang and Allenby 2003).

The model fit statistics (both in-sample and out-of-sample) indicate the following. First, car choices are interdependent. Model 1 is the worst-fitting model, and all attempts to incorporate geographic and/or demographic interdependence in the model lead to improved in-sample and out-of-sample fit. A comparison of the fit statistics between

Table CS2.2 Model fit comparison

Model	Specification[a]	In-sample fit[b]	Out-of-sample fit[c]
1 Standard probit model	$z_i = x_i^{1'}\beta^1 + x_i^{2'}\beta^2 + x_i^{3'}\beta^3 + x_i^{4'}\beta^4 + \varepsilon_i$	−237.681	0.177
2 Random coefficients model	$z_{ik} = x_i^{1'}\beta_k^1 + x_i^{2'}\beta^2 + x_i^{3'}\beta^3 + x_i^{4'}\beta^4 + \varepsilon_{ik}$, $\beta_k^1 = \Gamma[x_k^2, x_k^3] + \eta_k$	−203.809	0.158
3 Autoregressive model with geographic neighboring effect	$z_i = x_i^{1'}\beta^1 + x_i^{2'}\beta^2 + x_i^{3'}\beta^3 + x_i^{4'}\beta^4 + \theta_i + \varepsilon_i$, $\theta_i = \rho\sum_{j=1}^{m} w_{ij}^{g1}\theta_j + u_i$	−146.452	0.136
4 Autoregressive model with geographic neighboring effect	$z_i = x_i^{1'}\beta^1 + x_i^{2'}\beta^2 + x_i^{3'}\beta^3 + x_i^{4'}\beta^4 + \theta_i + \varepsilon_i$, $\theta_i = \rho\sum_{j=1}^{m} w_{ij}^{g2}\theta_j + u_i$	−148.504	0.135
5 Autoregressive model with demographic neighboring effect	$z_i = x_i^{1'}\beta^1 + x_i^{2'}\beta^2 + x_i^{3'}\beta^3 + x_i^{4'}\beta^4 + \theta_i + \varepsilon_i$, $\theta_i = \rho\sum_{j=1}^{m} w_{ij}^{d}\theta_j + u_i$	−151.237	0.139
6 Autoregressive model with geographic and demographic neighboring effect	$z_i = x_i^{1'}\beta^1 + x_i^{2'}\beta^2 + x_i^{3'}\beta^3 + x_i^{4'}\beta^4 + \theta_i + \varepsilon_i$, $\theta_i = \rho\sum_{j=1}^{m} w_{ij}\theta_j + u_i$, $w_{ij} = \phi w_{ij}^{g2} + (1-\phi)w_{ij}^{d}$	−133.836	0.127

[a] $y_i = 1$ if a Japanese brand car is purchased (i indexes person and k indexes zip code), $x_i^1 = [1$, difference in price, difference in option], $x_i^2 =$ [age, income, ethnic group, education], $x_i^3 =$ [longitude, latitude], $x_i^4 =$ [group1, group2, ..., group19]. Note that for other subgroups, estimation of these group-level effects is not feasible because of the perfect or close to perfect correlation between the group dummy variable and y.

[b] In-sample model fit is measured by the log marginal density calculated using the importance sampling method of Newton and Raftery.

[c] Out-of-sample fit is measured by mean absolute deviation of estimated choice probability and actual choice.

model 2 and models 3–6 indicates that there is stronger evidence of interdependence in the autoregressive models than in a random effects model based on zip codes. Moreover, adding quadratic and cubic terms for longitude and latitude results in a slight improvement in in-sample fit (from −203.809 to −193.692) and out-of-sample fit (from 0.158 to 0.154). This supports the view that people have similar preferences not only because they share similar demographic characteristics that may point to similar patterns of resource allocation (an exogenous explanation), but also because there is an endogenous interdependence among people. Furthermore, models 3 and 4 produce very similar fit statistics, which indicates that a geographic neighbor based weighting matrix performs a good approximation to a geographic distance based weighting matrix. The introduction of both geographic and demographic referencing schemes in model 6 leads to an improvement in the fit, showing that both reference groups are important in influencing the individual's preference.

Parameter estimates for the six models are reported in Table CS2.3. We note that, in general, the coefficient estimates are largely consistent across the six models, indicating that all of the models are somewhat successful in reflecting the data structure. Since

Table CS2.3 Posterior mean of coefficient estimates

Coefficient	Model 1	Model 2	Model 3	Model 4	Model 5	Model 6
Intercept	−0.799	1.015	1.135	1.489	1.116	1.274
Price	−0.004	−0.003	**−0.037**	**−0.039**	**−0.042**	**−0.038**
Option	0.010	0.005	0.031	0.034	−0.028	−0.015
Age	−0.014	**−0.014**	**−0.126**	**−0.119**	**−0.153**	**−0.080**
Income	**0.014**	**0.001**	**0.142**	**0.146**	**0.141**	**0.167**
Ethnic	**2.271**	**2.042**	**7.513**	**7.679**	**7.382**	**9.670**
Education	−0.279	**−0.032**	**−3.662**	**−4.003**	**−3.914**	**−3.425**
Latitude	**−2.901**	**−5.778**	**−6.397**	**−6.965**	**−6.231**	**−6.841**
Longitude	**−1.739**	**−2.575**	**−3.216**	**−3.537**	**−3.330**	**−3.619**
Group1	−1.069	−1.214	−0.897	−0.765	−0.682	−0.552
Group2	−1.591	−1.123	−1.159	−1.264	−1.378	−1.432
Group3	−1.142	−1.256	−0.569	−0.512	−0.165	−0.273
Group4	−1.207	−0.954	−3.159	−3.641	−2.331	−4.076
Group5	−0.499	−0.421	−2.315	−1.967	−1.578	2.523
Group6	−0.703	−0.946	−2.213	−2.786	−1.922	2.165
Group7	−0.604	−0.787	−1.887	−1.365	−2.134	2.632
Group8	−0.238	−0.113	1.459	1.781	0.967	0.957
Group9	−0.445	−0.335	2.646	2.893	1.384	2.484
Group10	−0.597	−0.667	0.132	0.197	0.268	0.169
Group11	**−1.207**	**−1.016**	**−3.376**	**−3.132**	**−2.711**	**−2.954**
Group12	**−1.634**	**−1.147**	**−4.105**	**−3.198**	**−3.452**	**−4.103**
Group13	−0.333	−0.178	2.139	2.164	2.551	2.872
Group14	**−1.669**	**−1.124**	2.778	2.115	2.160	2.949
Group15	−0.630	−0.421	1.986	1.468	1.235	1.836
Group16	**−0.981**	**−0.563**	**−2.071**	**−2.569**	**−2.135**	**−3.088**
Group17	−0.577	−0.289	1.468	1.923	1.497	2.313
Group18	−1.018	−0.956	0.912	0.813	0.569	0.783
Group19	−0.856	−0.433	1.876	1.579	1.692	0.964
σ^2	NA	NA	**16.784**	**16.319**	**15.691**	**18.235**
ρ	NA	NA	**0.473**	**0.462**	**0.487**	**0.512**
α	NA	NA	NA	NA	NA	**0.598**

Note: Bold face indicates that 0 lies outside the 95 % highest posterior density interval of the estimate.

model 6 yields the best in-sample and out-sample fit, we focus our discussion on its parameter estimates. The estimates indicate that price, age, income, ethnic origin, longitude and latitude are significantly associated with car purchases, with Asians, younger people, people with high incomes, and with more southern and more western zip codes preferring Japanese makes of car. The ethnic variable turns out to have a very large coefficient, indicating that Asian people will have a significantly higher likelihood of choosing a Japanese-made car. Furthermore, ρ is significantly positive, indicating a positive correlation among consumer preferences. Yang and Allenby (2003) also consider a mixture of the demographic and geographic reference models (model 6). Model 6 is a mixture of a model based on geographic reference groups with the model based on demographic reference groups. The mixing probability parameter is written as $\exp(\alpha)/(1+\exp(\alpha))$. $\alpha > 0$ implies that the mixture weight on the geographic reference group model is greater than the demographic reference group model as shown in the last line of Table CS2.3.

Figure CS2.2 displays estimates of the elements of the augmented parameter θ against the longitude and latitude of each observation in sample. Most of the estimates have posterior distributions away from zero, providing evidence of interdependent choices. An analysis of the difference in covariates for the two groups reported in Table CS2.4 ($\theta > 0$ and $\theta \leq 0$) does not reveal any statistically significant differences except for the longitude variable. That is, the augmented parameters that capture the endogenous nature of preferences are not simply associated with all the covariates in the analysis.

This empirical application demonstrates that: (i) there is a preference interdependence among individual consumers that reflects conformity ($\rho > 0$); (ii) the preference interdependence is more likely to have an endogenous influence structure than a simple exogenous structure; (iii) the geographically defined network is more important in

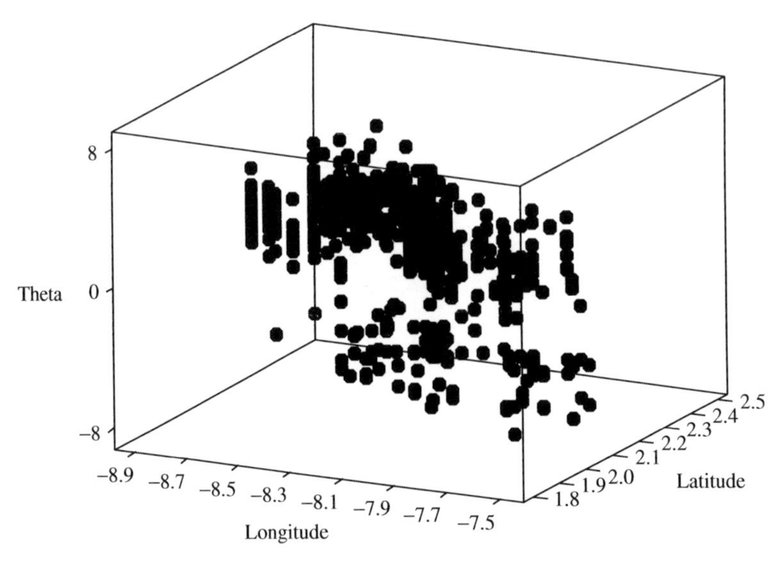

Figure CS2.2 Estimated θ for each individual (note that the origin for longitude is -110 and for latitude is 30; all the longitude and latitude information presented here is related to this origin specification)

Table CS2.4 Profiling the two groups

Average	$\theta > 0$	$\theta \leq 0$	p-value
Age	48.297	48.357	0.962
Income	69.835	71.934	0.418
Ethnic	0.116	0.094	0.275
Education	0.372	0.379	0.673
Latitude	4.152	4.158	0.502
Longitude	−8.281	−8.186	0.031

explaining individual consumer behavior than the demographic network. However, since our data is cross-sectional, we are unable to identify the true cause of the interdependence.

DISCUSSION

Choice models have been used extensively in the analysis of marketing data. In these applications, most of the analysis depends on the assumption that the individual forms his or her own preferences and makes a choice decision independent of other people's preferences. However, people live in a world where they are interconnected, information is shared, recommendations are made, and social acceptance is important. Interdependence is therefore a more realistic assumption in models of preference heterogeneity.

The model can be applied and extended in several ways. Opinion leaders, for example, are individuals who exert a high degree of influence on others, and could be identified with extreme realizations of the augmented parameter, θ_i. Aspiration groups that affect others, but are not themselves affected, could be modeled with an autoregressive matrix W that is asymmetric. Temporal aspects of influence, including word of mouth, could be investigated with longitudinal and cross-sectional data with the autoregressive matrix W defined on both dimensions. Such time series data would help us to identify the source and nature of interdependence. Finally, the model can be extended to apply to multinomial response data to investigate the extent of interdependent preference in brand purchase behavior, or extended to study the interdependence in β coefficients across people. These applications and extensions will contribute to our understanding of extended product offerings and the appropriateness of 'iid' heterogeneity assumptions commonly made in models of consumer behavior.

R IMPLEMENTATION

R We have implemented the method of Yang and Allenby in an R function, `rinterprobit`, which is available on the website for this book. The draw of θ requires taking the Cholesky root of an $N \times N$ matrix (here $N = 857$). While this is computationally feasible, larger N may require specialized inversion routines or the procedure of Smith and LeSage (2000).

Case Study 3 Overcoming Scale Usage Heterogeneity

Questions that use a discrete ratings scale are commonplace in survey research. Examples in marketing include customer satisfaction measurement and purchase intention. Survey research practitioners have long commented that respondents vary in their usage of the scale; common patterns include using only the middle of the scale or using the upper or lower end. These differences in scale usage can impart biases to correlation and regression analyses. In order to capture scale usage differences, Rossi *et al.* (2001) develop a model with individual scale and location effects and a discrete outcome variable. The joint distribution of all ratings scale responses is modeled, rather than specific univariate conditional distributions as in the ordinal probit model. The model is applied to a customer satisfaction survey where it is shown that the correlation inferences are much different once proper adjustments are made for the discreteness of the data and scale usage. The adjusted or latent ratings scale is also more closely related to actual purchase behavior.

BACKGROUND

Customer satisfaction surveys, and survey research in general, often collect data on discrete rating scales. Figure CS3.1 shows a sample questionnaire of this type from Maritz Marketing Research Inc., a leading customer satisfaction measurement firm. In this sample questionnaire a five-point scale (excellent to poor) is used, while in other cases seven- and ten-point scales are popular. Survey research practitioners have found that respondents vary in their usage of the scale. In addition, it has been observed that there are large cultural or cross-country differences in scale usage, making it difficult to combine data across cultural or international boundaries. These different usage patterns, which we term 'scale usage heterogeneity', impart biases to many of the standard analyses

Bayesian Statistics and Marketing P. E. Rossi, G. M. Allenby and R. McCulloch

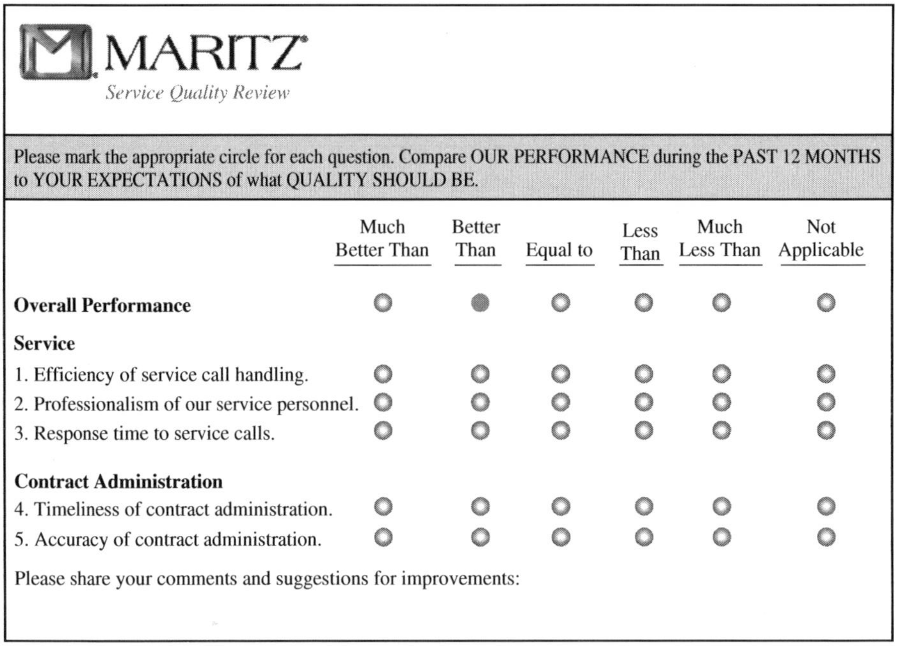

MARITZ
Service Quality Review

Please mark the appropriate circle for each question. Compare OUR PERFORMANCE during the PAST 12 MONTHS to YOUR EXPECTATIONS of what QUALITY SHOULD BE.

	Much Better Than	Better Than	Equal to	Less Than	Much Less Than	Not Applicable
Overall Performance	○	●	○	○	○	○
Service						
1. Efficiency of service call handling.	○	○	○	○	○	○
2. Professionalism of our service personnel.	○	○	○	○	○	○
3. Response time to service calls.	○	○	○	○	○	○
Contract Administration						
4. Timeliness of contract administration.	○	○	○	○	○	○
5. Accuracy of contract administration.	○	○	○	○	○	○

Please share your comments and suggestions for improvements:

Figure CS3.1 Example of customer satisfaction survey questionnaire

conducted with ratings data, including regression and clustering methods as well as the identification of individuals with extreme views.

The standard procedure for coping with scale usage heterogeneity is to center each respondent's data by subtracting the mean over all questions asked of the respondent and dividing by the standard deviation. The use of respondent means and standard deviations assumes that the response data is continuously distributed and from an elliptically symmetric distribution. Furthermore, the estimates of the individual location and scale parameters obtained by computing the mean and standard deviation over a small number of questions are often imprecise.

In order to choose an appropriate modeling strategy for ratings data, we must consider the types of analyses that will be conducted with this data as well as the basic issues of what sort of scale information (ratio, interval or ordinal) is available in this data. To facilitate this discussion, assume that the data is generated by a customer satisfaction survey; however, the methods developed here apply equally well to any data in which a ratings scale is used (examples include purchase intentions and psychological attitude measurement). In the typical customer satisfaction survey, respondents are asked to give their overall satisfaction with a product as well as assessments of satisfactions with various dimensions of the product or service. Ratings are made on five-, seven- or ten-point scales. We will focus on two major uses of customer satisfaction ratings data: measurement of the relationship between overall satisfaction and satisfaction with specific product attributes; and identification of customers with extreme views. Scale usage heterogeneity can substantially bias analyses aimed at either use.

For example, if some respondents tend to use either the low or high end of the scale, this will tend to bias upward any measure of correlation between two response

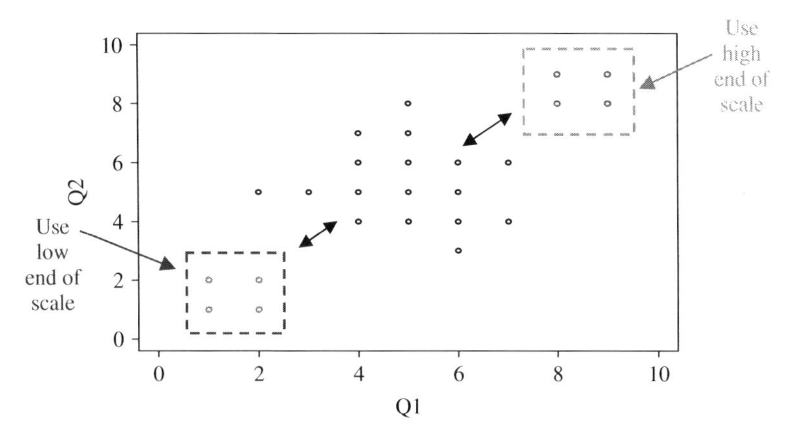

Figure CS3.2 Scale usage heterogeneity and upward correlation bias

items. The middle group of points in Figure CS3.2 represents a hypothetical situation in which all respondents have the same scale usage. If some respondents use the upper or lower end of the scale, this will move points outward from the middle grouping, creating a higher but spurious correlation. Thus, any covariance-based analysis of rating scale data such as regression or factor analysis can be substantially biased by scale usage heterogeneity, aside from the problems associated with using discrete data in methods based on the assumption of continuous elliptically symmetric distributions. In addition, any cluster analysis or filtering of the data for respondents with extreme views will tend to identify a group of nay or yea sayers whose true preferences may not be extreme or even similar.

Practitioners have long been aware of scale usage patterns and often center the data. We observe N respondents answering M questions on a discrete rating scale consisting of the integers 1 to K; the data array is denoted $X = \{x_{ij}\}$, an $N \times M$ array of discrete responses $x_{ij} = \{k\}$, $k = 1, \ldots, K$. Centering would transform the X array by subtracting row means and dividing by the row standard deviation:

$$X^* = \left[\left(x_{ij} - \bar{x}_i\right)/s_i\right].$$

After the data are centered, standard correlation and regression methods are used to examine the relationship between various questions. To identify extreme respondents or to cluster respondents, it is more typical to use the raw response data.

To select the appropriate analysis method, it is important to reflect on the nature of the scale information available in ratings data. Our perspective is that the discrete response data provides information on underlying continuous and latent preference/satisfaction. Clearly, the ratings provide ordinal information in the sense that a higher discrete rating value means higher true preference/satisfaction. It is our view that ratings data can also provide interval information, once the scale usage heterogeneity has been properly accounted for. However, we do not believe that even properly adjusted ratings data can provide ratio-level information. For example, if a respondent gives only ratings at the top end of the scale, we cannot infer that he/she is extremely satisfied. We can only infer that the level of relative satisfaction is the same across all items for this respondent.

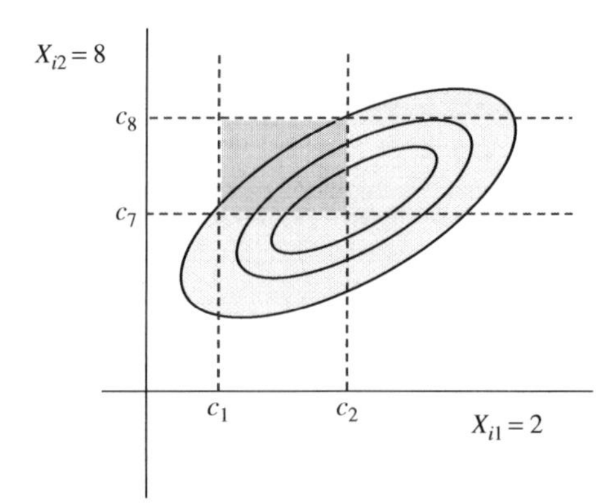

Figure CS3.3 Computing the multinomial probabilities

Centering acknowledges this fundamental identification problem, but results in imprecise row mean and row standard deviation estimates which introduce considerable noise into the data. In most cases, fewer than 20 questions are used to form respondent means and standard deviation estimates. Furthermore, the use of centered data in correlation, regression or clustering analyses ignores the discrete aspect of this data (some transform the data prior to centering, but no standard transformation can change the discreteness of the data). In the next section, we develop a model which incorporates both the discrete aspects of the data and scale usage heterogeneity.

MODEL

Our model is motivated by the basic view that the data in the X response array is a discrete version of underlying continuous data. For $i = 1, \ldots, N$ and $j = 1, \ldots, M$, let y_{ij} denote the latent response of individual i to question j. Let $y_i' = [y_{i1}, \ldots, y_{iM}]$ denote the latent response of respondent i to the entire set of M questions. Assume there are $K + 1$ common and ordered cut-off points $\{c_k : c_{k-1} \leq c_k, k = 1, \ldots, K\}$, where $c_0 = -\infty$, $c_K = +\infty$, such that for all i, j and k,

$$x_{ij} = k \text{ if } c_{k-1} \leq y_{ij} \leq c_k \tag{CS3.1}$$

and

$$y_i \sim N(\mu_i^*, \Sigma_i^*). \tag{CS3.2}$$

The interpretation of the model in (CS3.1)–(CS3.2) is that the observed responses are iid multinomial random variables, where the multinomial probabilities are derived from an underlying continuous multivariate normal distribution. The set of cut-offs $[c_0, \ldots, c_K]$ discretizes the normal variable y_{ij}.

The probability that $x_i' = [x_{i1}, \ldots, x_{iM}]$ takes on any given realization (a vector of M integers between 1 and K) is given by the integral of the joint normal distribution of y over the appropriately defined region. For example, if $M = 2$, and $x_i' = [2, 8]$, then Figure CS3.3 depicts this integral of a bivariate normal distribution over a rectangle defined by the appropriate cut-offs.

The above-proposed model is different in nature from some latent variable models used in Bayesian analyses of discrete data. Most such models deal with grouped data in the form of contingency tables, where the prior distribution of multinomial probabilities is usually taken as Dirichlet over multidimensional arrays. There, the problem of interest is usually of modeling probabilities forming certain patterns of statistical dependence. Here we are interested in modeling individual responses to make individual measurements comparable for the sake of correlation and regression analysis.

It should be noted that the model in (CS3.1) is not a standard ordinal probit model. We have postulated a model of the *joint* discrete distribution of the responses to all M questions in the survey. Standard ordinal probit models would focus on the conditional distribution of one discrete variable given the values of another set of variables. Obviously, since our model is of the joint distribution, we can make inferences regarding various conditional distributions so that our model encompasses the standard conditional approach. In analysis of ratings survey data, we are required to have the capability of making inferences about the marginal distribution of specific variables as well as conditional distributions so that a joint approach would seem natural.

The model in (CS3.1)–(CS3.2) is overparameterized since we have simply allowed for an entirely different mean vector and covariance matrix for each respondent. In order to allow for differences in respondent scale usage without overparameterization, the y vector is written as a location and scale shift of a common multivariate normal variable:

$$y_i = \mu + \tau_i \iota + \sigma_i z_i, \qquad z_i \sim N(0, \Sigma). \tag{CS3.3}$$

We have allowed for a respondent-specific location and scale shift to generate the mean and covariance structure in (CS3.2) with $\mu_i^* = \mu + \tau_i \iota$ and $\Sigma_i^* = \sigma_i^2 \Sigma$. Both the analysis of the joint distribution of questions and the identification of customers with extreme views are based on the set of model parameters $\{z_i\}$, μ, Σ.

The model (CS3.3) accommodates scale usage via the (τ_i, σ_i) parameters. For example, a respondent who uses the top end of the scale would have a large value of τ and a small value of σ. It is important to note that this model can easily accommodate lumps of probability at specific discrete response values. For example, in many customer satisfaction surveys, a significant fraction of respondents will give only the top value of the scale for all questions in the survey. In the model outlined in (CS3.1)–(CS3.3), we would simply have a normal distribution centered far out (via a large value of τ) so that there is a high probability that y will lie in the region corresponding to the top rating. We can also set τ to zero and make σ_i large to create 'piling up' at both extremes. However, our model cannot create lumps of probability at two different non-extreme values (to accommodate a respondent who uses mostly 2 s and 9 s on a ten-point scale).

As a modeling strategy, we have chosen to keep the cut-offs c common across respondents and shift the distribution of the latent variable. Another strategy would be to keep a common latent variable distribution for all respondents and shift the cut-offs to induce different patterns of scale usage. The problem here would be choosing a flexible but not too overparameterized distribution for the cut-offs. Given that we often have a small number of questions per respondent (M), a premium should be placed on parsimony. There is also a sense in which the model in (CS3.3) can be interpreted as a model with respondent-specific cut-offs. If we define

$$c_i^* = \tau_i + \sigma_i c$$

where c is the vector of cut-offs, then we have the same model with respondent-specific cut-offs and a common invariant distribution of latent preferences.

The model specified in (CS3.3) is not identified. The entire collection of τ_i parameters can be shifted by a constant, and a compensating shift can be made to μ without changing the distribution of the latent variables. Similarly, we can scale all σ_i and make a reciprocal change to Σ. As discussed below, we solve these identification problems by imposing restrictions on the hierarchical model. $(\tau_i, \ln \sigma_i)$ are assumed to be bivariate normal:

$$\begin{bmatrix} \tau_i \\ \ln \sigma_i \end{bmatrix} \sim N(\phi, \Lambda). \tag{CS3.4}$$

The model in (CS3.4) allows for a correlation between the location and scale parameters. For example, if there is a subpopulation which uses the high end of the scale, then we would expect an inverse relationship between τ and σ. In most applications of hierarchical models, the location and scale parameters are assumed to be independent.

We achieve identification of τ_i by imposing the restriction $\mathrm{E}[\tau_i] = 0$. Since the distribution of τ_i is symmetric and unimodal, we are also setting the median and mode of this distribution to zero. Greater care must be exercised in choosing the identification restriction for σ_i, as this distribution is skewed. One logical choice might be to set $\mathrm{E}[\sigma_i^2] = 1$. This imposes the identification restriction, $\phi_2 = -\lambda_{22}$. However, as the dispersion parameter (λ_{22}) is increased, the distribution of σ_i becomes concentrated around a mode smaller than 1.0 with a fat right tail. Our view is that this is an undesirable family of prior distributions. The right panel of Figure CS3.4 illustrates the prior on σ_i for small and large values of λ_{22}.

A more reasonable approach is to restrict the mode of the prior on σ_i to be 1. This imposes the restriction $\phi_2 = \lambda_{22}$. The left panel of Figure CS3.4 shows this family of distributions. As λ_{22} increases, these distributions retain the bulk of their mass around 1, but achieve greater dispersion by thickening the right tail.[1] Thus, we employ two identification restrictions:

$$\phi_1 = 0 \quad \text{and} \quad \phi_2 = \lambda_{22}. \tag{CS3.5}$$

Even with the cut-offs $\{c_k\}$ assumed fixed and known, the model in (CS3.1) and (CS3.3) is a very flexible model which allows for many possible discrete outcome distributions. In particular, the model allows for 'piling up' of probability mass at either or both endpoints of the ratings scale – a phenomenon frequently noted in CSM data. For further flexibility, we could introduce the cut-offs as free parameters to be estimated. However, we would have to recognize that identification restrictions must be imposed on the cut-off parameters since shifting all cut-offs by a constant or scaling all cut-offs is redundant with appropriate changes in μ and Σ. For this reason, we impose the following identification restrictions:

$$\begin{aligned} \sum_k c_k &= m_1, \\ \sum_k c_k^2 &= m_2. \end{aligned} \tag{CS3.6}$$

[1] In Rossi *et al.* (2001) the strategy of setting the mean to 1 is used for identification.

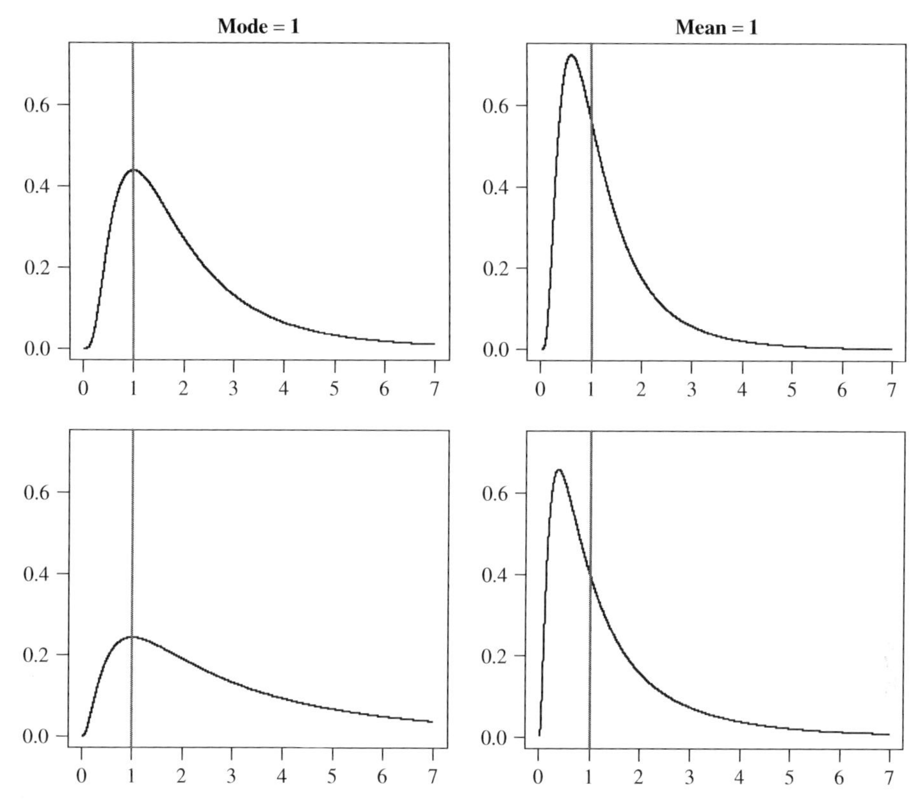

Figure CS3.4 Alternative prior distributions for σ_i

The model in (CS3.1)–(CS3.6) is fully identified but introduces $K-2$ free cut-off parameters. In order to make the model more parsimonious, we will consider further restrictions on the c_k parameters. If we impose equal spacing of the cut-offs, then by (CS3.6) there would be no free c_k parameters. Once the identification restrictions in (CS3.6) are imposed, the only sort of flexibility left in the cut-offs is to introduce skewness or nonlinear spread in the values. In order to allow for nonlinear spread while keeping the number of parameters to a minimum, we impose the further restriction that the cut-off values satisfy a quadratic equation:

$$c_k = a + bk + ek^2, \qquad k = 1, \ldots, K-1. \tag{CS3.7}$$

For example, consider the case of a ten-point scale with $a = 5.5$, $b = 1$ and $e = 0$; then $c_1 = 1.5, c_2 = 2.5, \ldots, c_9 = 9.5$. Johnson and Albert (1999) review univariate ordinal probit models in which the cut-offs are not parameterized and are estimated with a diffuse prior subject to different identification conditions.

Given the identification restrictions in (CS3.6) and the parameterization in (CS3.7), e is the only free parameter; that is, given m_1, m_2, and e, we can solve for a and b by substituting for c_k in (CS3.6) using (CS3.7). In our implementation, m_1 and m_2 are selected so that when $e = 0$, we obtain a standard equal spacing of cut-off values, with each centered around the corresponding scale value. This means that for a ten-point scale, $m_1 = \sum_{k=1}^{K-1}(k+0.5) = 49.5$ and $m_2 = \sum_{k=1}^{K-1}(k+0.5)^2 = 332.25$. The mapping from e to the quadratic coefficients in (CS3.7) is only defined for a bounded range of e values.

The role of e is to allow for a skewed spreading of the cut-off values as shown in Figure CS3.5. A positive value of e spreads out the intervals associated with high scale ratings and compresses the intervals on the low end. This will result in massing of probability at the upper end.

PRIORS AND MCMC ALGORITHM

To complete the model, we introduce priors on the common parameters:

$$\pi(\mu, \Sigma, \phi, \Lambda, e) = \pi(\mu)\pi(e)\pi(\Sigma)\pi(\phi)\pi(\Lambda), \tag{CS3.8}$$

with

$$\begin{aligned} &\pi(\mu) \propto \text{constant}, \\ &\pi(e) \propto \text{ Unif}\,[-0.2, 0.2], \\ &\Sigma \sim \text{IW}(\nu_\Sigma, V_\Sigma), \\ &\Lambda \sim \text{IW}(\nu_\Lambda, V_\Lambda). \end{aligned} \tag{CS3.9}$$

That is, we are using flat priors on the means and the cut-off parameter and standard Wishart priors for the inverse of the two covariance matrices. We note that the identification restriction in (CS3.5) means that the prior on Λ induces a prior on ϕ. Figure CS3.5 shows that the range of e in our uniform prior is more than sufficient to induce wide variation in the patterns of skewness in the cut-offs. We use diffuse but proper settings

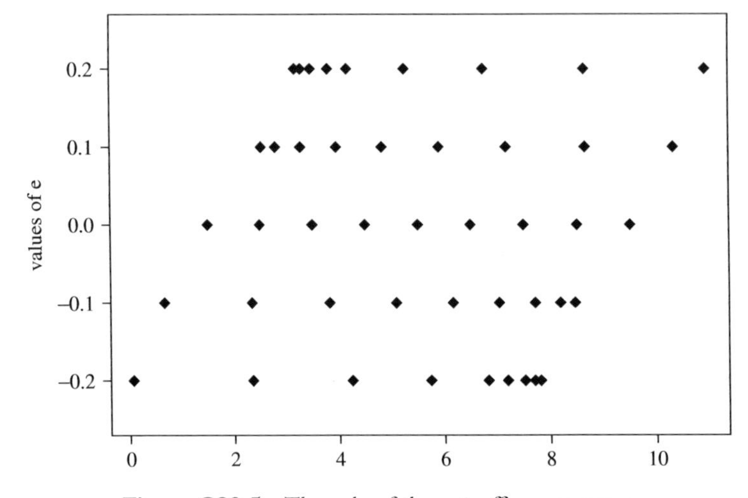

Figure CS3.5 The role of the cut-off parameter, e

for the priors on Σ^{-1} and Λ^{-1}. These parameter values center the prior on Σ over the identity matrix

$$\nu_\Sigma = \dim(\Sigma) + 3 = K + 3, \qquad V_\Sigma = \nu_\Sigma I. \tag{CS3.10}$$

The prior on Λ influences the degree of shrinkage in the τ_i, σ_i estimates. Our hierarchical model adapts to the information in the data regarding the distribution of τ_i, σ_i, subject to the influence of the prior on the hyperparameter Λ. There will rarely be more than a small number of questions on which to base estimates of τ_i, σ_i. This means that the prior on Λ may be quite influential. In most hierarchical applications, there is a subset of units for which a good deal of information is available. This subset allows for determination of Λ via adaptive shrinkage. However, in our situation, this subset is not available and the prior on Λ has the potential to exercise more influence than normal. For these reasons, we will exercise some care in the choice of the prior on Λ. We will also consider and recommend prior settings somewhat tighter than typically used in hierarchical contexts.

The role of the prior on Λ is to induce a prior distribution on τ_i, σ_i. To examine the implications for choice of the prior hyperparameters, we will compute the marginal prior on τ_i, σ_i via simulation. The marginal prior is defined by

$$\pi(\tau, \sigma) = \int p(\tau, \sigma|\Lambda)\pi(\Lambda|\nu_\Lambda, V_\Lambda)d\Lambda. \tag{CS3.11}$$

To assess ν_Λ, V_Λ, we consider a generous range of possible values for τ_i, σ_i, but restrict prior variation to not much greater than this permissible range. For τ, we consider the range ± 5 to be quite generous in the sense that this encompasses much of a ten-point scale. For σ, we must consider the role of this parameter in restricting the range of possible values for the latent variable. Small values of σ correspond to respondents who only use a small portion of the scale, while large values would correspond to respondents who use the entire scale. Consider the ratio of the standard deviation of a 'small' scale range user (e.g. someone who only uses the bottom or top three scale numbers) to the standard deviation of a 'mid-range' user who employs a range of five points on the ten-point scale. This might correspond to a 'small' value of σ. The ratio for a respondent who uses endpoints and the 'middle' of the scale (e.g. 1, 5, 10) – a 'large range' user – to the 'mid-range' user could define a 'large value' of σ. These computations suggest that a generous range of σ values would be (0.5, 2). We employ the prior settings corresponding to a relatively informative prior on τ_i, σ_i:

$$\nu_\Lambda = 20, \qquad V_\Lambda = (\nu_\Lambda - 2 - 1)\bar{\Lambda}, \qquad \bar{\Lambda} = \begin{bmatrix} 4 & 0 \\ 0 & 0.5 \end{bmatrix}. \tag{CS3.12}$$

The settings in (CS3.12) ensure that $E[\Lambda] \doteq \bar{\Lambda}$. If ν_Λ is set to 20, then the resulting marginal prior on τ_i, σ_i provides coverage of the relevant range without admitting absurdly large values, as illustrated in Figure CS3.6.

The model defined in equations (CS3.1)–(CS3.4) with priors given by (CS3.9) and identification restrictions (CS3.5)–(CS3.7) is a hierarchical model that can be analyzed with some modifications to the standard Gibbs sampler. Our interest centers not only on common parameters but also on making inferences about the respondent-specific scale usage and latent preference parameters, ruling out the use of classical statistical methods. Four problems must be overcome to construct the sampler.

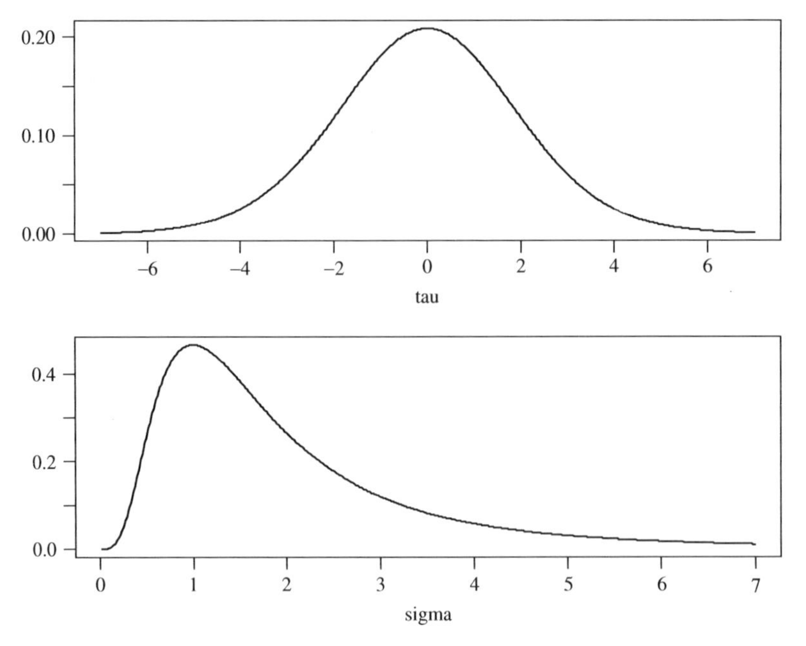

Figure CS3.6 Marginal priors on τ and σ

1. Data augmentation requires a method for handling truncated multivariate normal random variables as in the multinomial probit model (see Chapter 4).
2. One of the conditional distributions in the Gibbs sampler requires the evaluation of the integral of multivariate normal random variables over a rectangle. We use the GHK simulation method (see Chapter 2).
3. The random effects model and priors are not always conditionally conjugate.
4. We accelerate the Gibbs sampler by integrating out some of the latent variables to block e and $\{y_i\}$.

We refer the reader to the appendix of Rossi *et al.* (2001) for more details.

DATA

To illustrate our method, we examine a customer satisfaction survey done in a business-to-business context with an advertising product. This dataset can be loaded once the package **R** `bayesm` package has been installed using the R command `data(customerSat)`. A total of 1811 customers were surveyed as to their satisfaction with overall product performance, aspects of price (three questions) and various dimensions of effectiveness (six questions). Figure CS3.7 lists the specific questions asked. All responses are on a 10 point ratings scale.

On a scale from 1 to 10 where 10 means an "Excellent" performance and 1 means a "Poor" performance, please rate BRAND on the following items:

Q1. Overall value

Price:

Q2. Setting competitive prices.

Q3. Holding price increases to a reasonable minimum for the same ad as last year.

Q4. Being appropriately priced for the amount of customers attracted to your business.

Effectiveness:

Q5. Demonstrating to you the potential effectiveness of your advertising purchase.

Q6. Attracting customers to your business through your advertising.

Q7. Reaching a large number of customers.

Q8. Providing long-term exposure to customers throughout the year.

Q9. Providing distribution to the number of households and/or business your business needs to reach.

Q10. Proving distribution to the geographic areas your business needs to reach.

Figure CS3.7 List of questions

Scale Usage Heterogeneity

Figure CS3.8 plots the median over the ten questions versus the range of responses for each of the 1811 respondents. Since all responses are integer, the points are 'jittered' slightly so that the number of respondents at any given combination of range (0–9) and median (1–10) can be gauged. Figure CS3.8 shows considerable evidence of scale usage heterogeneity. A number of respondents are using only the top end of the scale which is represented by points in the lower right-hand corner of the figure. In fact, a reasonably large number give only the top rating response (10) to all questions. On the other hand, there are very few customers who use the lower end of the scale (lower left-hand corner) and a large number who use a much of the scale.

Our hierarchical model should capture the scale usage heterogeneity in the distribution of (τ_i, σ_i). Figure CS3.9 provides histograms of the N posterior means of (τ_i, σ_i). Both τ_i and σ_i display a great deal of variation from respondent to respondent, indicating a high degree of scale usage heterogeneity. Figure CS3.9 shows results for three prior settings, ranging from highly diffuse (top) to tight (bottom). For $\nu = 5$ and $\nu = 20$, the posteriors of both τ_i and σ_i are very similar, indicating little sensitivity to the prior. The hierarchical prior adapts to the information in the data, centering on values of τ_i and σ_i that are shrunk quite a bit. When $\nu = 100$, the prior is reasonably tight about the mean value. This prior has a greater influence and reduces the extent of shrinkage. For example, when $\nu = 100$, there appears to be a small mode in the τ_i distribution around 3. This mode accommodates those respondents who always give the highest rating (10) in answering all questions. More diffuse priors induce more shrinkage from the information in the data, and the τ_i distributions for those respondents that always

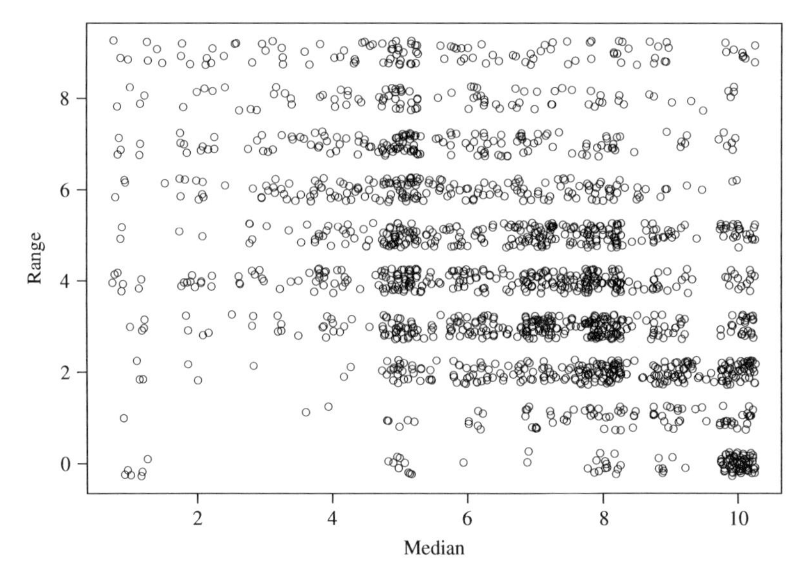

Figure CS3.8 Respondent range versus median (jittered values)

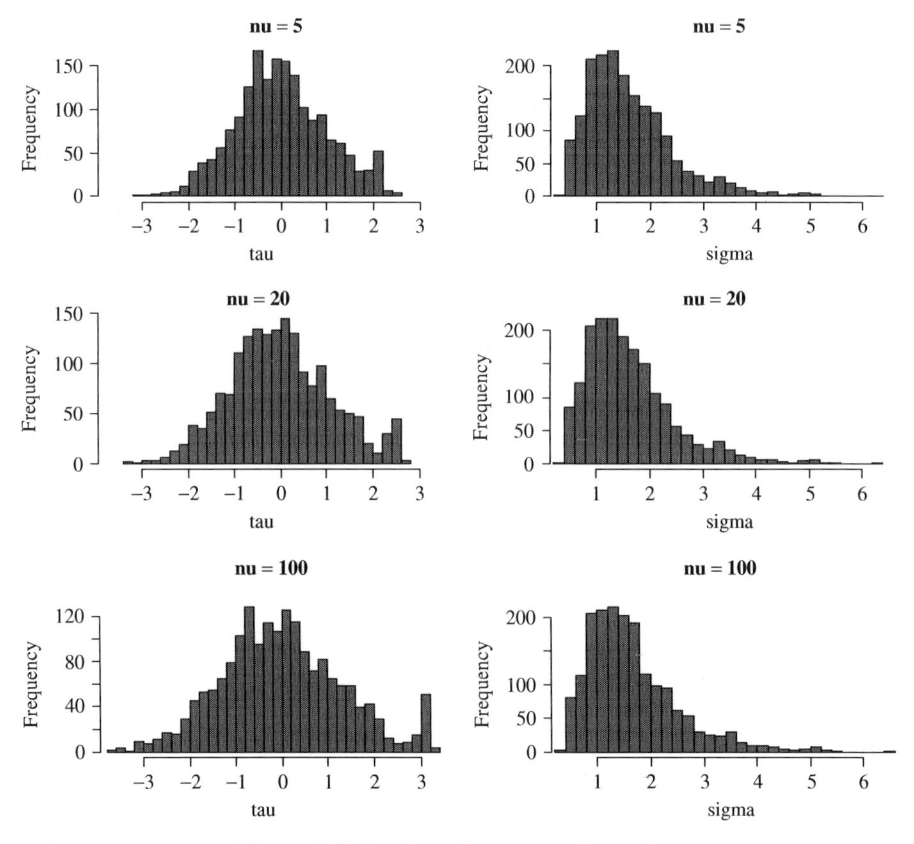

Figure CS3.9 Posterior means of τ_i and σ_i

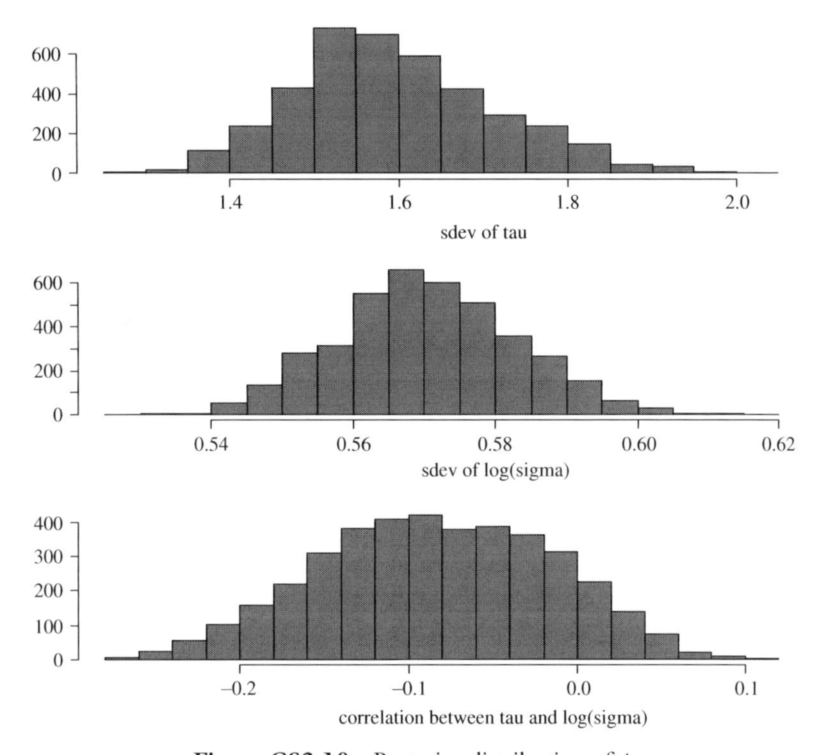

Figure CS3.10 Posterior distribution of Λ

respond with the highest rating are shrunk toward zero. With the higher prior, there is less of this shrinkage.

Figure CS3.10 displays the posterior distribution of Λ. The top panel shows the posterior distribution of $\sqrt{\lambda_{11}}$, the standard deviation of τ_i. This distribution is reasonably tight around 1.55 or so and puts virtually no mass near 2 which corresponds to the location of the prior. Even more striking is the posterior distribution of $\sqrt{\lambda_{22}}$, the standard deviation of $\log(\sigma_i)$. This distribution is centered tightly about 0.57, indicating that the more than 1000 respondents are quite informative about this scale usage parameter. Finally, the posterior distribution of the correlation between τ_i and $\log(\sigma_i)$ puts most of its mass on small negative values, indicating that there is a tendency for those who give high ratings to use less of the scale.

Our model also differs from the standard centering and the standard normal approaches in that we explicitly recognize the discrete nature of the outcome data. Moreover, the discrete outcomes do not appear to be merely a simple filtering of the underlying latent data in which latent data is simply rounded to integers using equal size intervals. The posterior of the quadratic cut-off parameter, e, has a mean of 0.014 with a posterior standard deviation of 0.0039. This implies a skewed set of cut-offs.

Correlation Analysis

One of the major purposes of these surveys is to learn about the relationship between overall satisfaction and various dimensions of product performance. The presence of scale

Table CS3.1 Raw Data means, together with Covariance (lower triangle) and Correlation (upper triangle) matrix

Q.	Mean										
1	6.06	6.50	0.65	0.62	0.78	0.65	0.74	0.59	0.56	0.44	0.45
2	5.88	4.38	7.00	0.77	0.76	0.55	0.49	0.42	0.43	0.35	0.35
3	6.27	4.16	5.45	7.06	0.72	0.52	0.46	0.43	0.46	0.38	0.40
4	5.55	5.36	5.43	5.16	7.37	0.64	0.67	0.52	0.52	0.41	0.40
5	6.13	4.35	3.83	3.62	4.53	6.84	0.69	0.58	0.59	0.49	0.46
6	6.05	4.82	3.29	3.15	4.61	4.61	6.49	0.59	0.59	0.45	0.44
7	7.25	3.64	2.70	2.73	3.42	3.68	3.66	5.85	0.65	0.62	0.60
8	7.46	3.28	2.61	2.79	3.23	3.51	3.41	3.61	5.21	0.62	0.62
9	7.89	2.41	1.99	2.18	2.39	2.72	2.47	3.20	3.02	4.57	0.75
10	7.77	2.55	2.06	2.33	2.42	2.67	2.51	3.21	2.95	3.54	4.89

Table CS3.2 Posterior mean of latent responses (z), together with covariance (lower triangle) and correlation (upper triangle) matrix (Σ); standard deviations in parentheses

Q.	Mean (μ)	Covariance/Correlation (Σ)									
1	6.50	2.46	0.59	0.53	0.73	0.57	0.66	0.45	0.40	0.24	0.26
	(0.08)	(0.29)									
2	6.23		2.98	0.77	0.76	0.47	0.36	0.25	0.26	0.14	0.15
	(0.08)		(0.31)								
3	6.55			3.33	0.69	0.43	0.34	0.27	0.30	0.20	0.23
	(0.08)			(0.32)							
4	6.08				3.28	0.56	0.57	0.37	0.36	0.20	0.21
	(0.08)				(0.35)						
5	6.53					2.82	0.63	0.47	0.47	0.32	0.32
	(0.08)					(0.30)					
6	6.55						2.39	0.49	0.45	0.26	0.26
	(0.08)						(0.30)				
7	7.46							2.93	0.63	0.62	0.60
	(0.08)							(0.28)			
8	7.56								2.43	0.61	0.58
	(0.08)								(0.23)		
9	7.90									2.64	0.78
	(0.08)									(0.25)	
10	7.82										2.76
	(0.08)										(0.26)

usage heterogeneity prevents meaningful use of the raw data for correlation purposes. Scale usage heterogeneity will bias the correlations upward. Table CS3.1 provides the means and correlations for the survey data. The correlations are uniformly positive and in the (0.4, 0.7) range. Table CS3.2 provides posterior means (standard deviations) and correlations of the standardized latent variable, z. Table CS3.2 provides a dramatically

Table CS3.3 Means of standard responses $(x - \bar{x})/s_x$, together with covariance (lower triangle) and correlation (upper triangle) matrix

Q.	Mean										
1	−0.29	0.66	−0.07	−0.13	0.03	−0.14	0.06	−0.11	−0.16	−0.24	−0.21
2	−0.42	−0.05	0.82	0.35	0.20	−0.19	−0.36	−0.32	−0.25	−0.26	−0.27
3	−0.18	−0.10	0.31	0.93	0.14	−0.21	−0.33	−0.33	−0.24	−0.24	−0.22
4	−0.60	0.02	0.14	0.11	0.62	−0.23	−0.17	−0.24	−0.20	−0.26	−0.28
5	−0.28	−0.09	−0.15	−0.18	−0.16	0.76	0.04	−0.07	−0.01	−0.10	−0.11
6	−0.32	0.04	−0.28	−0.27	−0.12	0.03	0.74	0.03	0.03	−0.12	−0.14
7	0.33	−0.08	−0.23	−0.26	−0.16	−0.05	0.02	0.67	0.01	0.06	0.05
8	0.46	−0.09	−0.16	−0.17	−0.12	−0.01	0.02	0.01	0.56	0.01	−0.04
9	0.68	−0.14	−0.17	−0.18	−0.16	−0.07	−0.08	0.04	0.00	0.58	0.31
10	0.61	−0.14	−0.20	−0.18	−0.18	−0.08	−0.10	0.03	−0.02	0.19	0.67

different view of the correlation structure behind the data. In general, the correlations adjusted for scale usage are smaller than those computed on the raw data. This accords with the intuition developed in the discussion of Figure CS3.2. In particular, the correlations between question 2 (price) and questions 9 and 10 (effectiveness in reach) are estimated to be one-half the size of the raw data after adjusting for scale usage heterogeneity. Intuitively, it seems reasonable that those who are very satisfied with (low) price should not also think that the advertising reach is also high. Table CS3.3 provides the correlation matrix of the centered data. Centering the data changes the uniformly positive correlations to mostly negative correlations. It does not seem intuitively reasonable that questions probing very similar aspects of the product should have zero to small negative correlations.

DISCUSSION

There are two challenges facing an analyst of ratings scale data: dealing with the discrete/ordinal nature of the ratings scale; and overcoming differences in scale usage across respondents. In the psychometrics literature, various optimal scaling methods have been proposed for transforming discrete/ordinal data into data that can be treated in a more continuous fashion. However, it is difficult to adapt these methods to allow for respondent-specific scale usage. We adopt a model-based approach in which an underlying continuous preference variable is subjected to respondent-specific location and scale shifts. This model can easily produce data of the sort encountered in practice where some respondents are observed to only use certain portions of the scale (in the extreme, only one scale value). Standard centering methods are shown to be inferior to our proposed procedure in terms of the estimation of relationships between variables.

Our analysis demonstrates that scale usage heterogeneity can impart substantial biases to estimation of the covariance structure of the data. In particular, scale usage heterogeneity causes upward bias in correlations and can be at the source of the collinearity problems observed by many in the analysis of survey data. The covariance structure is the key input to many different forms of analysis, including identification of segments via clustering and the identification of relationships through covariance-based structural modeling. Our procedures provide unbiased estimates of the covariance structure which can then be used as input to subsequent analysis.

While it is well documented that scale usage heterogeneity is prevalent in ratings scale data, not much is known about the determinants of this behavior. Important questions for future research include what sorts of respondents tend to exhibit a high degree of scale usage heterogeneity and how questionnaires and items can be designed to minimize scale usage heterogeneity. Our view is that it is desirable to reduce the magnitude of this phenomenon so as to increase the information content of the data. Whether it is possible to design survey instruments for customer satisfaction that are largely free of this problem is open to question. Ultimately, the answer must come from empirical applications of our model that will detect the extent of this problem.

R IMPLEMENTATION

R The R implementation of this method is embodied in the function, `rscaleUsage`, which is included in *bayesm*. We made a few changes to the basic algorithm in Rossi *et al.* (2001). First, we did not use an independence Metropolis step for drawing σ_i, instead we used a relatively fine grid over a relatively large interval. Second, we used a uniform prior for e on a grid on the interval $(-0.1, 0.1)$ and we used a random walk on this grid (moving to left and right grid points with equal probability except at the boundary where we propose the next innermost point with probability one). We use 100 replications to compute the GHK estimates of normal probabilities. We use 500 as the default number of grid points for the griddy Gibbs parameter draws (σ_i, e, Λ). We use the default prior settings given in (CS3.9), (CS3.10), and (CS3.12).

Case Study 4
A Choice Model with Conjunctive Screening Rules

This case study considers the situation where a consumer chooses from among a finite set of alternatives. When there are a large number of alternatives, deciding which option to choose becomes a formidable task. Consumers have been shown to use various heuristics to simplify their decisions. The choice model discussed here is consistent with two-stage decision processes wherein a subset of alternatives is screened from the universal set, and the final choice is from the reduced set. The presence of a first-stage screening rule leads to discontinuities in the likelihood function. This is common in many models of consumer behavior and leads to computational difficulties for non-Bayesian procedures based on maximization. Gilbride and Allenby (2004) show that a latent variable approach using Bayesian data augmentation leads to a simple and elegant solution to the problem.

BACKGROUND

Assume the consumer chooses from among J choice alternatives ($j = 1, \ldots, J$) characterized by a vector of attributes x_j with M elements, $\{x_{j1}, \ldots, x_{jm}, \ldots, x_{jM}\}$. The first stage of the choice model identifies the alternatives for further consideration using an indicator function, $I(x_{jm} > \gamma_m)$, that equals one when the value of attribute m for alternative j is greater than the cutoff γ_m. An alternative is an element of the set of screened alternatives, referred to as the choice set, if the indicator function is equal to one for all M attributes, that is, alternative j is in the choice set if

$$I(x_j, \gamma) = \prod_m I(x_{jm} > \gamma_m). \tag{CS4.1}$$

For example, if price is used to screen the alternatives, so that only offerings above a lower threshold are considered, then $I(x_j, \gamma) = I(\text{price}_j > \gamma_{\text{price}})$. If an upper threshold for price is present, then negative price can be used as an argument, $I(x_j, \gamma) = I(-\text{price}_j > \gamma_{\text{price}})$. Here, γ is a parameter to be estimated. This formulation allows for dynamic choice sets when an attribute (such as price) changes over the choice history.

The second stage of the model uses a standard latent variable approach:

$$z = X\beta + \varepsilon, \qquad \varepsilon \sim N(0, 1). \tag{CS4.2}$$

A standard choice model assumes that consumers select the alternative with the maximum latent variable z. In this model, consumers select the alternative with the maximum z for those alternatives in the choice set, that is, those that have passed the first-stage screening process.

Equation (CS4.1) is known as a conjunctive, or an 'and', decision rule. Gilbride and Allenby (2004) also consider the disjunctive ('or') and compensatory decision rule. Since (CS4.1) and (CS4.2) combine to determine the probability that an alternative is chosen, the presence of screening rules leads to discontinuities in the likelihood surface. Small changes in γ lead to abrupt changes in choice probabilities. For example, if we increase γ_{price}, then we may observe choice probabilities jump to zero for small changes in this parameter as items fall out of the consideration set.

MODEL

If an alternative is in the choice set, then its choice probability is determined relative to the other offerings in the set. If an alternative does not satisfy the screening rule, then its choice probability is zero. The choice model can be written hierarchically as sequence of conditional distributions:

$$\begin{aligned} &y|z, \gamma, x, \\ &z|\beta, x, \\ &\gamma, \\ &\beta, \end{aligned} \tag{CS4.3}$$

where y indicates the alternative chosen, $z_j = x_j'\beta + \varepsilon_j$.

Estimation proceeds by drawing iteratively from the conditional distributions

$$z_j|\beta, \gamma, y, x, z_{-j}, \tag{CS4.4}$$

$$\gamma_m|y, z, x, \gamma_{-m}, \tag{CS4.5}$$

$$\beta|z, x \tag{CS4.6}$$

where $-j$ means 'all elements of the vector except j'.

The conditional distribution in (CS4.4) is a truncated normal distribution for those alternatives (j) in the choice set. That is, the element of z for the chosen alternative is greater than the elements of z that correspond to the alternatives not chosen, given that they are in the choice set. If an alternative is not in the choice set, which is determined by the indicator function $I(x, \gamma)$, then z_j is drawn from a non-truncated distribution.

The conditional distribution for the cutoff parameter, γ_m, in (CS4.5) is dependent on the observed data, y, and the augmented parameter, z. The conditioning arguments simply restrict the support of the distribution of γ_m, that is, the posterior is proportional to the prior on the restricted support. Permissible values of the cutoff parameters are those that lead to a choice set where the maximum of the augmented variable, z, for those in the choice set, corresponds to the observed choice. If an alternative was chosen it must not have been screened out, and an alternative with a latent z larger than the chosen alternative must have been screened out. This can happen because some values of the augmented parameter in (CS4.4) are drawn from a truncated distribution while others are not. Variation in z results in different choice sets by defining the permissible values of the cutoffs.

The conditional distribution in (CS4.6) corresponds to a standard Bayesian analysis of the regression model, as discussed in Chapter 2.

The Markov chain defined by (CS4.4)–(CS4.6) generates draws from the full conditional distribution of model parameters despite the irregularity of the likelihood surface. Given the cutoffs, γ, the model (CS4.1) becomes a standard discrete choice model. Given the set of augmented variables, $\{z\}$, some values of the cutoff are acceptable and others are not. Across multiple draws of the augmented variables, however, the acceptable range of γ varies, and the Markov chain converges in distribution to the full conditional distribution of all model parameters. The augmented variable z reflects a single error term that gives rise to the joint probability of the consideration set and the resulting choice. Gilbride and Allenby (2004) provide the exact form of these conditional distributions and the estimation algorithm.

In addition to the form of the screening rule, researchers can specify that the cutoff parameters, γ, take on a limited range of values. If an attribute, x, is continuous (such as price), then γ can also be continuous. However, if the attribute takes on a restricted number of values, then the cutoff is not well identified over an unrestricted range. In a conjoint analysis, for example, there are rarely more than four or five levels of an attribute. In these cases one can ensure that the cutoff parameter takes on a small number of finite values by a prior assumption that γ is multinomially distributed over a grid of values, and estimating the posterior distribution of the multinomial mass probabilities.

DATA

The screening rule model is illustrated in the context of a discrete choice conjoint study of consumer preferences for the features of a new camera format, the Advanced Photo System (APS). This photographic system offers several potential advantages over standard 35 mm film systems, including mid-roll film change, camera operation feedback, and a magnetic layer on the film that offers the ability to annotate the picture with titles and dates.

Respondents were screened and recruited by telephone, and interviewed at field locations in nine metropolitan areas. A total of 302 respondents participated. The screen was standard in the industry and was designed to exclude individuals who had little or no interest in photography, and retained approximately 55 % of the population contacted. This compares to approximately 67 % of the population that owns still cameras. Individuals participating in the study were therefore assumed to be good prospects for cameras in the price range of the new APS offerings, and representative of the APS camera market.

Respondents were asked to evaluate 14 buying scenarios. The order of the buying scenarios was randomized. Each scenario contained full profile descriptions of three compact 35 mm cameras, three new APS cameras, and a seventh no-buy option. Table CS4.1 lists the attributes that describe each of the cameras, and identifies the attributes that are new and specific to APS cameras. The order of the attributes was the same in each scenario. Table CS4.1 also displays restrictions on the availability of certain attributes. For example, the 4× zoom lens was available only on the 'high' body style. Prices ranged from \$41 to \$499, the average price for APS cameras was \$247, and the average price of the compact 35 mm cameras was \$185.

Each buying scenario requires the respondent to engage in a complex decision. With six offerings described on eight attributes, plus the no-buy option, we consider this to be a good data set to investigate the use of screening rules. The no-buy option was selected 14.7 % of the time or an average of 2.1 times in the 14 choice tasks. One of the APS camera offerings was chosen 39.9 % of the time, and one of the 35 mm cameras was selected 45.4 % of the time. In the analysis that follows, 12 choice tasks per respondent were used to calibrate the model and the final two retained for predictive testing. This results in 3624 observations for model estimation and 604 for hold-out validation.

Five models are fitted to the data. In each model, we make the assumption that the no-buy option is always in the choice set. 'Not buying' is always an option both in the empirical conjoint study we use and in real markets.

The first model is a heterogeneous probit model that assumes all choice options are considered by the respondent. For identification, effects-level coding is used for the non-price attributes, with the lowest level of each attribute set to zero, except for body style. Price is entered in log units. For the no-buy option, all attributes are set to zero (including the natural logarithm of price). The effects-level coding results in 18 parameters, 17 corresponding to binary indicator variables and the 18th corresponding to ln(price). In order to fix the scale in the probit model, the error term is assumed to be independent and identically distributed across choice tasks with unit variance. The latent variable formulation of our model is given by:

$$\Pr(j)_{hi} = \Pr(z_{hij} > z_{hik}\ \forall k), \tag{CS4.7}$$

$$z_{hi} = X_{hi}\beta_h + \varepsilon_{hi}, \qquad \varepsilon_{hi} \sim N(0, I), \tag{CS4.8}$$

$$\beta_h \sim N(\bar{\beta}, \Sigma_\beta), \tag{CS4.9}$$

where h indexes the respondent, i is the index for the buying scenario, and j and k indicate the alternatives within the buying scenario.

The second model assumes that the choice set comprises alternatives with the deterministic value of utility ($V_{hi} = X_{hi}\beta_h$) greater than a threshold value, γ_h. That is, the selected alternative has greatest utility of the options for which $V_{hik} > \gamma_h$:

$$\Pr(j)_{hi} = \Pr(z_{hij} > z_{hik}\ \forall k \text{ such that } I(V_{hik} > \gamma_h) = 1). \tag{CS4.10}$$

This model is sometimes referred to as the 'compensatory' model.

Table CS4.1 Camera attributes and levels

Attribute/Level	Available in 'Low' Body Style	Available in 'Medium' Body Style	Available in 'High' Body Style
A Basic body style and standard features			
A1 Low	×		
A2 Medium		×	
A3 High			×
B Mid-roll change[a]			
B1 None	×	×	×
B2 Manual	×		
B3 Automatic	×	×	×
C Annotation[a]			
C1 None	×	×	×
C2 Pre-set list	×	×	×
C3 Customized list	×	×	×
C4 Custom input method 1			×
C5 Custom input method 2		×	×
C6 Custom input method 3			×
D Camera operation feedback[a]			
D1 No	×	×	×
D2 Yes			×
E Zoom			
E1 None	×	×	×
E2 2×		×	×
E3 4×			×
F Viewfinder			
F1 Regular	×	×	×
F2 Large			×
G Camera settings feedback			
G1 None	×		
G2 LCD	×	×	×
G3 Viewfinder		×	×
G4 Both LCD and viewfinder		×	×

[a]Unique to APS camera at time of the study; all other features available on both APS and 35 mm cameras at the time of the study.
× Indicates attributes available on corresponding body style.

The third model assumes that the choice set is formed using a conjunctive screening rule over the M attributes for each alternative:

$$\Pr(j)_{hi} = \Pr(z_{hij} > z_{hik}\ \forall k \text{ such that } \prod_{m} I(x_{hikm} > \gamma_{hm}) = 1). \qquad \text{(CS4.11)}$$

x_{hikm} is the level of the attribute for respondent h in buying scenario i for alternative k and attribute m. γ_{hm} is a respondent-level parameter representing the threshold, or acceptable level, of attribute m for respondent h. Two alternative heterogeneity distributions are employed for the threshold parameters. When attribute m is continuously distributed (for example, price), then we employ a normal distribution to allow for heterogeneity:

$$\gamma_{hm} \sim N(\overline{\gamma}_m, \sigma_m^2). \tag{CS4.12}$$

When an attribute takes on a small number of discrete levels we assume that the threshold parameter is multinomially distributed. For example, consider the zoom lens attribute in Table CS4.1 taking on values of none, 2× and 4×. We recode the attribute levels (say, 0,1,2) and specify a grid of possible cutoffs (say, −0.5, 0.5, 1.5, 2.5) where the lowest cutoff value indicates that all attribute levels are acceptable, and the highest level indicates that none of the levels are acceptable. For nominally scaled attribute levels, where a prior ordering is not obvious (such as color), the grid has three points (−0.5, 0.5, 1.5). The full set of attributes, levels, and cutoffs is contained in Table CS4.5. The distribution of heterogeneity of the cutoff parameters with discrete levels is:

$$\gamma_{hm} \sim \text{Multinomial}(\theta_m), \tag{CS4.13}$$

where θ_m is the vector of multinomial probabilities associated with the grid for attribute m.

The fourth model assumes that the choice set is formed using a disjunctive decision rule:

$$\Pr(j)_{hi} = \Pr(z_{hij} > z_{hik}\ \forall k \text{ such that } \sum_m I(x_{hikm} > \gamma_{hm}) \geq 1). \tag{CS4.14}$$

The upper and lower grid points specify the role of the attribute in determining the choice set, depending on the decision rule. An attribute determines membership in the choice set only if it changes the value of the alternative indicator function. For the conjunctive model, if the threshold for attribute k is equal to −0.5, then $I(x_{hijk} > \gamma_{hk})$ always equals one and the value of $\Pi_m I(x_{hijm} > \gamma_{hm})$ does not change. Since different levels of the attribute do not affect choice sets, we conclude that attribute k is not being used to screen alternatives. For the disjunctive model, if $\gamma_{hk} = 2.5$, then $I(x_{hijk} > \gamma_{hk})$ always equals zero and the value of $\Sigma_m I(x_{hijm} > \gamma_{hm})$ does not change across different levels of attribute k. For the disjunctive model, the upper value of the threshold implies no screening based on that attribute. Figure CS4.1 illustrates the procedure and the interpretation.

The fifth model assumes that consumers are either using a conjunctive (CS4.11) or a disjunctive (CS4.14) screening rule. That is, we allow for the presence of structural heterogeneity (Yang and Allenby, 2000):

$$\begin{aligned}\Pr(j)_{hi} = {} & \phi \Pr(z_{hij} > z_{hik}\ \forall k \text{ such that } \prod_m I(x_{hikm} > \gamma_{hm}) = 1) \\ & + (1-\phi)\Pr(z_{hij} > z_{hik}\ \forall k \text{ such that } \sum_m I(x_{hikm} > \gamma_{hm}) \geq 1),\end{aligned} \tag{CS4.15}$$

where ϕ is the portion of the sample using the conjunctive decision rule and $1 - \phi$ is the portion using the disjunctive decision rule. A concise listing of the models is provided in Table CS4.2.

Attribute	Levels	Recoded Values (x)
Zoom Lens	None	0
	2×	1
	4×	2

Possible Cut-offs	Probability of each cut-off
$\gamma_{(1)} = -0.5$	$\theta_{(1)}$
$\gamma_{(2)} = 0.5$	$\theta_{(2)}$
$\gamma_{(3)} = 1.5$	$\theta_{(3)}$
$\gamma_{(4)} = 2.5$	$\theta_{(4)}$

Scenario #1:

Screening rule: $I(x > \gamma) = 1$

If $\gamma = -0.5$: $I(0 > -0.5) = 1$, $I(1 > -0.5) = 1$, $I(2 > -0.5) = 1$ — *All attribute levels are acceptable*

For the conjunctive model, if $\gamma = -0.5$ then zoom lens is not being used to screen alternatives.

Scenario #2:

Screening rule: $I(x > \gamma) = 1$

If $\gamma = 0.5$: $I(0 > 0.5) = 0$, $I(1 > 0.5) = 1$, $I(2 > 0.5) = 1$ — *Camera must have 2× zoom lens to be considered*

Scenario #3:

Screening rule: $I(x > \gamma) = 1$

If $\gamma = 1.5$: $I(0 > 1.5) = 0$, $I(1 > 1.5) = 0$, $I(2 > 1.5) = 1$ — *Camera must have 4× zoom lens to be considered*

Scenario #4:

Screening rule: $I(x > \gamma) = 1$

If $\gamma = 2.5$: $I(0 > 2.5) = 0$, $I(1 > 2.5) = 0$, $I(2 > 2.5) = 0$ — *None of the attribute levels are acceptable*

For the disjunctive model, if $\gamma = 2.5$ then zoom lens is not being used to screen alternatives.

Figure CS4.1 Example for specifying cut-offs for conjunctive and disjunctive Model discrete attributes

Table CS4.2 Model comparisons

Model	In-Sample	Predictive	
	Log marginal density	Hit probability	Hit frequency
HB probit	−3468.8	0.391	266
Identified Compensatory	−3449.6	0.393	267
Conjunctive	***−2990.8***	***0.418***	***276***
Disjunctive	−3586.2	0.39	267
Structural Heterogeneity	−3012.9	0.417	277

RESULTS

The results indicate that the conjunctive model fits the data best. In-sample and predictive fits of the five models are presented in Table CS4.2. The log marginal density is computed using the GHK estimator (see Chapter 2) and the importance sampling method of Newton and Raftery (see Chapter 6). Hit probability is defined as the average probability associated with the observed choices. Hit frequency is the number of times the predicted choice matches the actual choice. The parameters reported below for the conjunctive model result in 30 % of the alternatives screened out of the final choice set, which averaged five (out of seven) alternatives including the no-buy option.

Fit results for the other models are as follows. The fit for the disjunctive model is slightly worse than the standard probit model, with parameters that lead to only 2 % of the alternatives screened from final consideration. The slight decrease in the log marginal density for the disjunctive model is from the increase in the dimension of the model due to the cutoff parameters. Although the compensatory model screened out 23 % of the alternatives to form the choice set and has a slightly better in-sample fit than the probit model, its predictive accuracy is nearly identical. The structural heterogeneity model assigned nearly all the respondents (99 %) to the conjunctive screening rule model as opposed to the disjunctive model, so the fit statistics closely match those of the conjunctive model.

The results provide evidence that the conjunctive model is superior to both the standard probit and non-conjunctive models. The conjunctive model has the largest value of the log marginal density. Despite the addition of 11 cutoff parameters per respondent, the conjunctive model performs as well as or better than more parsimonious models in the hold-out sample. This establishes the predictive validity of the choice model with a conjunctive screening rule.

Tables CS4.3 and CS4.4 report part-worth estimates for the probit and conjunctive models. Table CS4.3 displays the mean and covariance matrix of the part-worths for the probit model, and Table CS4.4 reports the same statistics for the conjunctive model. Substantive differences exist in the part-worth estimates for body style, zoom, camera operation feedback, and price. The conjunctive model estimates are much smaller for these attributes, with the average price coefficient moving to zero. The reason for this difference is discussed below.

The cutoff parameters are reported in Table CS4.5. For discrete attributes, the cutoffs are reported in terms of multinomial point mass probabilities associated with the grid described in Figure CS4.1. The probabilities indicate the fraction of respondents with a cutoff parameter (γ) that screens out choice alternatives. For the conjunctive model, $\theta_1 = 0.598$ indicates that 59.8 % of the respondents are not screening alternatives based on 'body style', $\theta_2 = 0.324$ implies that 32.4 % of the respondents screen out offerings with the low body style, $\theta_3 = 0.059$ implies that 5.9 % of the respondents screen out offerings with low and medium body style, and $\theta_4 = 0.015$ implies that 1.5 % of the respondents screen on all alternative levels. Estimates of θ equal to approximately 2 % reflect the influence of the prior distribution and can be disregarded for practical purposes.

The use of screening rules to form choice sets is pervasive among the respondents in this research. Analysis of the individual-level cutoff values indicates that 92 % of respondents form choice sets using the conjunctive screening rule; 58 % are forming choice sets based on a single attribute, 33 % are using two attributes, and 2 % are using three attributes. A strength of the Bayesian approach outlined in this chapter is the ability to make individual-level inferences and estimate not only what attributes are important in forming choice sets but also who is using which screening rule.

Consider, for example, an aggregate analysis of the effect of the price threshold, estimated to be equal to -6.269. This corresponds to an average price threshold of \$527. Since the maximum price of an APS camera is \$499, this implies that, on average, respondents are not screening on price. However, such analysis masks the tail behavior of the distribution of cutoff values. As displayed in Figure CS4.2, 10 % of the respondents will only consider cameras under $276, 20 % will only consider cameras under $356, etc. In all, 41 % of the respondents are screening alternatives based on price.

Figure CS4.3 displays the proportion of respondents screening on each of the attributes. Body style, zoom, camera settings feedback and price are the attributes most often used to screen alternatives. These attributes are available on both 35 mm and APS cameras, and as a result respondents are more fully aware of their benefits. In contrast, none of the new APS attributes (for example, mid-roll film change) are being used to screen alternatives. In addition, the attributes used to form choice sets have substantively different part-worths in Tables CS4.3 and CS4.4. Our results indicate that once the choice set is formed, the price and body style do not play a role in the final decision. The zoom and camera feedback features, however, continue to play a role in the final selection of a camera beyond just screening.

The participants in the survey were recruited and qualified by industry standards so that they were a prospect for cameras in the price range of the APS offering. Despite this prequalification, some 41 % of respondents used price to screen alternatives and excluded those cameras above a particular threshold. Exploratory analysis revealed no relationship between the posterior estimate of the price threshold value and demographic variables, including income. However, replies to the question 'Typically, when you attend informal social gatherings away from your home, how often do you take a camera along with you?' were associated with the price threshold (p-value $= 0.0362$). Those respondents who indicated they 'almost always' take their camera (44 % of respondents) have an average price threshold $68 higher than respondents who take their camera less frequently.

Table CS4.3 Part-worth estimates for the probit model

Attribute	*Level*	*Beta*	*Posterior Mean*
Body style	Low body	1	0.325
	Medium body	2	**2.843**
	High body	3	**2.224**
Mid-roll change	Manual change	4	0.059
	Automatic change	5	**0.178**
Annotation	Pre-set list	6	**0.362**
	Customized list	7	**0.678**
	Input method 1	8	**−1.036**
	Input method 2	9	**0.767**
	Input method 3	10	**−0.985**
Camera operation feedback	Operation feedback	11	**0.357**
Zoom	2× Zoom	12	**1.122**
	4× Zoom	13	**1.477**
Viewfinder	Large viewfinder	14	**−0.158**
Settings feedback	LCD	15	**0.759**
	Viewfinder	16	**0.757**
	LCD and viewfinder	17	**0.902**
Price	ln(price)	18	**−0.71**

Table CS4.3 *(continued)*

Covariance matrix Σ_β

	1	2	3	4	5	6	7	8	9	10	11	12	13	14	15	16	17	18
1	**19.605**																	
2	**17.952**	**20.258**																
3	**15.303**	**17.674**	**17.845**															
4	−0.548	−0.906	−1.218	**1.803**														
5	**−2.077**	**−2.232**	**−2.153**	**0.507**	**1.4**													
6	−0.575	−0.309	−0.333	0.097	0.171	**0.629**												
7	−0.584	−0.366	−0.356	0.147	**0.231**	**0.255**	**0.685**											
8	**−8.252**	**−7.975**	**−6.735**	0.371	**1.134**	0.3	0.387	**5.392**										
9	**−1.921**	**−1.52**	**−1.285**	0.139	**0.395**	**0.237**	**0.26**	**0.974**	**1.096**									
10	**−5.553**	**−5.628**	**−4.753**	0.432	**0.854**	0.238	0.224	**2.734**	**0.726**	**3.315**								
11	−0.808	−0.67	−0.665	0.061	0.173	0.019	0.033	0.494	**0.225**	0.347	**1.055**							
12	−0.207	0.651	0.895	−0.009	0.022	0.106	0.121	0.134	0.167	−0.056	0.115	**1.42**						
13	−0.811	0.54	1.08	−0.225	−0.08	0.225	0.173	0.418	0.251	−0.021	−0.008	**1.07**	**2.21**					
14	**−1.394**	**−1.054**	−0.859	−0.011	0.149	0.051	−0.021	**0.592**	**0.231**	**0.487**	0.064	0.122	0.13	**0.947**				
15	−0.434	0.06	0.324	−0.075	0.04	0.106	0.12	0.167	0.181	0.04	0.092	**0.343**	**0.454**	0.141	**0.966**			
16	−0.289	0.035	0.161	0.011	0.044	0.083	0.067	0.129	0.179	0.083	0.123	**0.298**	0.336	0.119	**0.525**	**0.976**		
17	−0.472	−0.042	0.12	0.019	0.057	0.109	0.118	0.255	0.173	0.084	0.106	**0.321**	**0.427**	0.074	**0.526**	**0.538**	**0.937**	
18	**−2.789**	**−3.459**	**−3.278**	0.144	**0.287**	−0.045	−0.054	**1.22**	0.12	**0.923**	0.046	**−0.412**	**−0.52**	0.142	−0.216	−0.206	−0.192	**1.079**

Estimates in bold have more than 95 % of the posterior mass away from 0.

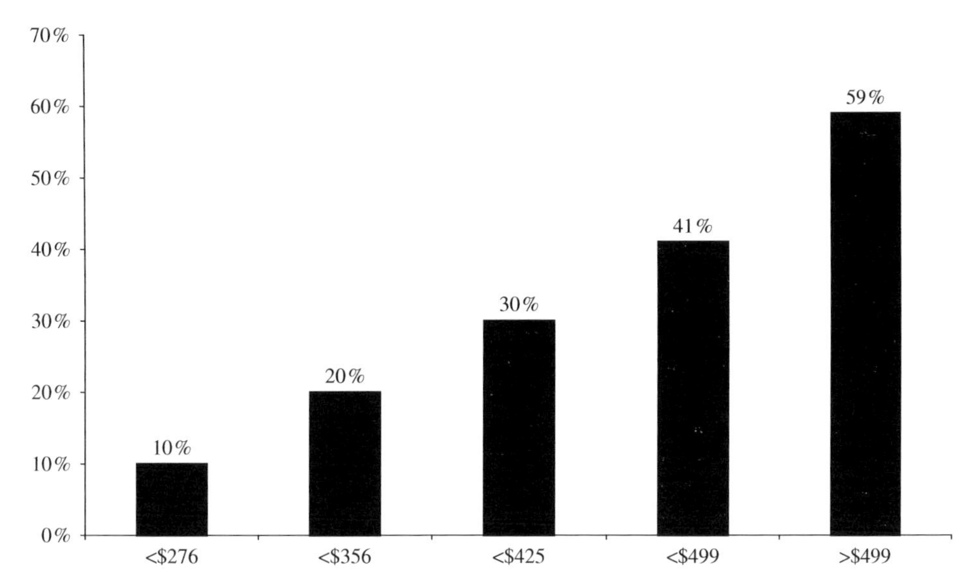

Figure CS4.2 Distribution of price threshold, conjunctive model

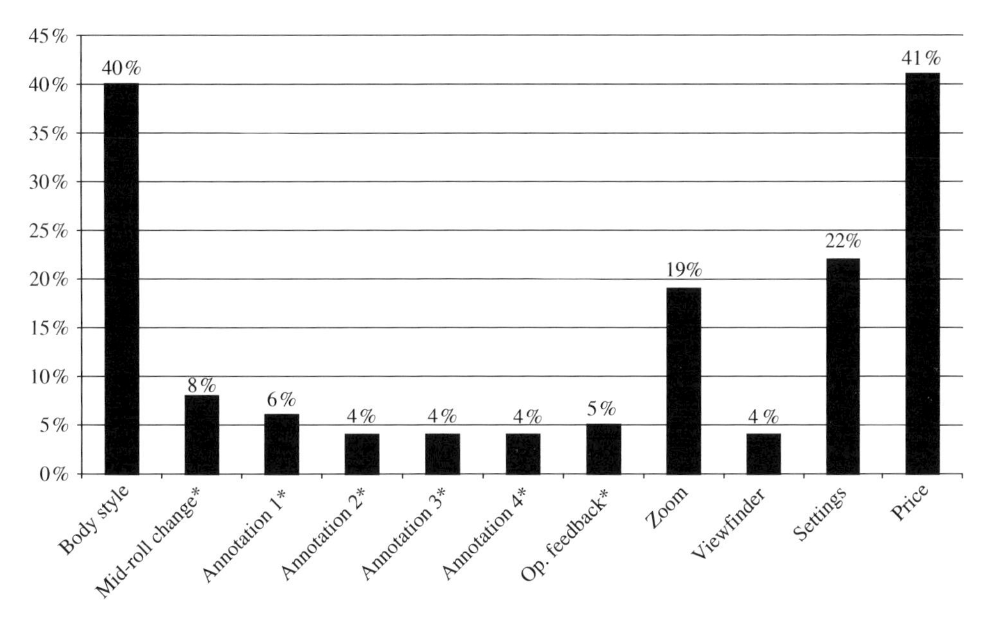

Figure CS4.3 Proportion of respondents screening on each attribute, conjunctive model (* indicates attributes only available on APS cameras at the time of this study)

Although exploratory in nature, this analysis shows that other variables correlate with the price threshold and may be used in conjunction with the model to better identify prospects in the marketplace.

The empirical results support the conjunctive screening rule model. Better in-sample and out-of-sample fit despite the large increase in the number of parameters argues for the use of the screening rule model over the more parsimonious probit. Consumers tend to screen alternatives using attributes they are already familiar with and an exploratory analysis was conducted to demonstrate how the price threshold can be associated with other variables in a posterior analysis. After controlling for their effect in forming choice sets, the importance of the screening attributes in the final choice among acceptable alternatives is smaller than that indicated by a standard probit model.

DISCUSSION

The use of screening rules in this study is pervasive, with 92 % of respondents using this heuristic to manage the complexity of the choice problem. The results also show that consumers screen alternatives using attributes that are well known as opposed to the new and novel attributes also included in this study. Some attributes appear to be only used in forming choice sets, while others are used in both screening and final product choice. The proposed model can apply to a variety of screening rules and handle problems of realistic size. The ability to capture a non-compensatory screening process and make attribute-level inferences for individuals is due to the advantages of modern Bayesian methods.

A key element in our model is accurately navigating the likelihood of the data where only the final choice is observed. Past researchers have attempted to average over consideration sets to create a smooth likelihood surface. This approach requires a complex partition of the error space corresponding to the $2^J - 1$ consideration sets and the choice probability within each set. Alternatively, the consideration set probability

Table CS4.4 Part-worth estimates for the conjunctive model

Attribute	*Level*	*Beta*	*Posterior Mean*
Body style	Low body	1	−0.34
	Medium body	2	0.577
	High body	3	0.024
Mid-roll change	Manual change	4	0.257
	Automatic change	5	**0.169**
Annotation	Pre-set list	6	**0.289**
	Customized list	7	**0.604**
	Input method 1	8	−0.369
	Input method 2	9	**0.655**
	Input method 3	10	**−0.619**
Camera operation feedback	Operation feedback	11	**0.413**
Zoom	2× Zoom	12	**0.841**
	4× Zoom	13	**1.231**
Viewfinder	Large viewfinder	14	−0.139
Settings Feedback	LCD	15	**0.374**
	Viewfinder	16	**0.373**
	LCD and viewfinder	17	**0.52**
Price	ln(price)	18	0.056

Table CS4.4 (*continued*)

Covariance matrix Σ_β

	1	2	3	4	5	6	7	8	9	10	11	12	13	14	15	16	17	18
1	**16.183**																	
2	**15.035**	**15.918**																
3	**14.839**	**15.176**	**16.419**															
4	−1.35	−1.218	−1.398	**1.88**														
5	**−1.903**	−1.821	**−1.854**	**0.625**	**1.523**													
6	−0.637	−0.527	−0.591	0.117	0.148	**0.632**												
7	−0.667	−0.531	−0.604	0.198	0.194	**0.241**	**0.742**											
8	**−6.744**	**−6.458**	**−6.301**	0.653	**0.956**	0.349	0.399	**4.828**										
9	**−2.246**	**−2.069**	**−2.022**	0.319	**0.481**	**0.229**	**0.26**	**1.191**	**1.269**									
10	**−5.42**	**−5.306**	**−5.148**	0.662	**0.876**	0.318	0.305	**2.752**	**1.073**	**3.892**								
11	−0.641	−0.479	−0.589	0.1	0.169	0.007	0.009	0.451	0.205	0.287	**1.122**							
12	0.443	0.499	0.491	0.035	−0.051	−0.005	0.006	−0.119	−0.048	−0.161	0.126	**1.056**						
13	0.106	0.427	0.387	−0.101	−0.179	0.112	0.08	0.084	0.014	−0.234	−0.015	0.557	**1.653**					
14	**−1.292**	−1.256	−1.289	0.057	0.154	0.065	−0.047	0.616	0.281	**0.695**	0.022	−0.049	−0.112	**1.213**				
15	−0.648	−0.611	−0.494	0.005	0.07	0.045	0.079	0.28	0.125	0.186	0.087	0.097	0.139	0.102	**0.864**			
16	−0.65	−0.745	−0.736	0.125	0.104	0.043	0.059	0.341	0.143	0.302	0.065	0.105	0.056	0.065	**0.365**	**0.903**		
17	−0.662	−0.717	−0.704	0.15	0.112	0.088	0.113	0.381	0.141	0.285	0.077	0.122	0.161	0.014	**0.38**	**0.422**	**0.878**	
18	**−2.338**	**−2.463**	**−2.464**	0.137	0.233	0.048	0.042	**0.957**	0.258	**0.813**	0.057	−0.184	−0.25	0.209	0.007	0.011	0.026	**0.774**

Estimates in bold have more than 95 % of the posterior mass away from 0

Table CS4.5 Threshold estimates for the conjunctive model posterior means

Attribute	Levels	Recoded values	Possible cut-offs	Probability of each cut-off
Body style	Low	0	$\gamma_{1(1)} = -0.5$	$\theta_1 = 0.598$
	Medium	1	$\gamma_{1(2)} = 0.5$	$\theta_2 = 0.324$
	High	2	$\gamma_{1(3)} = 1.5$	$\theta_3 = 0.059$
			$\gamma_{1(4)} = 2.5$	$\theta_4 = 0.015$
Mid-roll change	None	0	$\gamma_{2(1)} = -0.5$	$\theta_5 = 0.926$
	Manual	1	$\gamma_{2(2)} = 0.5$	$\theta_6 = 0.027$
	Auto	2	$\gamma_{2(3)} = 1.5$	$\theta_7 = 0.028$
			$\gamma_{2(4)} = 2.5$	$\theta_8 = 0.020$
Annotation 1	None	0	$\gamma_{3(1)} = -0.5$	$\theta_9 = 0.940$
	Pre-set	1	$\gamma_{3(2)} = 0.5$	$\theta_{10} = 0.022$
	Custom	2	$\gamma_{3(3)} = 1.5$	$\theta_{11} = 0.018$
			$\gamma_{3(4)} = 2.5$	$\theta_{12} = 0.020$
Annotation 2	None	0	$\gamma_{4(1)} = -0.5$	$\theta_{13} = 0.960$
	Input #2	1	$\gamma_{4(2)} = 0.5$	$\theta_{14} = 0.020$
			$\gamma_{4(3)} = 1.5$	$\theta_{15} = 0.020$
Annotation 3	None	0	$\gamma_{5(1)} = -0.5$	$\theta_{16} = 0.961$
	Input #3	1	$\gamma_{5(2)} = 0.5$	$\theta_{17} = 0.019$
			$\gamma_{5(3)} = 1.5$	$\theta_{18} = 0.020$
Annotation 4	None	0	$\gamma_{6(1)} = -0.5$	$\theta_{19} = 0.961$
	Input #4	1	$\gamma_{6(2)} = 0.5$	$\theta_{20} = 0.019$
			$\gamma_{6(3)} = 1.5$	$\theta_{21} = 0.020$
Operation feedback	None	0	$\gamma_{7(1)} = -0.5$	$\theta_{22} = 0.947$
	Feedback	1	$\gamma_{7(2)} = 0.5$	$\theta_{23} = 0.034$
			$\gamma_{7(3)} = 1.5$	$\theta_{24} = 0.020$
Zoom lens	None	0	$\gamma_{8(1)} = -0.5$	$\theta_{25} = 0.808$
	2×	1	$\gamma_{8(2)} = 0.5$	$\theta_{26} = 0.140$
	4×	2	$\gamma_{8(3)} = 1.5$	$\theta_{27} = 0.032$
			$\gamma_{8(4)} = 2.5$	$\theta_{28} = 0.019$
Viewfinder	Regular	0	$\gamma_{9(1)} = -0.5$	$\theta_{29} = 0.961$
	Large	1	$\gamma_{9(2)} = 0.5$	$\theta_{30} = 0.019$
			$\gamma_{9(3)} = 1.5$	$\theta_{31} = 0.019$
Settings feedback	None	0	$\gamma_{10(1)} = -0.5$	$\theta_{32} = 0.783$
	LCD	1	$\gamma_{10(2)} = 0.5$	$\theta_{33} = 0.157$
	Viewfinder	2	$\gamma_{10(3)} = 1.5$	$\theta_{34} = 0.021$
	Both	3	$\gamma_{10(4)} = 2.5$	$\theta_{35} = 0.020$
			$\gamma_{10(4)} = 3.5$	$\theta_{36} = 0.019$
Continuous Attribute:	-ln(price)		$\gamma_{11} = -6.269$	

All estimates have more than 95% of the posterior mass away from 0.

could be estimated by treating each of the $2^J - 1$ partitions as a model parameter. By parameterizing the screening rules, using data augmentation to accurately partition the error space, and using the Markov chain to navigate the posterior distribution of all parameters, we can estimate a model that has a discontinuous likelihood function.

Managers frequently define markets based on particular attributes (such as body style) or price points, recognizing that certain consumers will only consider offerings with specific attribute levels, regardless of other elements of the product offering. Past methods of identifying these attributes and levels have tended to rely on compensatory models that may not be a good representation of the decision process. While compensatory models can often predict as well as nonlinear models, inferences for strategic decisions are often biased. The proposed method model overcomes the statistical and practical problems of estimating discontinuous choice models with attribute-level screening rules.

R IMPLEMENTATION

R The book website contains both data and R code for this case study. The R functions are defined in the file `Screening Rules.R` and the data is read in and processed using the command `run Screening Rules.R`. These functions implement the conjunctive screening rule in (CS4.1). Users will find that the algorithm converges to a stationary distribution quickly, within the first 100 or so iterations. The code is designed specifically for the APS conjoint application. In particular, it makes the following assumptions:

- A no-choice option is present and coded as the last (seventh) choice alternative.
- There are $M - 1 = 10$ discrete attributes and one continuously scaled attribute, with the continuous attribute coded as the mth attribute.
- No parameters are associated with the no-choice option. Instead, all levels of body style (the first attribute) are estimated. For the remaining discrete attributes, $M - 1$ attribute levels are identified by the data, and dummy variable coding is used.

Users should be aware of these conditions when modifying the code for their use.

Case Study 5 Modeling Consumer Demand for Variety

Consumers are often observed to purchase more than one variety of a product on a given shopping trip. The simultaneous demand for varieties is observed not only for packaged goods such as yogurt or soft drinks but also in many other product categories such as movies, music CDs, and apparel. Linear utility models used to justify a random utility logit or probit choice specification cannot accommodate interior solutions with more than one variety (alternative) chosen, and standard demand models in the economics literature exhibit only interior solutions. The reason is that models with mixtures of interior and corner solutions are computationally demanding, particularly when heterogeneity is incorporated into the model specification. Modern Bayesian methods are effective at reducing this computational burden, facilitating the estimation of models with nonstandard utility specifications.

In this case study, we examine the translated additive utility structure of Kim *et al.* (2002) that nests the linear utility structure, while allowing for the possibility of interior solutions where more than one variety is selected. The model can be used, for example, to undertake calculations to value assortments by asking how much prices must be lowered to compensate households (in utility terms) for the removal of flavors or to directly compute the monetary value stemming from the contribution of a given flavor to the assortment. This sort of computation gets at the fundamental question of the value of variety.

BACKGROUND

Standard choice models are based on a linear utility structure in which one and only one variety will be selected at each purchase occasion. This is clearly at variance with data on the purchase of multiple varieties of the same product. We develop a new

Bayesian Statistics and Marketing P. E. Rossi, G. M. Allenby and R. McCulloch

utility-based model and estimation procedure that can accommodate a mixture of corner and interior solutions.

In order to allow for simultaneous demand for multiple varieties, each variety can be thought of as an imperfect substitute for the other varieties. The proposed demand model is based on a translated, nonlinear, but additive utility structure. This model provides a parsimonious specification that allows both interior and corner solutions as well as diminishing marginal utility, while nesting the standard linear utility structure. The likelihood function for this model is derived from normal random errors in marginal utility. Evaluation of the likelihood involves high-dimensional integrals of normal distributions over rectangular regions and, for this reason, has not been used in either the economics or marketing literature. Coupled with a simulation approach to evaluate the likelihood, a Bayesian hierarchical model of household heterogeneity is used.

The model is applied to data on purchases of varieties of yogurt. In this data set, all purchases involve corner solutions and households frequently purchase more than one variety on the same shopping trip. Estimates from the hierarchical model reveal differences between varieties in base preference as well as different rates of diminishing marginal utility. Households are found to differ greatly in their preferences for varieties, with some households showing extreme preferences for particular flavors. Finally, we briefly discuss use of the model to investigate the value households place on particular varieties by computing compensating values at the household level – the monetary equivalent of the household's loss in utility.

MODEL

As introduced in Section 4.4.3, we define utility over the $i = 1, \ldots, m$ varieties as

$$\overline{U}(x) = \sum_i \psi_i (x_i + \gamma_i)^{\alpha_i}, \qquad \text{(CS5.1)}$$

where x is the vector of quantity demanded with elements x_i, and ψ_i, γ_i, and α_i are parameters of the utility function. The utility function in (CS5.1) is an additive but nonlinear utility function. Equation (CS5.1) defines a valid utility function under the restrictions that $\psi_i > 0$ and $0 < \alpha_i \leq 1$. An additive utility structure is used because the products we consider are not jointly consumed. For example, there is no interaction in utility from consuming two flavors together.

The utility in (CS5.1) is a family of translated utility functions where γ_i controls the translation and α_i influences the rate of diminishing marginal returns. If γ_i is zero, then there will be no corner solutions as the indifference curves are tangent to the axes. However, for positive γ, the indifference curves will have finite nonzero slope at the axes, creating the possibility of a corner solution. As pointed out in Chapter 4, the likelihood function for the variety model is based on the Kuhn–Tucker conditions which involve marginal utility. Marginal utility for the variety model is given by

$$\overline{U}_i = \psi_i \alpha_i (x_i + \gamma_i)^{\alpha_i - 1}, \qquad \text{(CS5.2)}$$

which suggests a more natural parameterization for estimation: $\beta_i = \ln(\alpha_i \psi_i)$ and $\delta_i = \alpha_i - 1$. In this parameterization, β_i is estimated unrestricted and δ_i is constrained to lie in the interval $(-1, 0)$. A uniform prior distribution over $(-1, 0)$ is assumed for the δ_i parameters.

Heterogeneity is introduced into the model by specifying a random effects distribution for the β parameters:

$$\beta_h = \ln(\alpha_h \psi_h) \sim N(\bar{\beta}, V_\beta), \quad \text{(CS5.3)}$$

where $h = 1, \ldots, H$ indexes the households. The multivariate normal distribution of heterogeneity for β_h is flexible as it does not restrict individual parameters to a specific support and allows elements to covary across the population. The model in (CS5.3) could easily be modified to include a set of covariates in the mean function, $\bar{\beta} = \Delta z_h$, as discussed in Chapter 5. Marketing mix variables (other than price) such as display and feature are also candidates for inclusion. Display and feature activity are the same for all varieties of the same brand, which implies these covariates will drop out of the resulting demand functions used in our empirical analysis.

A random effects distribution is not specified for the other model parameters because there does not exist sufficient information in the data. For example, allowing households to exhibit different curvature parameters (α_i) would require long purchase histories that we do not observe. However, there are no conceptual problems specifying a random effects distribution for all model parameters.

We specify the standard conditionally conjugate priors for $\bar{\beta}$, V_β:

$$\begin{aligned} \bar{\beta} | V_\beta &\sim N(\bar{\bar{\beta}}, V_\beta \otimes A^{-1}), \\ V_\beta &\sim \text{IW}(\nu, V). \end{aligned} \quad \text{(CS5.4)}$$

In this case, A is a scalar. Recognizing that the $\{\beta_h\}$ are conditionally independent, we can define a hybrid MCMC algorithm for this model which is suggested by the following conditional distributions:

$$\beta_h | \delta, X_h, P_h, \bar{\beta}, V_\beta, \quad \text{(CS5.5)}$$

$$\delta | \{\beta_h\}, X, P, \quad \text{(CS5.6)}$$

$$\bar{\beta}, V_\beta | \{\beta_h\}. \quad \text{(CS5.7)}$$

X is a matrix of the demand for all products, P is the matrix of observed prices. X_h, P_h are the submatrices corresponding to demand and price for household h. (CS5.5) can be achieved via H RW Metropolis draws, one for each household. Data for all households can be pooled and another RW Metropolis draw can be used for the common δ parameters. Given $\{\beta_h\}$ draws, (CS5.7) can be accomplished by a standard one-for-one draw, as illustrated for hierarchical models in Chapters 3 and 5. The likelihood function required to implement the Metropolis steps in (CS5.5) and (CS5.6) is developed in Chapter 4.

DATA

The model is estimated using scanner-panel data from the yogurt category. Five flavors of Dannon yogurt (blueberry, mixed berry, pina colada, plain, and strawberry) in the popular 8 oz size are examined. In these data, there is virtually no brand-switching between the Dannon and Yoplait brands of yogurt. Less than 2 % of the observations involve purchase of both brands. Therefore, attention is restricted to the dominant market share Dannon brand varieties. However, there is nothing conceptually different about applying our utility specification to multiple varieties from multiple brands. The data are from 332 households and include 2380 purchase occasions.

Table CS5.1 presents information on the frequency of corner and interior solutions. A corner solution is defined as a purchase occasion on which only one variety is purchased. Table CS5.1 shows the problem with application of standard choice models since more than 20 % of the purchase occasions involve the simultaneous purchase of more than one variety, but never all varieties. We can eliminate the possibility that household 'portfolio' effects are the only possible explanation for the simultaneous purchase of multiple varieties. The 69 single-person households in our data set have 464 purchase occasions with the same frequency of interior solutions (26 %).

Tables CS5.2 and CS5.3 provide information on the quantity of yogurt purchased. Table CS5.2 reports the distribution of purchase quantity. More than 50 % of the purchase occasions involve quantities of at least two units, indicating that the data

Table CS5.1 Frequency of corner and interior solutions in the yogurt data

	Purchase incidence	Corner solution	Interior solution
Strawberry	989	571	418
Blueberry	545	309	236
Pina colada	570	281	289
Plain	361	338	23
Mixed berry	570	325	245
Total	2380	1824 (76.6 %)	556 (23.4 %)

Table CS5.2 Distribution of purchase quantity

Purchase quantity	Frequency	%
1	1140	47.90
2	745	31.30
3	228	9.58
4	133	5.59
5	42	1.76
6	60	2.52
7	8	0.34
8	17	0.71
9	1	0.04
10	3	0.13
11	1	0.04
12	1	0.04
21	1	0.04

Table CS5.3 Frequency of corner and interior solutions for multiple unit purchases

	Observations	Corner Solution	Interior Solution
Strawberry	336	208 (0.62)	128 (0.38)
Blueberry	207	129 (0.62)	78 (0.38)
Pina colada	183	99 (0.54)	84 (0.46)
Plain	117	108 (0.92)	9 (0.08)
Mixed berry	198	140 (0.71)	58 (0.29)

is capable of providing information about flavor satiation. Table CS5.3 displays the incidence of corner and interior solutions for purchase occasions in which at least two units are demanded. This table shows that plain and mixed berry yogurt are more often purchased in isolation, while the other flavors are typically not. One possible explanation for the differing behavior of plain and mixed berry varieties is that they are strongly preferred by some households. Alternatively, consumers may not tire of these varieties as rapidly as the others. The model discussed above has both features and will allow us to distinguish between these effects.

RESULTS

Table CS5.4 presents parameter estimates. As expected, there is considerable evidence of nonlinear utility and different rates of satiation for different flavors. The flavor satiation parameters $\delta = 1 - \alpha$ range from a maximum of -0.02 for plain, indicating little curvature in the utility function, to a minimum of -0.57 for strawberry that is associated with greater curvature. Estimates of the mean of the random effects distribution for β are all negative, indicating that strawberry has the highest marginal utility when $x = 0$, and plain has the lowest marginal utility, on average, when $x = 0$. Estimates of the elements of the covariance matrix, V_β, indicate substantial heterogeneity in the population, with the most diverse preferences for plain.

DISCUSSION

Demand systems derived from valid utility functions can be used to make policy recommendations in pricing and assortment. Without a valid utility structure, reduced-form models must be used in which the demand for yogurt is linked in an ad hoc manner to the assortment of yogurts available. Even if agreement could be reached on the nature of the link between yogurt expenditure and variety as well as on a measure of variety, estimation of such a reduced form would require time series variation in variety offered, which is not present in our data. A utility-based model allows us to analyze counterfactual experiments such as what demand would be in the absence of a particular variety. Direct utility calculations can be used to determine the utility loss and compensating value of deleting a flavor from the assortment.

Table CS5.4 Parameter estimates (posterior standard deviations)

	Common parameters	
	$\bar{\beta}$	$\delta = 1 - \alpha$
Strawberry	0.00[a]	−0.57 (0.07)
Blueberry	−0.85 (0.13)	−0.40 (0.06)
Pina colada	−0.70 (0.13)	−0.46 (0.07)
Plain	−2.66 (0.33)	−0.02 (0.21)
Mixed berry	−0.67 (0.12)	−0.33 (0.06)

	Covariance/*Correlation* matrix (V_β)			
Blueberry	2.89 (0.44)	*0.21*	*0.18*	*0.31*
Pina colada	0.53 (0.29)	2.15 (0.33)	*0.30*	*0.46*
Plain	1.12 (0.74)	1.59 (0.62)	13.33 (2.55)	*0.28*
Mixed berry	0.70 (0.26)	0.90 (0.23)	1.36 (0.69)	1.76 (0.25)

[a]Fixed for identification.
Note: : $\beta_i = \ln(\alpha_i \psi_i)$.

Compensating Values

The value of an assortment can be determined by computing the compensating value of adding or removing an offering from the product line. As the assortment is altered, the utility attainable for a fixed level of expenditure changes. The compensating value is the amount that the budgetary allotment would have to increase or decrease to yield the same level of utility as that attained prior to any change in the assortment. As utility is measured on an arbitrary scale, it is converted to the monetary scale of compensating value for interpretability purposes.

The removal of a flavor from a product line will result in a decrease in the attainable utility of all consumers, and is affected by factors such as whether other offerings are considered good substitutes and the marginal utility of consumption. These factors involve more than one of the model parameters and depend on the current level of expenditure. The compensating value depends on the set of substitutes available.

Compensating value is computed by numerically evaluating the indirect utility function and computing the increase in expenditures required to attain the level of utility derived from the full assortment. Since prices and total grocery expenditure vary across purchase occasions, computations are undertaken observation by observation and then summed for each household.

Define for each observation the indirect utility function

$$V_{ht}(p_{ht}, E_{ht}) = \max_{x} E_{\beta_h,\delta|\text{data}}[U(x|\beta_h, \delta)] \quad \text{such that } p'_{ht}x = E_{ht}, \tag{CS5.8}$$

where U is defined over all five flavors. E_{ht} is the total yogurt expenditure for household h at purchase occasion t. Note that we integrate or average utility with respect to the posterior distribution of (β_h, δ).

It is also possible to include an outside good in addition to the yogurt varieties (as in Kim *et al.* 2002). Some would argue that it is important to include an outside good to allow for the substitution possibility of consuming more of the outside good when a flavor is removed. Failure to include the outside good may overstate the value of an assortment. However, Kim *et al.* (2002) had difficulty estimating the parameters in a model with the outside good. There is also some ambiguity as to the exact definition of the outside good. Kim *et al.* use the total expenditure on grocery products as a measure of the outside good. Given the large expenditures on this definition of the outside good relative to yogurt expenditures, the outside good will be viewed as exhibiting little satiation and the yogurt flavors will have high satiation parameters. This will tend to push the parameters to the boundaries of their admissible region. For this reason, we will confine attention here to the model with only inside goods.

To find the compensating value (CV) for deletion of flavor i, we use the indirect utility function defined in (CS5.8). CV is the solution to

$$V_{ht}(p_{ht}, E_{ht}) = V_{ht}^{(i)}(p_{ht}, E_{ht} + \mathrm{CV}_{ht}^{(i)}), \tag{CS5.9}$$

where $V_{ht}^{(i)}$ is defined by

$$V_{ht}^{(i)}(p_{ht}, E_{ht}) = \max_{x} E_{\beta_h,\delta|\text{data}}[U(x|\beta_h, \delta)]$$
$$\text{such that } p'_{ht}x = E_{ht} \text{ and } x_i = 0. \tag{CS5.10}$$

The indirect utility functions can be evaluated by numerical optimization methods (e.g., **R** R routine `constrOptim`). These utility calculations are made conditional on estimates of household-level parameters that are not available in non-Bayesian models.

$\mathrm{CV}_{ht}^{(i)}$ is defined as the amount by which expenditure must be increased to compensate household h on purchase occasion t so that utility will remain unaffected by the deletion of flavor i. If a flavor has unique value with poor substitutes, then the compensating value will be high. In addition, some households may have an extreme preference for a given flavor which may also cause the CV to be large. We can sum up the $\mathrm{CV}_{ht}^{(i)}$ over purchase occasions to the household level, $\mathrm{CV}_{h}^{(i)} = \sum_{t=1}^{T_h} \mathrm{CV}_{ht}^{(i)}$.

As discussed in Kim *et al.* (2002), compensating values for the deletion of particular flavors can be large because of the presence of heterogeneous preferences. Some households have strong preference for one flavor. These households require large compensating values when their favorite flavor is deleted. Calculations based on plugging in the average household preference parameter, $\bar{\beta}$, underestimate the compensating value of flavors.

Implications for Pricing Policy

As a policy matter, retailers face space constraints and do not offer a complete assortment of products. If a retailer were to delete a flavor from a store inventory, utility losses might drive customers away from the yogurt category and away from the store. The retailer would need to compensate customers for reduced variety by lowering prices either store-wide or in the yogurt category. This strategy has been adopted by warehouse format competitors such as Wal-Mart; less variety is offered with lower prices. The indirect utility function offers a means of computing the price reductions necessary to compensate for loss in variety.

To compute the price reductions necessary to compensate for loss in variety, we again use the indirect utility functions defined above. To calculate utility-compensating price reductions, we solve for the percentage reduction in prices required to restore aggregate utility to the level present prior to its deletion from the product line. In performing this computation, we integrate over the observed purchases and posterior distribution of model parameters, including the distribution of heterogeneity. We can perform two types of calculations: we will compute the amount by which we must reduce the price of the remaining yogurts to compensate for loss of a flavor; and we will calculate the amount by which all store prices (outside good as well as yogurt) must be reduced to compensate for loss of variety. Here we assume that overall grocery expenditure will not be affected by deletion of a yogurt flavor. We will only comment on the first calculation; the reader is referred to Kim *et al.* for discussion of the second.

We compute price reductions necessary to compensate for the loss of a flavor, using only yogurt prices. That is, the retailer lowers prices of yogurts alone to compensate for loss of variety without changing the prices of other goods in the store (as captured by the outside good). We solve the following problem to determine the level by which we must reduce yogurt prices, $r^*_{i,\text{yogurt}}$, to compensate for deletion of flavor i:

$$\text{find } r^*_{i,\text{yogurt}} \text{ such that } \overline{V}^{(i)}(r^*_{i,\text{yogurt}}) = \overline{V},$$

where

$$\begin{aligned} \overline{V} &= \sum_{h=1}^{H}\sum_{t=1}^{T_h} V_{ht}, \\ V_{ht} &= \max E_{\beta_h,\delta|\text{data}}[U(x|\beta_h,\delta)] \text{ such that } x'p_{ht} = E_{ht} \end{aligned} \qquad \text{(CS5.11)}$$

and

$$\begin{aligned} \overline{V}^{(i)}(r^*_{i,\text{yogurt}}) &= \sum_{h=1}^{H}\sum_{t=1}^{T_h} V^{(i)}_{ht}(r^*_{i,\text{yogurt}}), \\ V^{(i)}_{ht}(r^*_{i,\text{yogurt}}) &= \max E_{\beta_h,\delta|\text{data}}[U(x|\beta_h,\delta)] \\ &\quad \text{such that } (x'p_{ht})r^*_i = E_{ht} \text{ and } x_i = 0. \end{aligned} \qquad \text{(CS5.12)}$$

Our computations take into account the distribution of flavor preferences by integrating over the distribution of household parameters. These results show that deletion of all flavors (except for mixed berry) result in losses in utility that require substantial price reductions (15–21 %). Plain yogurt requires the highest price cut even though, on average, households do not like plain yogurt. The reason is that there is a subset of households that place a high value on plain yogurt. In a traditional retailing environment, the only way the retailer can compensate for this is to lower prices on the remaining yogurts substantially.

These price reduction results all hinge on the distribution of household heterogeneity. On average, households regard the yogurt flavors as quite substitutable, but this statement is very misleading. Many households have a decided preference for one flavor which drives these utility and price reduction results. To illustrate this point, we consider the price reduction experiment conditional on the average value of the utility parameter, $\bar{\beta}$. Here we see that, conditional on $\bar{\beta}$, there are only negligible reductions in utility, which means that we do not have to lower prices at all to compensate for the loss of a flavor. In models with nonlinear utility functions, any sort of welfare calculation such as the compensating value computation undertaken here is very sensitive to how household heterogeneity is handled. We cannot simply 'plug in' the mean value without obtaining very misleading results. This illustrates the importance of integrating over the distribution of parameters in a full decision-theoretic approach.

If it were possible to customize the assortment to each household, the utility loss incurred by reduction in the size of the assortment would be reduced. In web retailing, this is a very real possibility. A web retailer can have the full assortment of varieties, but displaying this information to the buyer may be costly in terms of navigation of the website and purchase decisions. For example, a web retailer such as Peapod could offer its customers the full array of 25 or more Dannon yogurt flavors. The danger here is that the customers incur a larger search and ordering cost than a customer facing a much more limited variety at the standard bricks-and-mortar retailer. One way of avoiding these information-processing mental costs is to customize the assortment based on past purchase behavior.

R IMPLEMENTATION

R We have implemented the MCMC algorithm in the R function `rhierVarietyRw`, which is available on the book web site. At the core of this function is the routine to compute the log-likelihood for the variety model. As discussed in Chapter 4, this requires integration of a multivariate normal density over a half-plane whenever there is a 'zero' in the vector of demands. In addition, if there is more than one nonzero demand, we must compute a Jacobian term. This function implements a partial vectorization of the computation of the variety likelihood for observations with one or two nonzero demands.

Care should be taken to ensure that there has been an adequate burn-in for this procedure. Our experience suggests than at least 5000 draws are required. Also, some experimentation with scaling of the RW increment matrix (the parameter `s`) may be required for optimal performance. We recommend making trial runs of shorter duration, **R** followed by relative numerical efficiency calculations (using `numEff`) on the resulting output to optimize the choice of the RW increment size.

Appendix A An Introduction to Hierarchical Bayes Modeling in R

In order to facilitate computation of the models in this book, we created a set of programs written in R. R is a general-purpose programming and statistical analysis system; it is free and available on the web. We have made our suite of programs into what is called an R 'package'. This package, called *bayesm*, is easy to download and install from within R and is thoroughly documented, including test examples.

This appendix provides an introduction to the R environment and *bayesm* using the example of a hierarchical binary logit model.

We start with the installation of the R statistical package and *bayesm*, provide a short introduction to the R language and programming, and conclude with a case study involving a heterogeneous binary logit model calibrated on conjoint data. We assume the user is working in a Windows environment. All R commands and objects are the same under Windows and Linux, but the install procedure and graphical user interface (GUI) are different.

A.1 SETTING UP THE R ENVIRONMENT

Obtaining R

Visit http://cran.r-project.org/ or google "R language". CRAN is a network of mirror sites that allow you to download precompiled binary versions of R or source.

Look under 'Precompiled Binary Distributions' and click on 'Windows'. Click on 'base' on the next page and download the rwXXXX.exe file (XXXX = 2011 at present) – right-click and 'Save Target As'. Double-click the file name under Windows Explorer and R will install itself in the usual fashion for Windows software.

You may also obtain the optimized BLAS version for Pentium 4 chips by visiting the 'contributed' link (for Windows users this link is labeled 'contrib' and is located directly below the 'base' link) and clicking on the ATLAS directory. Simply download the RBLAS.dll file for your chip (invariably Pentium 4, found in the directory P4/) and

Bayesian Statistics and Marketing P. E. Rossi, G. M. Allenby and R. McCulloch

replace the RBLAS.dll file in the bin directory (this will be something like C:\Program Files\R\rwXXXX\bin). Our experience is that this will double the speed of many common matrix operations.

Customizing the R Shortcut

The standard R installation will create a desktop shortcut to invoke R. This will start up R pointing to the rwXXXX directory. It is more useful to modify the shortcut to start up with R pointing to a directory which is 'closer' to the directories you plan on working in. To modify the shortcut, right-click on it and choose 'Properties', change the 'Start in:' value to any valid directory on your machine. You can also copy the shortcut and make a shortcut for each of your major 'projects'.

If you have more than 512 MB of memory (we highly recommend at least 1 GB memory), you may also want to add the option to increase the memory available on the shortcut. This is accomplished by adding a parameter to the command line which invokes R. Again, right-click the shortcut, choose 'Properties', and modify the command in the 'Target' line. For example:

"C:\Program Files\R\rwXXXX\bin\Rgui.exe"-- max-mem-size = 1900 Mb

Note the close quotation marks after Rgui.exe *before* the specification of the '--max-mem-size' parameter. You may also copy this shortcut to your taskbar – do not forget to 'unlock' it first!

Invoking R and Using the R GUI

Double-click on the shortcut to invoke R. Something like the screen shot in Figure A.1 will appear.

If you create graphics, a separate graphics window will appear within the RGUI main window. For example, if we create a histogram of 1000 normal random numbers by using the `hist` command (more below in graphics section), a window will automatically open with the plot, as shown in Figure A.2. (It is possible to create multiple graphics windows and 'paint' different graphics in each – see commands `dev.cur()`, `dev.list()`, `dev.set()`.) The contents of this graphics window can be 'clipped' into the clipboard for pasting into your favorite application by selecting the graph window and using Ctrl-W (not Ctrl-C) and then paste or Ctrl-V to paste into the application. Note: if you use Ctrl-W you will copy the graph as a Windows metafile graphic which is device-independent and will print at the highest resolution of the device. Ctrl-C copies the graph as a bitmap which can compromise resolution.

Obtaining Help in R

There are a number of different ways to obtain help in R. The 'Help' menu allows you to access the manuals in PDF or HTML form as well as to search for keywords. Note:

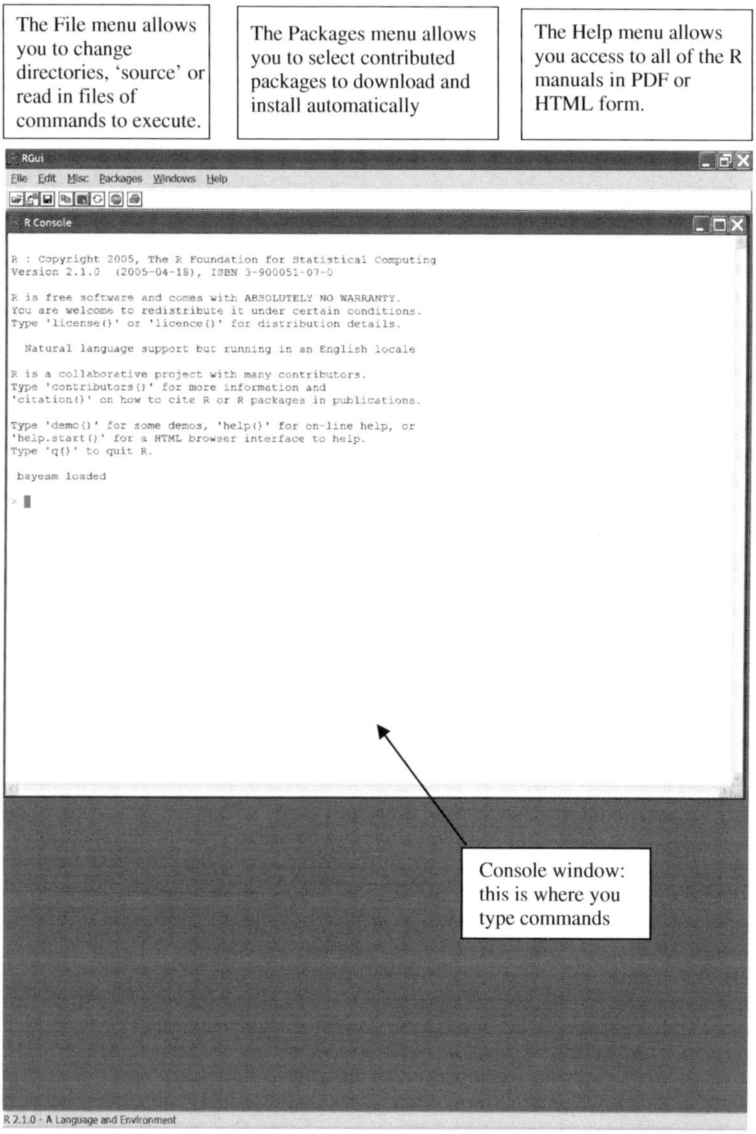

The File menu allows you to change directories, 'source' or read in files of commands to execute.

The Packages menu allows you to select contributed packages to download and install automatically

The Help menu allows you access to all of the R manuals in PDF or HTML form.

RGui

File Edit Misc Packages Windows Help

R Console

```
R : Copyright 2005, The R Foundation for Statistical Computing
Version 2.1.0  (2005-04-18), ISBN 3-900051-07-0

R is free software and comes with ABSOLUTELY NO WARRANTY.
You are welcome to redistribute it under certain conditions.
Type 'license()' or 'licence()' for distribution details.

  Natural language support but running in an English locale

R is a collaborative project with many contributors.
Type 'contributors()' for more information and
'citation()' on how to cite R or R packages in publications.

Type 'demo()' for some demos, 'help()' for on-line help, or
'help.start()' for a HTML browser interface to help.
Type 'q()' to quit R.

 bayesm loaded

>
```

Console window: this is where you type commands

R 2.1.0 - A Language and Environment

Figure A.1 The R window or graphical user interface

the Help menu does not appear in the R GUI unless the console window is 'active'. To make any window active, click on the button bar or the blue title bar at the top of the window.

You can also use the `help` and `help.search` commands in the R console windows. For example, in Figure A.2 we used two commands, `hist` and `rnorm`. `help(rnorm)` produces a window with the content shown in Figure A.3 (a shortcut is the command `?rnorm`).

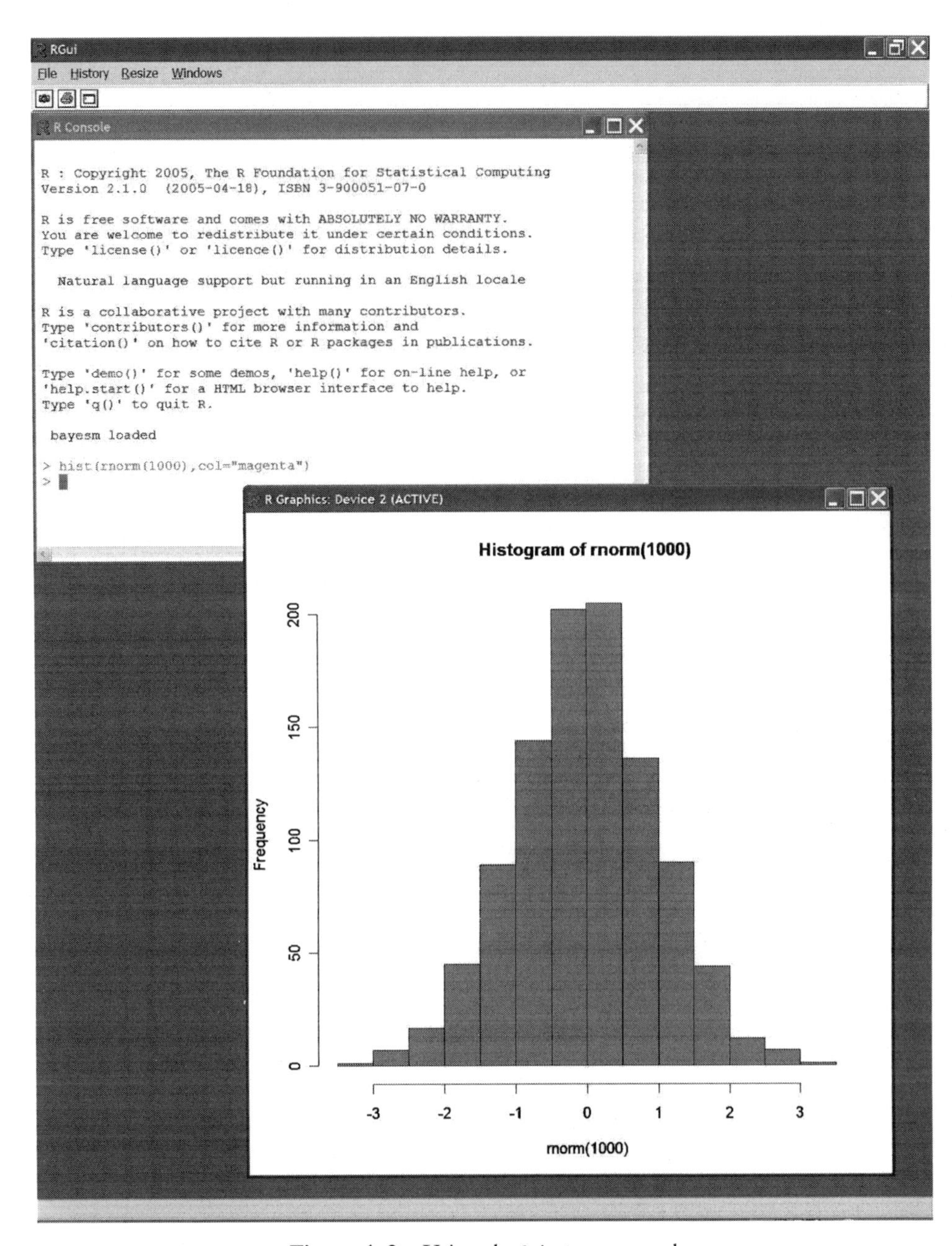

Figure A.2 Using the `hist` command

Each R Help window has the same sections:

Description

Usage

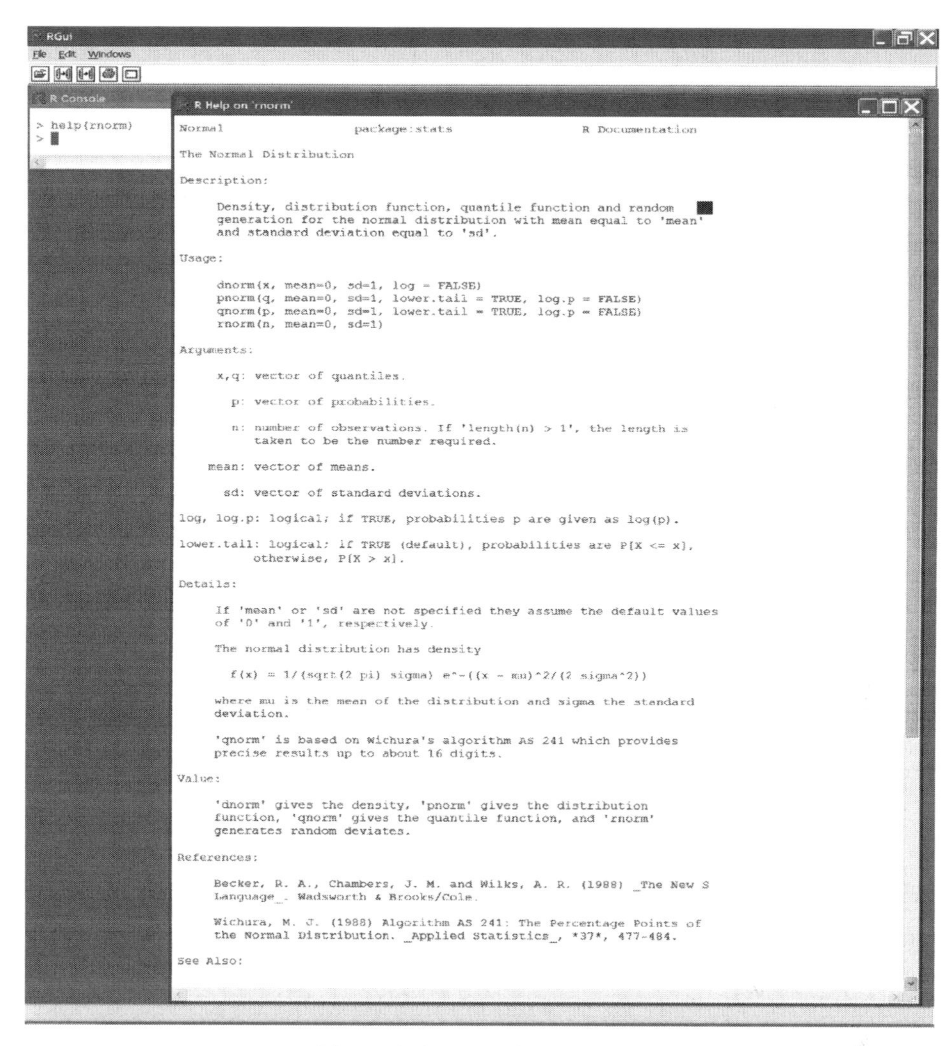

Figure A.3 R. Help on `rnorm`

Arguments

Details

Value

References

See Also

Examples (this is 'cut off' in the screen shot in Figure A.3).

Usage, Arguments and Examples are the most useful.

If you are not sure what command you need, `help.search("key word")` can be very useful.

Installing `bayesm`

bayesm can be installed from a CRAN mirror site by selecting Packages from the menu bar in R. Users must first set the CRAN mirror, then install *bayesm* directly from CRAN by simply clicking on 'bayesm' in the Packages window displayed. This is shown in the screen shot in Figure A.4.

Program Editing and R

For all but the simplest tasks, it is useful to edit a file with R commands in it. R syntax is sufficiently complex that it is difficult to write directly into the command window without making numerous syntax errors.

Open a text editor (Vim, an improved version of vi, is recommended; this is shareware, available at http://www.vim.org/); type in R commands and save the file. You can either cut the commands from the editor window and paste into the R console window to run or use the R command, `source` – also available from the File menu in the Windows GUI.

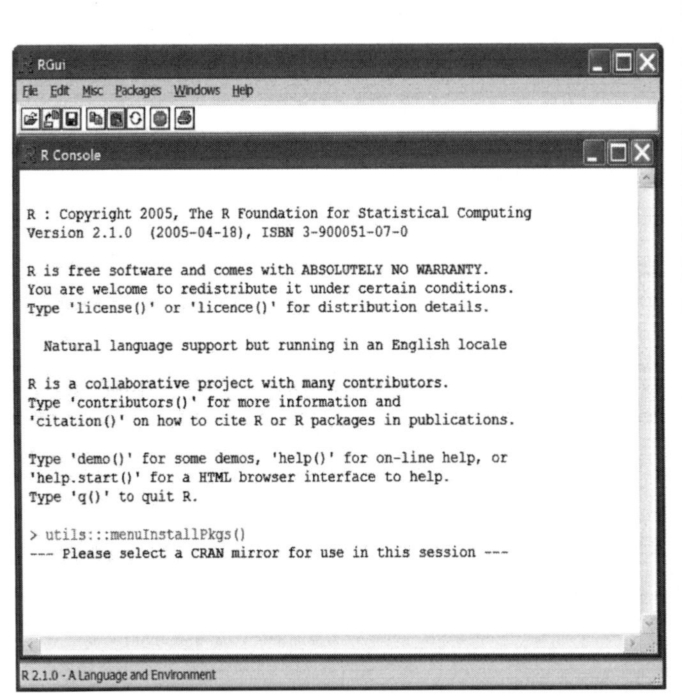

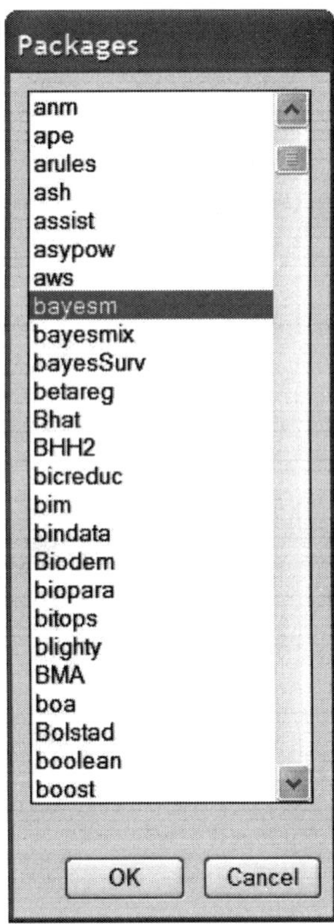

Figure A.4 Installing `bayesm`

A.2 THE R LANGUAGE

R is a functionally oriented language. All commands are functions which act upon objects of various types. All commands produce objects as well. The basic R command is of the form `object=function(object)`. Functions can be composed to produce powerful (but sometimes hard-to-read) expressions. Users can define their own functions. Writing these user functions constitutes R programming. Venables *et al.* (2001) provide an introduction to R. For a more advanced treatment, see Venables and Ripley (2000).

Let us start by reading in some data. Suppose we have a file in a spreadsheet with some regression data on several different units. The file has a `UNIT` variable to identify which unit the data comes from, a dependent variable Y, and two independent variables X_1, X_2.

```
UNIT      Y          X1                X2
 A        1       0.23815           0.4373
 A        2       0.55508           0.47938
 A        3       3.03399          -2.17571
 A        4      -1.49488           1.66929
 B       10      -1.74019           0.35368
 B        9       1.40533          -1.2612
 B        8       0.15628          -0.27751
 B        7      -0.93869          -0.0441
 B        6      -3.06566           0.14486
```

We write this data out of Excel by saving it as a text (tab-delimited file), data.txt (use the 'Save as' option on the File menu and choose 'text file' in the file type box). Note: there is no simple, direct way to read .xls files in R.

We can read this file into R using the `read.table` command:

```
> df=read.table("data.txt",header=TRUE)
> df
  UNIT  Y       X1       X2
1    A  1  0.23815  0.43730
2    A  2  0.55508  0.47938
3    A  3  3.03399 -2.17571
4    A  4 -1.49488  1.66929
5    B 10 -1.74019  0.35368
6    B  9  1.40533 -1.26120
7    B  8  0.15628 -0.27751
8    B  7 -0.93869 -0.04410
9    B  6 -3.06566  0.14486
```

The `read.table` function has two arguments: the name of the file, and the argument `header`. There are many other arguments, but they are optional and often have defaults. The default for the header argument is the value `FALSE`. Assigning the value `header=TRUE` tells the `read.table` function to expect that the first line of the file will contain (delimited by spaces or tabs) the names of each variable. `TRUE` and `FALSE` are examples of reserved values in R indicating a logical switch for true or false. Another useful reserved value is `NULL`, which is often used to create an object with nothing in it.

The command `df=read.table(...)` assigns the output of the `read.table` function to the R object named `df`. The object `df` is a member of a class or type of object called a ***data frame***. A data frame is preferred by R as the format for data sets. A data frame contains a set of observations on variables with the same number of observations in each variable. In this example, each of the variables, Y, X_1, and X_2, is of type numeric (R does not distinguish between integers and floating-point numbers), while the variable `UNIT` is character.

There are two reasons to store your data as a data frame: first, most R statistical functions require a data frame; and second, the data frame object allows the user to access the data either via the variables names or by viewing the data frame as a two-dimensional array:

```
> df$Y
[1]  1  2  3  4 10  9  8  7  6
> mode(df$Y)
[1] "numeric"
> df[,2]
[1]  1  2  3  4 10  9  8  7  6
```

We can refer to the Y variable in `df` by using the `df$XXX` notation (where `XXX` is the name of the variable). The `mode` command confirms that this variable is, indeed, numeric. We can also access the Y variable by using notation in R for subsetting a portion of an array. The notation `df[,2]` means the values of the second column of `df`. Below we will explore the many ways we can subset an array or matrix.

Using Built-in Functions: Running a Regression

Let us now use the built-in linear model function in R to run a regression of Y on X_1 and X_2, pooled across both units A and B:

```
> lmout=lm(Y ~ X1 + X2, data=df)
> names(lmout)
 [1] "coefficients"  "residuals"     "effects"       "rank"
 [5] "fitted.values" "assign"        "qr"            "df.residual"
 [9] "xlevels"       "call"          "terms"         "model"
> print(lmout)

Call:
lm(formula = Y ~ X1 + X2, data = df)

Coefficients:
(Intercept)          X1           X2
      5.084      -1.485       -2.221
> summary(lmout)

Call:
lm(formula = Y ~ X1 + X2, data = df)
```

```
Residuals:
    Min      1Q  Median      3Q     Max
-3.3149 -2.4101  0.4034  2.5319  3.2022

Coefficients:
            Estimate Std. Error t value Pr(>|t|)
(Intercept)   5.0839     1.0194   4.987  0.00248 **
X1           -1.4851     0.8328  -1.783  0.12481
X2           -2.2209     1.3820  -1.607  0.15919
---
Signif. codes:  0 '***' 0.001 '**' 0.01 '*' 0.05 '.' 0.1 ' ' 1

Residual standard error: 2.96 on 6 degrees of freedom
Multiple R-Squared: 0.3607,     Adjusted R-squared: 0.1476
F-statistic: 1.693 on 2 and 6 DF,  p-value: 0.2612
```

`lm` is the function in the package (`stats`) which fits linear models. Note that the regression is specified via a 'formula' that tells `lm` which are the dependent and independent variables. We assign the output from the `lm` function to the object, `lmout`. `lmout` is a special type of object called a 'list'. A list is simply an ordered collection of objects of any type. The `names` command will list the names of the elements of the list. We can access any element of the list by using the `$` notation:

```
> lmout$coef
(Intercept)          X1          X2
   5.083871   -1.485084   -2.220859
```

Note that we only need to specify enough of the name of the list component to uniquely identify it – for example, `lmout$coef` is the same as `lmout$coefficients`.

We can print the object `lmout` and get a brief summary of its contents. 'Print' is a generic command which uses a different 'print method' for each type of object. Print recognizes that `lmout` is a list of type `lm` and uses a specific routine to print out the contents of the list. A more useful summary of contents of `lmout` can be obtained with the `summary` command.

Inspecting Objects and the R Workspace

When you start up R, R looks for a file. Rdata is in the directory in which R is started (you can also double-click the file to start R). This file contains a copy of the R 'workspace', which is a list of R objects created by the user. For example, we just created two R objects in the example above: `df` (the data frame) and `lmout` (the `lm` output object).

To list all objects in the current workspace, use the command `ls()`:

```
> ls()
[1] "df"    "lmout"
```

This does not tell us very much about the objects. If you just type the object name at the command prompt and return, then you will invoke the default print method for this type of object as we saw above in the data frame example.

As useful command is the structure (str for short) command:

```
> str(df)
'data.frame':    9 obs. of  4 variables:
 $ UNIT: Factor w/ 2 levels "A","B": 1 1 1 1 2 2 2 2 2
 $ Y   : int  1 2 3 4 10 9 8 7 6
 $ X1  : num   0.238  0.555  3.034 -1.495 -1.740 ...
 $ X2  : num   0.437  0.479 -2.176  1.669  0.354 ...
```

Note that the str command tells us a bit about the variables in the data frame. The UNIT variable is of type factor with two levels. Type factor is used by many of the built-in R functions and is way to store qualitative variables.

The R workspace exists only in memory. You must either save the workspace when you exit (you will be prompted to do this) or you must re-create the objects when you start the program again.

Vectors, Matrices and Lists

From our point of view, the power of R comes from statistical programming at a relatively high level. To do so, we will need to organize data as vectors, arrays and lists. Vectors are ordered collections of the same type of object. If we access one variable from our data frame above, it will be a vector:

```
> df$X1
[1]  0.23815  0.55508  3.03399 -1.49488 -1.74019  1.40533
0.15628 -0.93869 -3.06566
> length(df$X1)
[1] 9
> is.vector(df$X1)
[1] TRUE
```

The function is.vector returns a logical flag as to whether or not the input argument is a vector. We can also create a vector with the c() command:

```
> vec=c(1,2,3,4,5,6)
> vec
[1] 1 2 3 4 5 6
> is.vector(vec)
[1] TRUE
```

A matrix is a two-dimensional array. Let us create a matrix from a vector:

```
> mat=matrix(c(1,2,3,4,5,6),ncol=2)
> mat
     [,1] [,2]
[1,]    1    4
[2,]    2    5
[3,]    3    6
```

matrix() is a command to create a matrix from a vector. The option ncol is used to create the matrix with a specified number of columns (see also nrow). Note that the matrix is created column by column from the input vector (the first subscript varies the fastest: mat[1,1], mat[2,1], mat[3,1], mat[1,2], mat[2,2], mat[3,2]). We can create a matrix row by row with the following command:

```
> mat=matrix(c(1,2,3,4,5,6),byrow=TRUE,ncol=2)
> mat
     [,1] [,2]
[1,]    1    2
[2,]    3    4
[3,]    5    6
```

We can also convert a data frame into a matrix:

```
> dfmat=as.matrix(df)
> dfmat
  UNIT Y    X1         X2
1 "A"  " 1" " 0.23815" " 0.43730"
2 "A"  " 2" " 0.55508" " 0.47938"
3 "A"  " 3" " 3.03399" "-2.17571"
4 "A"  " 4" "-1.49488" " 1.66929"
5 "B"  "10" "-1.74019" " 0.35368"
6 "B"  " 9" " 1.40533" "-1.26120"
7 "B"  " 8" " 0.15628" "-0.27751"
8 "B"  " 7" "-0.93869" "-0.04410"
9 "B"  " 6" "-3.06566" " 0.14486"
> dim(dfmat)
[1] 9 4
```

Note that all of the values of the resulting matrix are character-valued as one of the variables in the data frame (UNIT) is character-valued. Finally, matrices can be created from other matrices and vectors using the cbind (column bind) and rbind (row bind) commands:

```
> mat1
     [,1] [,2]
[1,]    1    2
[2,]    3    4
[3,]    5    6
> mat2
     [,1] [,2]
[1,]    7   10
[2,]    8   11
[3,]    9   12
> cbind(mat1,mat2)
     [,1] [,2] [,3] [,4]
[1,]    1    2    7   10
[2,]    3    4    8   11
[3,]    5    6    9   12
```

```
> rbind(mat1,mat2)
     [,1] [,2]
[1,]    1    2
[2,]    3    4
[3,]    5    6
[4,]    7   10
[5,]    8   11
[6,]    9   12
> rbind(mat1,c(99,99))
     [,1] [,2]
[1,]    1    2
[2,]    3    4
[3,]    5    6
[4,]   99   99
```

R supports multidimensional arrays as well. Below is an example of creating a three-dimensional array from a vector.

```
> ar=array(c(1,2,3,4,5,6),dim=c(3,2,2))
> ar
, , 1

     [,1] [,2]
[1,]    1    4
[2,]    2    5
[3,]    3    6

, , 2

    [,1] [,2]
[1,]    1    4
[2,]    2    5
[3,]    3    6
```

Again, the array is created by using the vector to fill out the array dimension by dimension, with the first subscript varying faster than the second and the second varying faster than the third as shown above. A $3 \times 2 \times 2$ array has 12 elements, not the six provided as an argument. R will repeat the input vector as necessary until the required number of elements are obtained.

A list is an ordered collection of objects of any type. It is the most flexible object in R that can be indexed. As we have seen in the `lm` function output, lists can also have names:

```
> l=list(1,"a",c(4,4),list(FALSE,2))
> l
[[1]]
[1] 1

[[2]]
[1] "a"
```

```
[[3]]
[1] 4 4

[[4]]
[[4]][[1]]
[1] FALSE

[[4]][[2]]
[1] 2

> l=list(num=1,char="a",vec=c(4,4),list=list(FALSE,2))
> l$num
[1] 1
> l$list
[[1]]
[1] FALSE

[[2]]
[1] 2
```

In the example, we created a list of a numeric value, character, vector and another list. We also can name each component and access them with the `$` notation.

Accessing Elements and Subsetting Vectors, Arrays, and Lists

To access an element of a vector, simply enclose the index of that element in square brackets.

```
> vec=c(1,2,3,2,5)
> vec[3]
[1] 3
```

To access a subset of elements, there are two approaches: either specify a vector of integers of the required indices; or specify a logical variable which is `TRUE` for the desired indices:

```
> index=c(3:5)
> index
[1] 3 4 5
> vec[index]
[1] 3 2 5
> index=vec==2
> index
[1] FALSE  TRUE FALSE  TRUE FALSE
> vec[index]
[1] 2 2
> vec[vec!=2]
[1] 1 3 5
```

`c(3:5)` creates a vector from the 'pattern' or sequence from 3 to 5. The `seq` command can create a wide variety of different patterns.

To properly understand the example of the logical index, it should be noted that = is an assignment operator while == is a comparison operator. `vec==2` creates a logical vector with flags for if the elements of `vec` are 2. The last example uses the 'not equal' comparison operator !=. We can also access the elements not in a specified index vector:

```
> vec[-c(3:5)]
[1] 1 2
```

To access elements of arrays, we can use the same ideas as for vectors but we must specify a set of row and column indices. If no indices are specified, we get all of the elements on that dimension. For example, earlier we used the notation `df[,2]` to access the second column of the data frame `df`.

We can pull off the observations corresponding to unit A from the matrix version of `dfmat` using the following commands:

```
> dfmat
  UNIT Y    X1         X2
1 "A"  " 1" " 0.23815" " 0.43730"
2 "A"  " 2" " 0.55508" " 0.47938"
3 "A"  " 3" " 3.03399" "-2.17571"
4 "A"  " 4" "-1.49488" " 1.66929"
5 "B"  "10" "-1.74019" " 0.35368"
6 "B"  " 9" " 1.40533" "-1.26120"
7 "B"  " 8" " 0.15628" "-0.27751"
8 "B"  " 7" "-0.93869" "-0.04410"
9 "B"  " 6" "-3.06566" " 0.14486"
> dfmat[dfmat[,1]=="A",2:4]
  Y    X1         X2
1 " 1" " 0.23815" " 0.43730"
2 " 2" " 0.55508" " 0.47938"
3 " 3" " 3.03399" "-2.17571"
4 " 4" "-1.49488" " 1.66929"
```

The result is a 4×3 matrix. Note that we are using the values of `dfmat` to index into itself. This means that R evaluates the expression `dfmat[,1] == "A"` and passes the result into the matrix subsetting operator `[ ]` which is a function that processes `dfmat`.

To access elements of lists, we can use the `$` notation if the element has a name or we can use a special operator `[[ ]]`. To see how this works, let us make a list with two elements, each corresponding to the observations for unit A and B. Note that the size of the matrices corresponding to each unit is different – unit A has four observations and unit B has five! This means that we cannot use a three-dimensional array to store this data (we would need a 'ragged' array):

```
>
ldata=list(A=dfmat[dfmat[,1]=="A",2:4],
   B=dfmat[dfmat[,1]=="B",2:4])
> ldata
```

```
$A
  Y     X1         X2
1 "  1" " 0.23815" " 0.43730"
2 "  2" " 0.55508" " 0.47938"
3 "  3" " 3.03399" "-2.17571"
4 "  4" "-1.49488" " 1.66929"

$B
 Y      X1         X2
5 "10" "-1.74019" " 0.35368"
6 " 9" " 1.40533" "-1.26120"
7 " 8" " 0.15628" "-0.27751"
8 " 7" "-0.93869" "-0.04410"
9 " 6" "-3.06566" " 0.14486"

> ldata[1]
$A
 Y      X1         X2
1 "  1" " 0.23815" " 0.43730"
2 "  2" " 0.55508" " 0.47938"
3 "  3" " 3.03399" "-2.17571"
4 "  4" "-1.49488" " 1.66929"

> is.matrix(ldata[1])
[1] FALSE
> is.list(ldata[1])
[1] TRUE
> ldata$A
  Y     X1         X2
1 "  1" " 0.23815" " 0.43730"
2 "  2" " 0.55508" " 0.47938"
3 "  3" " 3.03399" "- 2.17571"
4 "  4" "-1.49488" " 1.66929"
> is.matrix(ldata$A)
[1] TRUE
```

If we specify `ldata[1]`, we do not get the contents of the list element (which is a matrix) but we get a list! If we specify `ldata$A`, we obtain the matrix. If we have a long list or we do not wish to name each element, we can use the `[[ ]]` operator to access elements in the list:

```
> is.matrix(ldata[[1]])
[1] TRUE
> ldata[[1]]
  Y     X1         X2
1 "  1" " 0.23815" " 0.43730"
2 "  2" " 0.55508" " 0.47938"
3 "  3" " 3.03399" "-2.17571"
4 "  4" "-1.49488" " 1.66929"
```

Loops

As with all interpreted languages, loops in R are slow. That is, they typically take more time than if implemented in a compiled language. On the other hand, matrix/vector operations are typically faster in R than in compiled languages such as C and Fortran unless the optimized BLAS is called. Thus, wherever possible, 'vectorization' – writing expressions as only involving matrix/vector arithmetic – is desirable. This is more of an art than a science, however.

If a computation is fundamentally iterative (such as maximization or MCMC simulation), a loop will be required. A simple loop can be accomplished with the `for` structure. The syntax is of the form

```
for (var in range) { }
```

`var` is a numeric loop index. `range` is a range of values of `var`. Enclosed in the braces is any valid R expression. There can be more than one R statement in the R expression. The simplest example is a loop over a set of i values from 1 to N:

```
x = 0
for (i in 1:10)
   {
     x = x + 1
   }
```

Let us loop over both units and create a list of lists of the regression output from each:

```
> ldatadf=list(A=df[df[,1]=="A",2:4],B=df[df[,1]=="B",2:4])
> lmout=NULL
> for (i in 1:2) {
+     lmout[[i]]=lm(Y ~ X1+X2,data=ldatadf[[i]])
+     print(lmout[[i]])
+     }

Call:
lm(formula = Y ~ X1 + X2, data = ldatadf[[i]])

Coefficients:
(Intercept)           X1           X2
      4.494       -2.860       -3.180

Call:
lm(formula = Y ~ X1 + X2, data = ldatadf[[i]])

Coefficients:
(Intercept)           X1           X2
      9.309        1.051        1.981
```

Here we subset the data frame directly rather than the matrix created from the data frame to avoid the extra step of converting character to numeric values and so that we can use the `lm` function which requires data frame input. We can see that the same subsetting command that works on arrays will also work on data frames.

Implicit Loops

In many contexts, a loop is used to compute the results of applying a function to either the row or column dimensions of an array. For example, if we wish to find the mean of each variable in a data frame, we want to apply the function `mean` to each column. This can be done with the `apply()` function:

```
> apply(df[,2:4],2,mean)
          Y          X1          X2
 5.5555556 -0.2056211 -0.0748900
```

The first argument specifies the array, the second the dimension (1 = row, 2 = column), and the third the function to be applied. In R the `apply` function is simply an elegant loop, so do not expect to speed things up with this. Of course, we could write this as a matrix operation which would be much faster.

Matrix Operations

One of the primary advantages of R is that we can write matrix/vector expressions directly in R code. Let us review some of these operators by computing a pooled regression using matrix statements.

The basic functions needed are:

`%*%`	matrix multiplication, e.g. `X %*% Y`, where X or Y or both can be vectors
`chol(X)`	compute Cholesky (square) root of square positive definite matrix $X = U'U$ where U is `chol(X)`; U is upper triangular
`chol2inv(chol(X))`	compute inverse of square positive definite matrix using its Cholesky root
`crossprod(X,Y)`	`t(X) %*% Y` – very efficient
`diag`	extract diagonal of matrix or create diagonal matrix from a vector

Less frequently used are:

`%x%`	Kronecker product (to be used carefully as Kronecker products can create very large arrays)
`backsolve()`	used to compute inverse of a triangular array

The R statements are:

```
y=as.numeric(dfmat[,2])
X=matrix(as.numeric(dfmat[,3:4]),ncol=2)
```

```
X=cbind(rep(1,nrow(X)),X)
XpXinv=chol2inv(chol(crossprod(X)))
bhat=XpXinv%*%crossprod(X,y)
res=as.vector(y-X%*%bhat)
ssq=as.numeric(res%*%res/(nrow(X)-ncol(X)))
se=sqrt(diag(ssq*XpXinv))
```

The first two statements create y and X. Then we add a column of ones using the `rep()` function for the intercept and compute the regression using Cholesky roots. Note that we must convert `res` into a vector to use the statement res `%*% res`. We also must convert `ssq` into a scalar from a 1×1 matrix to compute the standard errors in the last statement. We note that the method above is very stable numerically, but some users would prefer the *QR* decomposition. This would be simpler, but our experience has shown that the method above is actually faster in R.

Other Useful Built-in R Functions

R has thousands of built-in functions and thousands more than can be added from contributed packages. Some functions used in the book include:

`rnorm`	draw univariate normal random variates
`runif`	draw uniform random variates
`rchisq`	draw chi-square random variates
`mean`	compute mean of a vector
`var`	compute covariance matrix given matrix input
`quantile`	compute quantiles of a vector
`optim`	general-purpose optimizer
`sort`	sort a vector
`if`	standard if statement (includes else clause)
`while`	while loop
`scan`	read from a file to a vector
`write`	write a matrix to a file
`sqrt`	square root
`log`	natural log
`%%`	modulo (e.g. `100 %%10 = 0`)
`round`	round to a specified number of sign digits
`floor`	greatest integer < argument
`.C`	interface to C and C++ code (more later)

User-Defined Functions

The regression example above is a situation for which a user-defined function would be useful. To create a function object in R, simply enclose the R statements in braces and assign this to a function variable:

```
myreg=function(y,X){
#
# purpose: compute lsq regression
#
# arguments:
#     y -- vector of dep var
#     X -- array of indep vars
#
# output:
#     list containing lsq coef and std errors
#
XpXinv=chol2inv(chol(crossprod(X)))
bhat=XpXinv%*%crossprod(X,y)
res=as.vector(y-X%*%bhat)
ssq=as.numeric(res%*%res/(nrow(X)-ncol(X)))
se=sqrt(diag(ssq*XpXinv))
list(b=bhat,std_errors=se)
}
```

The code above should be executed either by cutting and pasting into R or by sourcing a file containing this code. This will define an object called `myreg`:

```
ls()
[1]  "ar"   "bhat"           "df"      "dfmat"    "i"         "index"
[7]  "l"    "last.warning"   "ldata"   "ldatadf"  "ldataidf"  "lmout"
[13] "mat"  "mat1"           "mat2"    "myreg"    "names"     "res"
[19] "se"   "ssq"            "vec"     "X"        "XpXinv"    "y"
```

To execute the function, we simply type it in with arguments at the command prompt or in another source file:

```
> myreg(X=X,y=y)
$b
          [,1]
[1,]  5.083871
[2,] -1.485084
[3,] -2.220859

$std_errors
[1] 1.0193862 0.8327965 1.3820287
```

`myreg` returns a list with b and the standard errors.

Objects are passed by copy in R rather than by reference. This means that if we give the command `myreg(Z,d)`, a copy of Z will be assigned to the 'local' variable X in the function `myreg` and a copy of d to y. In addition, variables created in the function (e.g. `XpXinv` and `res` in `myreg`) are created only during the execution of the function and then erased when the function returns to the calling environment.

The arguments are passed and copied in the order supplied at the time of the call, so you must be careful. The statement `myreg(d,Z)` will not execute properly. However, if we explicitly name the arguments as in `myreg(X=Z,y=d)` then the arguments can be given in any order.

Many functions have default arguments and R has what is called 'lazy' function evaluation, which means that if an argument is not needed it is not checked. See Venables *et al.* (2001) for a more discussion on default and other types of arguments. If a local variable cannot be found while executing a function, R will look in the environment or workspace that the function was called from. This can be convenient, but it can also be dangerous!

Many functions are dependent on other functions. If a function called within a function is only used by that calling function and has no other use, it can be useful to define these utility functions in the calling function definition. This means that they will not be visible to the user of the function. For example:

```
Myfun= function(X,y) {
#
# define utility function needed
#
Util=function(X) { ... }
#
# main body of myfun
#
...
}
```

Debugging Functions

It is a good practice to define your functions in a file and 'source' them into R. This will allow you to recreate your set of function objects for a given project without having to save the workspace.

To debug a function, you can use the brute force method of placing print statements in the function. `cat()` can be useful here. For example, we can define a 'debugging' version of `myreg` which prints out the value of se in the function. The `cat` command prints out a statement reminding us of where the 'print' output comes from (note the use of `fill=TRUE` which ensures that a new line will be generated on the console):

```
myreg=function(y,X){
#
# purpose: compute lsq regression
#
# arguments:
#     y -- vector of dep var
#     X -- array of indep vars
#
# output:
#     list containing lsq coef and std errors
#
XpXinv=chol2inv(chol(crossprod(X)))
bhat=XpXinv%*%crossprod(X,y)
```

```
res=as.vector(y-X%*%bhat)
ssq=as.numeric(res%*%res/(nrow(X)-ncol(X)))
se=sqrt(diag(ssq*XpXinv))
cat("in myreg, se = ",fill=TRUE)
print(se)
list(b=bhat,std_errors=se)
}
```

When run, this new function will produce the output:

```
> myregout=myreg(y,X)
in myreg, se =
[1] 1.0193862 0.8327965 1.3820287
```

R also features a simple debugger. If you `debug` a function, you can step through the function and inspect the contents of local variables. You can also modify their contents.

```
> debug(myreg)
> myreg(X,y)
debugging in: myreg(X, y)
debug: {
   XpXinv = chol2inv(chol(crossprod(X)))
   bhat = XpXinv %*% crossprod(X, y)
   res = as.vector(y - X %*% bhat)
   ssq = as.numeric(res %*% res/(nrow(X) - ncol(X)))
   se = sqrt(diag(ssq * XpXinv))
   cat("in myreg, se = ", fill = TRUE)
   print(se)
   list(b = bhat, std_errors = se)
}
Browse[1]>
debug: XpXinv = chol2inv(chol(crossprod(X)))
Browse[1]> X
[1]  1  2  3  4 10  9  8  7  6
Browse[1]> #OOPS!
debug: bhat = XpXinv %*% crossprod(X, y)
Browse[1]> XpXinv
            [,1]
[1,] 0.002777778
Browse[1]> Q
> undebug(myreg)
```

If there are loops in the function, the debugging command `c` can be used to allow the loop to finish. `Q` quits the debugger. You must turn off the debugger with the undebug command! If you want to debug other functions called by `myreg`, you must `debug()` them first!

Elementary Graphics

Graphics in R can be quite involved as the graphics capabilities are very extensive. For some examples of what is possible issue the commands `demo(graphics)`, demo(image) and demo(persp). Let us return to our first example – a histogram of a distribution.

```
hist(rnorm(1000),breaks=50,col="magenta")
```

This creates a histogram with 50 bars and with each bar filled in the color magenta (type colors() to see the list of available colors). This plot can be improved by inclusion of plot parameters to change the x and y axis labels and as well as the 'title' of the plot (see Figure A.5):

```
hist(rnorm(1000),breaks=30,col="magenta",xlab="theta",
 ylab="",main="Nonparametric Estimate of Theta Distribution")
```

Three other basic plots are useful:

plot(x,y)	scatterplot of x vs y
plot(x)	sequence plot of x
matplot(X)	sequence plots of columns of X
acf(x)	autocorrelation function of time series in x

The col, xlab, ylab, and main parameters work on all of these plots. In addition, the parameters

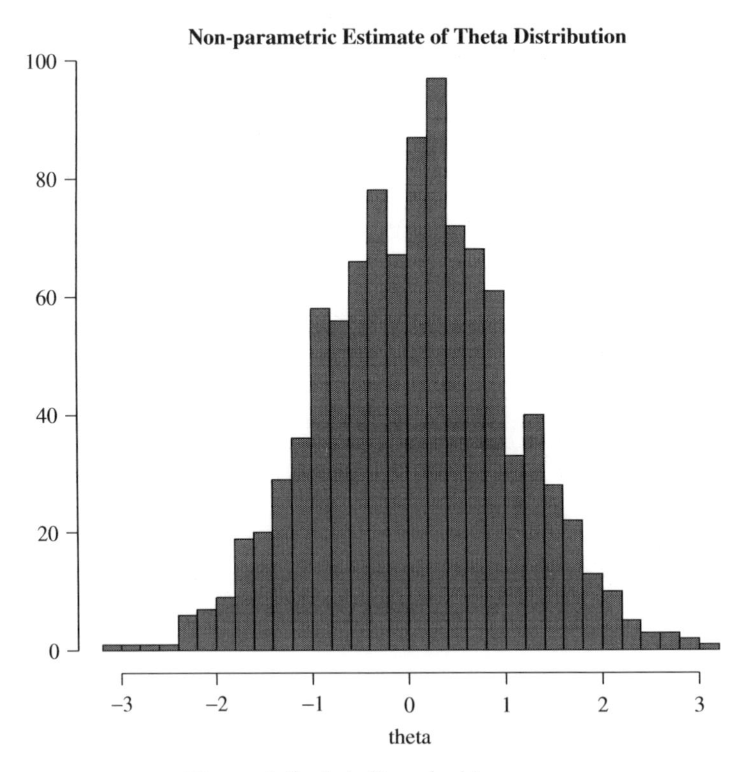

Figure A.5 Labelling the histogram

`type="l"`	connects scatterplot points with a lines
`lwd=x`	specifies the width of lines (1 is default, >1 is thicker)
`lty=x`	specifies type of line (e.g. solid vs dashed)
`xlim/ylim=c(z,w)`	specifies x/y axis runs from z to w

are useful. `?par` displays all of the graphic parameters available.

It is often useful to display more than one plot per page. To do this, we must change the global graphic parameters with the command `par(mfrow=c(x,y))`. This specifies an array of plots x by y plotted row by row.

```
par(mfrow=c(2,2))
X=matrix(rnorm(5000),ncol=5)
X=t(t(X)+c(1,4,6,8,10))

hist(X[,1],main="Histogram of 1st
   col",col="magenta",xlab="")
plot(X[,1],X[,2],xlab="col 1", ylab="col
   2",pch=17,col="red",
      xlim=c(-4,4),ylim=c(0,8))
title("Scatterplot")
abline(c(0,1),lwd=2,lty=2)
matplot(X,type="l",ylab="",main="MATPLOT")
acf(X[,5],ylab="",main="ACF of 5th Col")
```

`title()` and `abline()` are examples of commands which modify the current 'active' plot. Other useful functions are `points()` and `lines()` to add points and points connected by lines to the current plot. The commands above will produce Figure A.6.

System Information

The following commands provide information about the operating system:

`memory.limit()`	current memory limit
`memory.size()`	current memory size
`system.time(R expression)`	times execution of R expression
`proc.time()[3]`	current R session cpu usage in seconds
`getwd()`	obtain current working directory
`setwd()`	set current working directory
`Rprof(file="filename")`	turns on profiling and writes to filename
`Rprof("")`	turns off profiling
`summaryRprof(file="filename")`	summarizes output in profile file

Examples of usage are given below:

```
> memory.size()
[1] 191135504
> getwd()
[1] "C:/userdata/per/class/37904"
> x=matrix(rnorm(1e07),ncol=1000)
> memory.size()
[1] 332070456
> memory.limit()
```

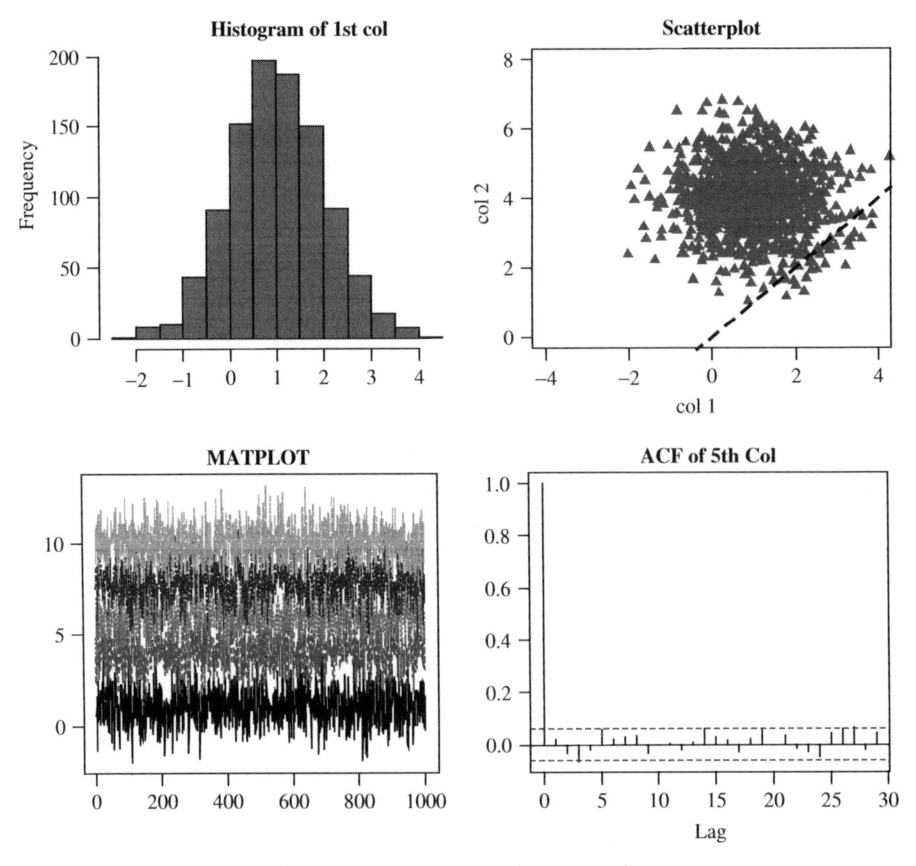

Figure A.6 Multiple plots example

```
[1] 1992294400
> begin=proc.time()[3]
> z=crossprod(x)
> end=proc.time()[3]
> print(end-begin)
[1] 6.59
> test=function(n){x=matrix(rnorm(n),ncol=1000);z=crossprod(x);
   cz=chol(z)}
> Rprof("test.out")
> test(1e07)
> Rprof()
> summaryRprof("test.out")
$by.self
           self.time self.pct total.time total.pct
rnorm           4.40     48.9       4.40      48.9
crossprod       4.16     46.2       4.16      46.2
matrix          0.22      2.4       4.72      52.4
.Call           0.12      1.3       0.12       1.3
as.vector       0.10      1.1       4.50      50.0
```

```
chol             0.00        0.0         0.12         1.3
test             0.00        0.0         9.00       100.0

$by.total
           total.time total.pct self.time self.pct
test             9.00     100.0      0.00      0.0
matrix           4.72      52.4      0.22      2.4
as.vector        4.50      50.0      0.10      1.1
rnorm            4.40      48.9      4.40     48.9
crossprod        4.16      46.2      4.16     46.2
.Call            0.12       1.3      0.12      1.3
chol             0.12       1.3      0.00      0.0

$sampling.time
[1] 9
```

The profile shows that virtually all of the time in the test function was in the generation of normal random numbers and in computing cross-products. The Cholesky root of a 1000×1000 matrix is essentially free! `crossprod` is undertaking 5 billion floating-point multiplies ($0.5 \times 10000 \times 1000 \times 1000$).

More Lessons Learned from Timing

If you are going to fill up an array with results, preallocate space in the array. Do not append to an existing array:

```
> n=1e04
> x=NULL
> zero=c(rep(0,5))
> begin=proc.time()[3]
> for (i in 1:n) {x=rbind(x,zero) }
> end=proc.time()[3]
> print(end-begin)
[1] 6.62
> x=NULL
> begin=proc.time()[3]
> x=matrix(double(5*n),ncol=5)
> end=proc.time()[3]
> print(end-begin)
[1] 0.07
```

A.3 HIERARCHICAL BAYES MODELING – AN EXAMPLE

You must install and load our R package, *bayesm*, in order to access the data and programs discussed in this section. See Appendix B for further details.

Data from the article by Allenby and Ginter (1995) are used to illustrate hierarchical Bayes analysis within the R environment. The context of the study was a bank offering a credit card to out-of-state customers, where the penalty associated with being an

Table A.1 Description of the data

Sample Size	946 Respondents
	14799 Observations
Attributes and Attribute Levels	
1. Interest rate	High, Medium, Low fixed, Medium variable
2. Rewards	The reward programs consisted of annual fee waivers or interest rebate reductions for specific levels of card usage and/or checking account balance. Four reward programs were considered.
3. Annual fee	High, Medium, Low
4. Bank	Bank A, Bank B, Out-of-State Bank
5. Rebate	Low, Medium, High
6. Credit line	Low, High
7. Grace period	Short, Long
Demographic Variables	Age (years) Annual Income ($000) Gender (=1 if female, =0 if male)

'unknown' brand, relative to other card characteristics, was to be assessed. Estimation of the disutility of having an unknown brand name would allow management to devise incentive programs that would make the credit card attractive to consumers. The attributes and attribute levels under study are presented in Table A.1.

Preferences were obtained from a trade-off study in which respondents were asked to choose between two credit cards that were identical in every respect except for two attributes. For example, respondents were asked to state their preference between the following offerings: the first card has a medium fixed annual interest rate and a medium annual fee, and the second card has a high fixed annual interest rate and low annual fee. Each respondent provided responses to between 13 and 17 paired comparisons involving a fraction of the attributes. A respondent trading off interest rates and annual fees, for example, did not choose between rebates and credit lines. As a result it was not possible to obtain fixed effect estimates of the entire vector of part-worths for any specific respondent. Moreover, even if all attribute levels were included for each respondent, constraints on the length of the questionnaire precluded collecting a sufficient number of trade-offs for accurate estimation of individual respondent part-worths. We will demonstrate that this data limitation is less important in random-effect models that pool information across respondents. In all, a total of 14,799 paired-comparisons from 946 respondents were available for analysis. Demographic information (age, income and gender) of the respondents were also available for analysis.

Choice probabilities for the binary logit model can be expressed in two ways:

$$\Pr(y_{hi} = 1) = \frac{\exp[x'_{hi1}\beta_h]}{\exp[x'_{hi1}\beta_h] + \exp[x'_{hi2}\beta_h]} = \frac{\exp[(x_{hi1} - x_{hi2})'\beta_h]}{1 + \exp[(x_{hi1} - x_{hi2})'\beta_h]}, \tag{A.1}$$

where h denotes the respondent, i denotes the paired-comparison, $\Pr(y_{hi} = 1)$ denotes the probability of selecting the first card, $\Pr(y_{hi} = 0)$ is the probability of selecting the second card, x is an attribute vector of dimension 14 with elements taking on values of 0 or 1 denoting absence or presence of the attribute level, and β is a vector of utility or part-worth coefficients to be estimated. Because of identification restrictions, the first level of each attribute in Table A.1 was set to zero, and the remaining attribute levels are dummy-coded to reflect the marginal utility associated with changes in the attribute levels from the first level.

The choice data file contains 14,799 rows and 16 columns. The first column is the respondent identifier (h), the second column is the choice indicator (y) and the remaining 14 columns contain the difference in the attribute vectors ($x_{hi1} - x_{hi2}$). The first 20 lines of the choice data file are displayed in Table A.2. The choice indicator takes on values of 1 or 0, with 1 indicating that the first card was selected by the respondent, and 0 indicating that the second card was selected. The elements of the differenced attribute vector take on values of 1, 0 or −1. Sample observations from the demographic data file are presented in Table A.3. Income is recorded in thousands of dollars ($000) and the gender variable is coded 1 for female.

R Both the choice/attribute and demographic data are available in the `bayesm` data set, `bank`. To access, this data use `data(bank)` (after installing and loading `bayesm`; see Appendix B). To obtain information about this dataset, use `help(bank)`.

Reading and Organizing Data for Analysis

The choice data are read into the R statistical package and organized for analysis with the hierarchical binary logit model using the following commands (see the examples section of the help file for the `bank` dataset):

```
data(bank)
choiceAtt=bank$choiceAtt
Z=bank$demo

## center demo data so that mean of random effects
## distribution can be interpreted as the average respondents
```

Table A.2 Sample choice observations

id	choice	d1	d2	d3	d4	d5	d6	d7	d8	d9	d10	d11	d12	d13	d14
1	1	1	0	0	−1	0	0	0	0	0	0	0	0	0	0
1	1	1	0	0	1	−1	0	0	0	0	0	0	0	0	0
1	1	1	0	0	0	1	−1	0	0	0	0	0	0	0	0
1	1	0	0	0	0	0	0	1	0	−1	0	0	0	0	0
1	1	0	0	0	0	0	0	1	0	1	−1	0	0	0	0
1	1	0	0	0	0	0	0	−1	1	−1	0	0	0	0	0
1	1	0	0	0	0	0	0	−1	1	1	−1	0	0	0	0
1	1	0	0	0	1	0	0	0	0	0	0	0	0	−1	0
1	0	0	0	0	−1	1	0	0	0	0	0	0	0	−1	0
1	1	0	0	0	0	−1	1	0	0	0	0	0	0	−1	0
1	1	0	0	0	−1	0	0	1	−1	0	0	0	0	0	0
1	1	0	0	0	1	−1	0	−1	0	0	0	0	0	0	0
1	0	0	0	0	−1	0	0	0	0	0	0	−1	0	0	0
1	1	0	0	0	−1	0	0	0	0	0	0	1	−1	0	0
1	1	0	0	0	0	0	0	0	0	−1	0	0	0	−1	0
1	0	0	0	0	0	0	0	0	0	1	0	0	0	−1	0
2	1	1	0	0	−1	0	0	0	0	0	0	0	0	0	0
2	1	1	0	0	1	−1	0	0	0	0	0	0	0	0	0
2	1	1	0	0	0	1	−1	0	0	0	0	0	0	0	0
2	1	0	0	0	0	0	0	1	0	−1	0	0	0	0	0

```
Z[,1]=rep(1,nrow(Z))
Z[,2]=Z[,2]-mean(Z[,2])
Z[,3]=Z[,3]-mean(Z[,3])
Z[,4]=Z[,4]-mean(Z[,4])
Z=as.matrix(Z)

hh=levels(factor(choiceAtt$id))
nhh=length(hh)
lgtdata=NULL
for (i in 1:nhh) {
     y=choiceAtt[choiceAtt[,1]==hh[i],2]
     nobs=length(y)
     X=as.matrix(choiceAtt[choiceAtt[,1]==hh[i],c(3:16)])
     lgtdata[[i]]=list(y=y,X=X)
            }

cat("Finished Reading data",fill=TRUE)
fsh()

Data=list(lgtdata=lgtdata,Z=Z)
Mcmc=list(R=20000,sbeta=0.2,keep=20)
out=rhierBinLogit(Data=Data,Mcmc=Mcmc)
```

The first and second lines above read the data into the R, and the last line of the file calls a subroutine to execute the Markov Chain Monte Carlo computations and return the results. The remaining lines of the file organize the data and set parameters for analysis.

Commands in the `bank` help file can be copied and pasted into the R console window. We begin our tutorial by cutting and pasting selected lines.

The R response to the commands is to return the prompt ">":

```
>data(bank)
>choiceAtt=bank$choiceAtt
>Z=bank$demo
>
```

indicating the command has been successfully executed. The objects `choiceAtt` and `Z` are now available for use. Both are data frames. R stores the data as an array whose dimension can be determined from the "dim" command:

```
> dim(choiceAtt)
[1] 14799    16
> dim(Z)
[1] 946   4
>
```

Table A.3 Sample demographic observations

id	age	income	gender
1	60	20	1
2	40	40	1
3	75	30	0
4	40	40	0
6	30	30	0
7	30	60	0
8	50	50	1
9	50	100	0
10	50	50	0
11	40	40	0
12	30	30	0
13	60	70	0
14	75	50	0
15	40	70	0
16	50	50	0
17	40	30	0
18	50	40	0
19	50	30	1
20	40	60	0
21	50	100	0

indicating that `choiceAtt` is comprised of 14,799 rows and 16 columns, and `Z` has 946 rows (one for each respondent) and four columns. More information about these variables can be obtained with the command `str`:

```
> str(choiceAtt)
'data.frame':   14799 obs. of  16 variables:
$ id    : int   1 1 1 1 1 1 1 1 1 1 ...
$ choice: int   1 1 1 1 1 1 1 1 0 1 ...
$ d1    : int   1 1 1 0 0 0 0 0 0 0 ...
$ d2    : int   0 0 0 0 0 0 0 0 0 0 ...
$ d3    : int   0 0 0 0 0 0 0 0 0 0 ...
$ d4    : int   -1 1 0 0 0 0 0 1 -1 0 ...
$ d5    : int   0 -1 1 0 0 0 0 0 1 -1 ...
$ d6    : int   0 0 -1 0 0 0 0 0 0 1 ...
$ d7    : int   0 0 0 1 1 -1 -1 0 0 0 ...
$ d8    : int   0 0 0 0 0 1 1 0 0 0 ...
$ d9    : int   0 0 0 -1 1 -1 1 0 0 0 ...
$ d10   : int   0 0 0 0 -1 0 -1 0 0 0 ...
$ d11   : int   0 0 0 0 0 0 0 0 0 0 ...
```

```
$ d12    : int   0 0 0 0 0 0 0 0 0 0 ...
$ d13    : int   0 0 0 0 0 0 0 -1 -1 -1 ...
$ d14    : int   0 0 0 0 0 0 0 0 0 0 ...
> str(Z)
'data.frame':    946 obs. of  4 variables:
$ id     : int  1 2 3 4 6 7 8 9 10 11 ...
$ age    : int  60 40 75 40 30 30 50 50 50 40 ...
$ income: int  20 40 30 40 30 60 50 100 50 40 ...
$ gender: int  1 1 0 0 0 0 1 0 0 0 ...
>
```

The data array, Z, contains demographic variables. We must mean-center these variables so that the mean of the random effects distribution can be interpreted as the average respondent's part-worths. To do this, we replace the first column of Z with the intercept vector and mean-center age, income, gender.

```
> Z[,1]=rep(1,nrow(Z))
> Z[,2]=Z[,2]-mean(Z[,2])
> Z[,3]=Z[,3]-mean(Z[,3])
> Z[,4]=Z[,4]-mean(Z[,4])
> Z=as.matrix(Z)
```

The data array, choiceAtt, contains a respondent identifier in column 1, the choice indicator in column 2, and a set of explanatory variables in columns 3–16. Each respondent provided between 13 and 17 responses to product descriptions, so it is not possible to format the data as a high-dimensional array. We instead form a data list by first using the commands:

```
> hh=levels(factor(choiceAtt$id))
> nhh=length(hh)
>
```

to identify the unique respondent indicators (hh) and to determine the number of respondents in the study. The output from the first of these two commands is a variable hh that denotes households containing a list of unique identifier codes:

```
> str(hh)
chr [1:946] "1" "2" "3" "4" "6" "7" "8" "9" "10" "11" "12" ...
> nhh
[1] 946
>
```

The quotes around the numbers indicate that the levels of the respondent identifiers are stored as a vector of characters, in contrast to the elements of Z that are stored as integers. The variable nhh is a scalar with value equal to 946, the number of respondents in the study.

The variables hh and nhh are used to create a data structure that allows easy access to each respondent's data. This is accomplished by creating a data list for each respondent that contains their choice vector (y_h) and matrix of attribute descriptors (X_h) whose lengths correspond to the number of responses. The data structure is named Data, and is created by iterating a set of commands using a loop:

```
> hh=levels(factor(choiceAtt$id))
> nhh=length(hh)
>
> lgtdata=NULL
>
> for (i in 1:nhh) {
+ y=choiceAtt[choiceAtt[,1]==hh[i],2]
+ nobs=length(y)
+ X=as.matrix(choiceAtt[choiceAtt[,1]==hh[i],c(3:16)])
+ lgtdata[[i]]=list(y=y,X=X)
+ }
> Data=list(lgtdata=lgtdata,Z=Z)
>
```

The NULL command initializes the variable lgtdata as a 'list' with zero length, and the loop beginning with the for command expands this list to contain respondents' data. Upon completion of the loop, the variable Data contains 946 elements, with each element containing the respondent's data. The first element of Data is comprised of two variables, y and X, lgtdata[[1]]$y and lgtdata[[1]]$X, as follows:

```
>Data$lgtdata[[1]]
 $y
  [1] 1 1 1 1 1 1 1 0 1 1 1 0 1 1 0

$X
   Med_FInt Low_FInt Med_VInt Rewrd_2 Rewrd_3 Rewrd_4 Med_Fee
       Low_Fee Bank_B Out_State
1     1     0     0    -1     0     0     0     0     0     0
2     1     0     0     1    -1     0     0     0     0     0
3     1     0     0     0     1    -1     0     0     0     0
4     0     0     0     0     0     0     1     0    -1     0
5     0     0     0     0     0     0     1     0     1    -1
6     0     0     0     0     0     0    -1     1    -1     0
7     0     0     0     0     0     0    -1     1     1    -1
8     0     0     0     1     0     0     0     0     0     0
9     0     0     0    -1     1     0     0     0     0     0
10    0     0     0     0    -1     1     0     0     0     0
11    0     0     0    -1     0     0     1    -1     0     0
12    0     0     0     1    -1     0    -1     0     0     0
```

```
13  0  0  0  -1  0  0  0  0   0  0
14  0  0  0  -1  0  0  0  0   0  0
15  0  0  0   0  0  0  0  0  -1  0
16  0  0  0   0  0  0  0  0   1  0
   Med_Rebate High_Rebate High_CredLine Long_Grace
1           0           0             0          0
2           0           0             0          0
3           0           0             0          0
4           0           0             0          0
5           0           0             0          0
6           0           0             0          0
7           0           0             0          0
8           0           0            -1          0
9           0           0            -1          0
10          0           0            -1          0
11          0           0             0          0
12          0           0             0          0
13         -1           0             0          0
14          1          -1             0          0
15          0           0            -1          0
16          0           0            -1          0
```

which corresponds to the data for the first respondent. Moreover, choices for the respondents are stored as vectors, and the (differenced) attribute levels (X) are stored as a matrix so that matrix operations (such as, matrix multiplication) can be performed:

```
> is.vector(Data$lgtdata[[1]]$y)
[1] TRUE
> is.matrix(Data$lgtdata[[1]]$X)
[1] TRUE
```

The remaining lines of code are used to write comments to the screen so that the user can monitor progress of the program, and to create other lists that are used within the subroutine that is called.

The list Mcmc is a list of parameters that control the Markov chain:

```
> Mcmc=list(R=20000,sbeta=0.2,keep=20)
>
```

where R is the number of iterations of the chain, sbeta is used to control the step size of the Metropolis–Hastings random-walk chain, and keep indicates the proportion of draw retained for analysis. If keep = 1, every draw is kept. If keep=20, every 20th draw is kept, and the other 19 draws are discarded from analysis. The keep variable is useful for reducing the memory requirements of the program. The contents of Mcmc can be checked by entering this variable name (Mcmc) at the command prompt:

```
> Mcmc
$R
[1] 20000

$sbeta
[1] 0.2

$keep
[1] 20

>
```

The last command above,

```
out=rhierBinLogit(Data,Mcmc)
```

calls a function named `rhierBinLogit` with arguments `Data` and `Mcmc` as defined above. The result is stored in the variable `out`. `out` will contain a list of variables that can be accessed using the `$` convention, as in `out$varname`.

Hierarchical Bayes Binary Logit R Code

The complete specification of the hierarchical Bayes binary logit model augments the likelihood in equation (A.1) with a random effects distribution of heterogeneity for β_h:

$$\beta_h \sim N(\Delta' z_h, V_\beta) \tag{A.2}$$

or

$$B = Z\Delta + U \qquad u_i \sim N(0, V_\beta), \tag{A.3}$$

where u_i and β_i are the ith rows of B and U.

The mean of the random effects distribution is dependent on the values of the demographic variables, z_h, and the estimated matrix of coefficients Δ. The prior distribution for Δ is

$$\text{vec}(\Delta | V_\beta) \sim N(\text{vec}(\bar{\Delta}), A^{-1} \otimes V_\beta), \tag{A.4}$$

and the prior on the covariance matrix is

$$V_\beta \sim \text{IW}(\nu, V_0). \tag{A.5}$$

The function `rhierBinLogit` extracts the relevant information from the calling arguments `Data` and `Mcmc`, initializes storage space for the draws, and then executes the computations associated with the Markov chain. Information contained in the calling arguments is copied into local variables.

To display the code in `rhierBinLogit`, type `rhierBinLogit` at the command prompt in R. This will display the code in the R console window.

The first part of `rhierBinLogit` checks for valid arguments and implements defaults. Here we are using the defaults for all prior settings.

The heart of the function starts with allocation of storage for the draws.

```
Vbetadraw=matrix(double(floor(R/keep)*nvar*nvar),ncol=nvar*nvar)
betadraw=array(double(floor(R/keep)*nlgt*nvar),
  dim=c(nlgt,nvar,floor(R/keep)))
Deltadraw=matrix(double(floor(R/keep)*nvar*nz),ncol=nvar*nz)
oldbetas=matrix(double(nlgt*nvar),ncol=nvar)
oldVbeta=diag(nvar)
oldVbetai=diag(nvar)
oldDelta=matrix(double(nvar*nz),ncol=nvar)
betad = array(0,dim=c(nvar))
betan = array(0,dim=c(nvar))
reject = array(0,dim=c(R/keep))
llike=array(0,dim=c(R/keep))
```

where `Vbetadraw` are the saved draws of the random effects covariance matrix (V_β), `betadraw` are the saved draws of the respondent-level coefficients (β_h), `Deltadraw` contains the draws of regression coefficients in the random effects distribution (Δ). A total of `R/keep` iterations are saved, and the draw of V_β is stored as a row vector of dimension nxvar^2.

Current draws of the variables are referred to as 'old' variables, and take on the names `oldbetas`, `oldVbeta`, `oldDelta`, and `oldVbetai`, the later of which is equal to V_β^{-1}. The variables `betad` and `betan` are the 'default' and 'new' draws of beta used in the Metropolis–Hastings step of the algorithm. The variables `reject` and `llike` are summary measures of the performance of the chain and will be discussed later.

The function `init.rmultiregfp` is then called to set up parameters for the multivariate regression:

```
Fparm=init.rmultiregfp(Z,ADelta,Deltabar,nu,V0)
```

Variables that monitor the speed of the chain, and are used to write out time-to-completion information to the screen, are initialized prior to the start of the first iteration:

```
itime=proc.time()[3]
cat("MCMC Iteration (est time to end - min)",fill=TRUE)
fsh()
```

and are found at the end of the iteration:

```
if((j%%100)==0)      {
ctime=proc.time()[3]
timetoend=((ctime-itime)/j)*(R-j)
cat(" ",j," (",round(timetoend/60,1),")",fill=TRUE)
fsh() }
```

'%%' is a 'mod' operator in R, so time-to-completion information will be written to the R window every 500th iteration.

The algorithm for estimating the model proceeds in two major steps for each of R iterations. The first step involves generating draws of β_h for each respondent, $h = 1, \ldots,$ nhh:

```
for (j in 1:R) {
      rej = 0
      logl = 0
      sV = sbeta*oldVbeta
      root=t(chol(sV))
      for (i in 1:nlgt) {

             betad = oldbetas[i,]
             betan = betad + root%*%rnorm(nvar)
             lognew = loglike(Data[[i]]$y,Data[[i]]$X,betan)
             logold = loglike(Data[[i]]$y,Data[[i]]$X,betad)
logknew = -.5*(t(betan)-Demo[i,]%*%oldDelta) %*% oldVbetai %*%
(betan-t(Demo[i,]%*%oldDelta))

logkold = -.5*(t(betad)-Demo[i,]%*%oldDelta) %*% oldVbetai %*%
(betad-t(Demo[i,]%*%oldDelta))

             alpha = exp(lognew + logknew - logold - logkold)
             if(alpha=="NaN") alpha=-1
             u = runif(n=1,min=0, max=1)
             if(u < alpha) {
                    oldbetas[i,] = betan
                    logl = logl + lognew } else {
                    logl = logl + logold
                    rej = rej+1  }
             }
```

A random-walk Metropolis–Hastings algorithm is used here, where the candidate vector of coefficients, β_h^n, is obtained by perturbing the existing vector of coefficients, β_h^d by a normal draw with mean zero and covariance proportional to the current draw of the covariance matrix, sbeta * V_β. In chapter 5 we provide an alternative method of tuning the RW chain. The method used here will work well for this sort of conjoint data with a small number of observations per respondent. The likelihood of the data is evaluated for both old and new coefficient vector, is multiplied by the density contribution from the distribution of heterogeneity, and used to form alpha, the acceptance probability for the Metropolis–Hastings algorithm. A uniform draw (u) is generated, and if u is less than alpha, the new vector of coefficients is accepted, otherwise it is rejected. The variable logl is used to accumulate an estimate of the log-likelihood of the data evaluated at the posterior draws, and the variable rej monitors the proportion of draws that are accepted by the algorithm. This set of computations is computed for each respondent on each iteration of the Markov chain.

The second major step generates the draws of the matrix of regression coefficients Δ and the covariance matrix V_β:

```
out=rmultiregfp(oldbetas,Z,Fparm)
oldDelta=out$B
oldVbeta=out$Sigma
oldVbetai=chol2inv(chol(oldVbeta))
```

which employs the function `rmultiregfp` which is part of our package, *bayesm*.

The remaining statements in the `rhierBinLogit` function are used to save the draws into the appropriate storage locations. The last line of the subroutine, which specifies the information passed back to the calling program, is a list:

```
return(list(betadraw=betadraw,Vbetadraw=Vbetadraw,
Deltadraw=Deltadraw, llike=llike,reject=reject))
```

which means that the saved draws are retrievable as `out$betadraw`, `out$Vbetadraw`, etc.

Obtaining Parameter Estimates

The Markov chain program is initiated by a call to `rhierBinLogit`. The call to `rhierBinLogit` will print out the problem and then give a trace of every 100th iteration along with the estimated time to completion.

A total of 20000 iterations of the Markov chain were completed in 130.69 minutes on Pentium 4 chip running at 1.78 GHz. This translates to about 150 iterations per minute for the 946 respondents and 14,799 observations. The structure of the variable out is as follows:

```
> str(out)
List of 5
$ betadraw : num [1:946, 1:14, 1:1000] 0.01251 -0.36642 ...
$ Vbetadraw: num [1:1000, 1:196] 0.0691 0.0963 0.1058 ...
$ Deltadraw: num [1:1000, 1:56] -0.0147 -0.0141 -0.0302 ...
$ llike    : num [, 1:1000] -9722 -9520 -9323 -9164 -9007 ...
$ reject   : num [, 1:1000] 0.621 0.614 0.607 0.605 0.608 ...
>
```

Figure A.7 displays a plot of the draws of the mean of the random effects distribution, Δ. The plot is of the 'intercept' elements of Δ:

```
> index=4*c(0:13)+1
> matplot(out$Deltadraw[,index],type="l",xlab="Iterations/20",
   ylab=" ",main="Average Respondent Part-Worths")
>
```

Similarly, Figure A.8 displays a plot of the diagonal elements of the covariance matrix:

```
> index=c(0:13)*15+1
> index
[1]   1  16  31  46  61  76  91 106 121 136 151 166 181 196
```

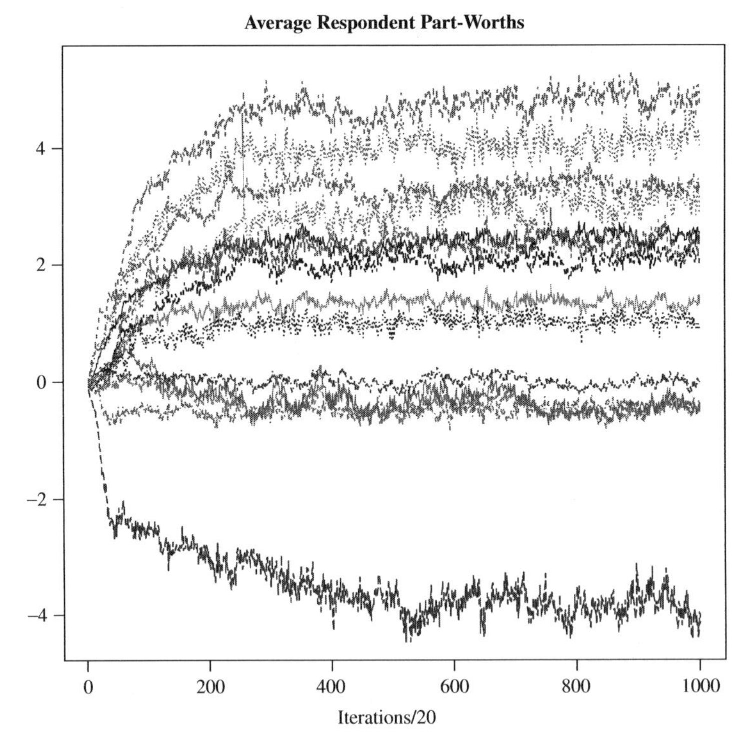

Figure A.7 Draws of the mean of the random effects distribution: every 20th draw was retained for analysis

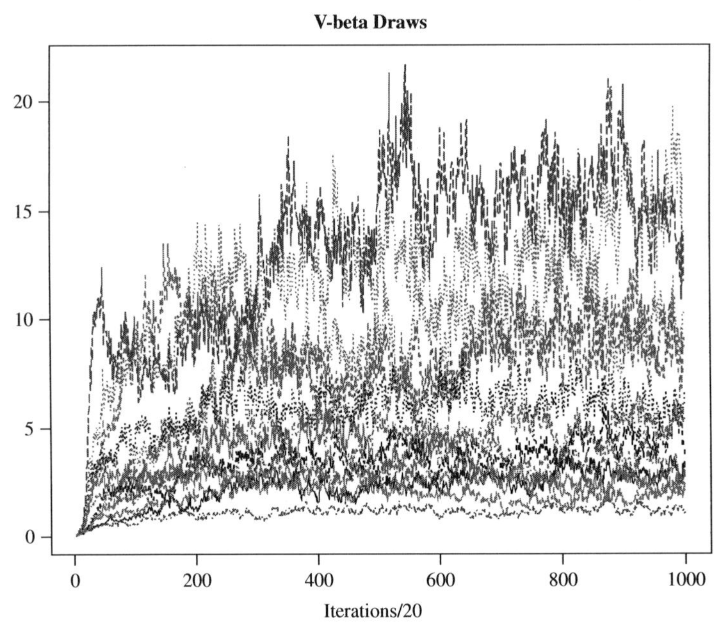

Figure A.8 Draws of the diagonal elements of the covariance matrix of random effects: every 20th draw retained for analysis

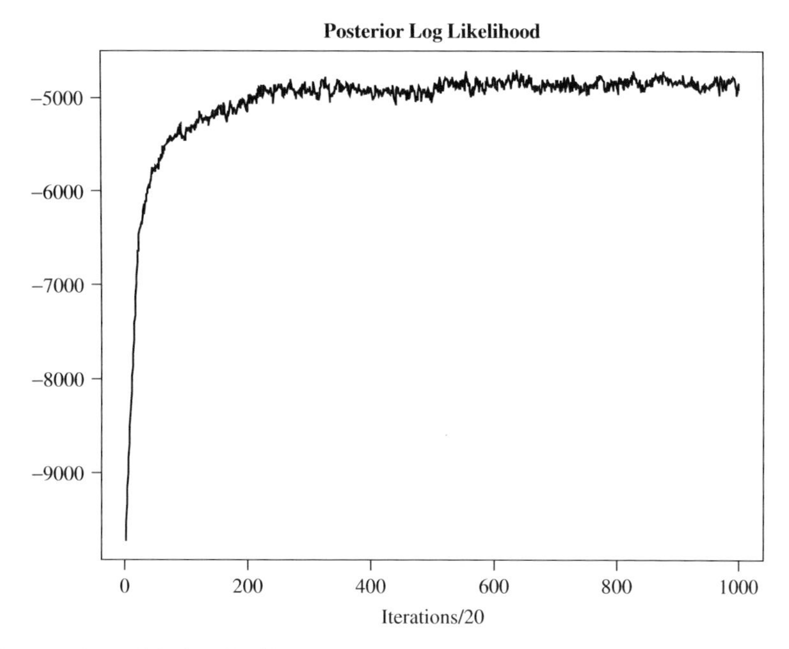

Figure A.9 Values of the log-likelihood of the data evaluated at posterior draws of individual-level part-worth estimates (β_h): every 20th draw retained for analysis

```
> matplot(out$Vbetadraw[,index],type="l",xlab="Iterations/20",
    ylab=" ",main="V-beta Draws")
>
```

Figure A.9 displays a plot of the log-likelihood values that are useful for assessing model fit:

```
> plot(out$llike,type="l",xlab="Iterations/20",ylab=" ",
main="Posterior Log Likelihood")
>
```

Figure A.10 displays a plot of the rejection rate of the Metropolis–Hastings algorithm:

```
> plot(out$reject,type="l",xlab="Iterations/20",ylab=" ",
main="Rejection Rate of Metropolis-Hastings Algorithm")
>
```

It is important to remember that the draws converge in distribution to the posterior distribution of the model parameters, in contrast to other forms of estimation (for example, maximum likelihood and method of moments) where convergence is to a point. Upon convergence, the draws can be used to construct summary measures of the distribution, such as histograms, confidence intervals and point estimates. Figure A.11 displays the distribution of heterogeneity for selected part-worths using draws of the individual respondent part-worths, $\{\beta_h, h = 1, \ldots, H\}$:

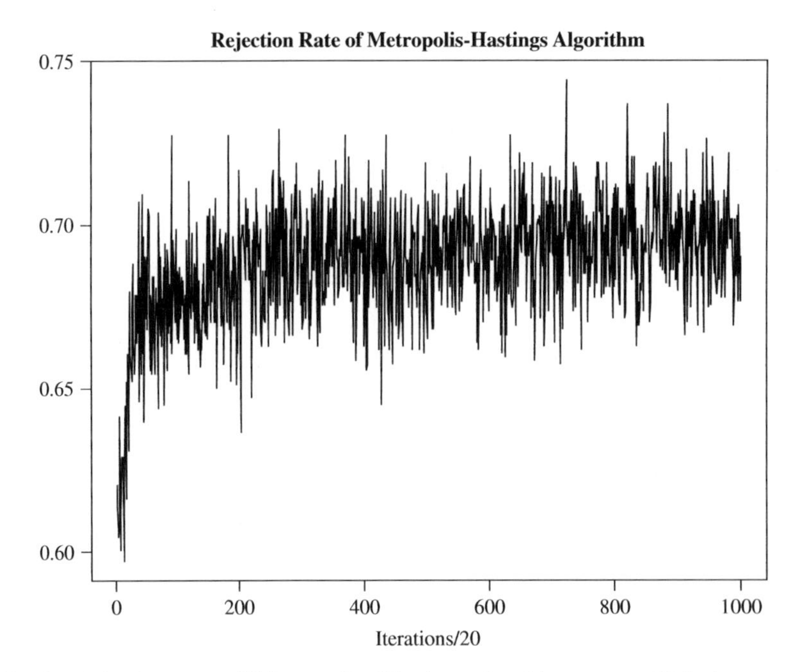

Figure A.10 Rejection rate of Metropolis–Hastings algorithm: every 20th draw retained for analysis

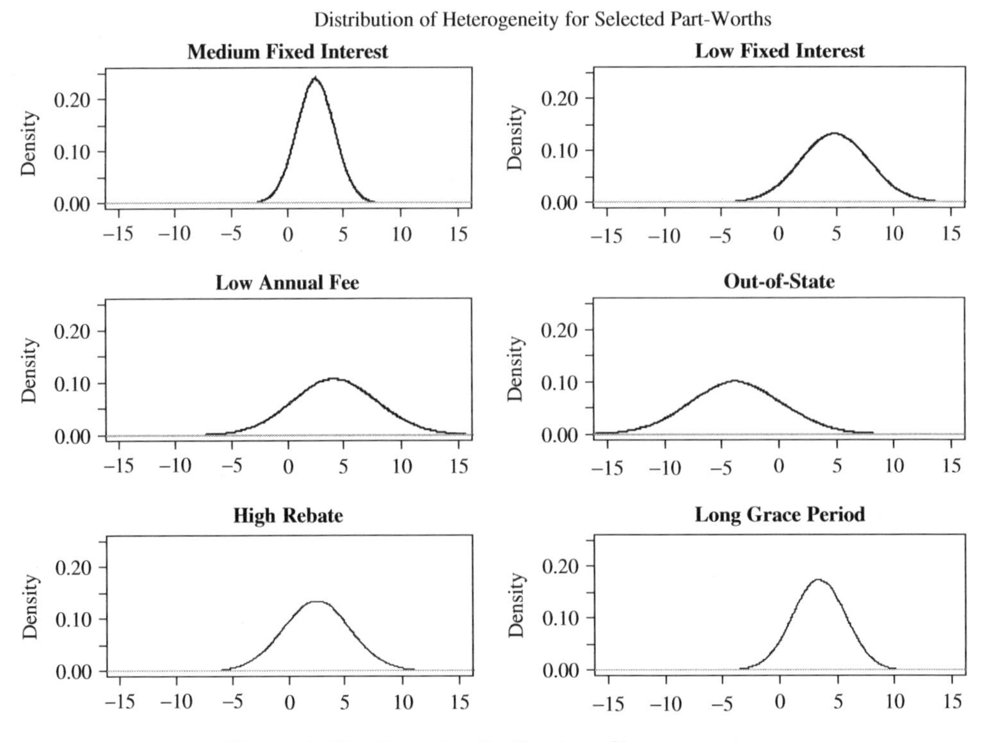

Figure A.11 Posterior distribution of heterogeneity

```
> par(mfrow=c(3,2),oma=c(2,0,3,0))
> plot(density(out$betadraw[,1,500:1000]),main="Medium
    Fixed Interest",xlab=" ",xlim=c(-15,15),ylim=c(0,.25))
> plot(density(out$betadraw[,2,500:1000]),main="Low Fixed
Interest",xlab=" ",xlim=c(-15,15),ylim=c(0,.25))
> plot(density(out$betadraw[,8,500:1000]),main="Low Annual
Fee",xlab=" ",xlim=c(-15,15),ylim=c(0,.25))
> plot(density(out$betadraw[,10,500:1000]),main="Out-of-
State",xlab=" ",xlim=c(-15,15),ylim=c(0,.25))
> plot(density(out$betadraw[,12,500:1000]),main="High
    Rebate",xlab=" ",xlim=c(-15,15),ylim=c(0,.25))
> plot(density(out$betadraw[,14,500:1000]),main="Long Grace
Period",xlab=" ",xlim=c(-15,15),ylim=c(0,.25))
> mtext("Distribution of Heterogeneity for Selected Part-
Worths",side=3,outer=TRUE,cex=1.3)
>
```

and Figure A.12 displays estimates of the individual-level posterior distributions for one respondent (no. 250) using $\{\beta_{250}\}$:

```
> par(mfrow=c(3,2),oma=c(2,0,3,0))
> plot(density(out$betadraw[250,1,500:1000]),main="Medium
    Fixed Interest",xlab=" ",xlim=c(-15,15),ylim=c(0,.35))
> plot(density(out$betadraw[250,2,500:1000]),main="Low Fixed
```

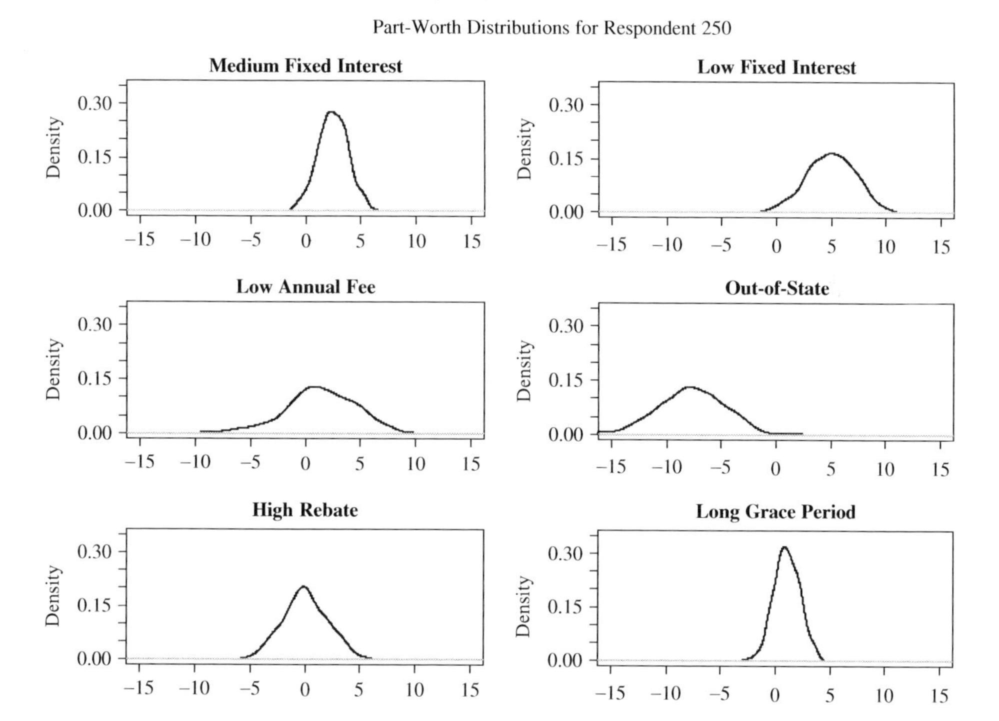

Figure A.12 Posterior distribution of part-worths for respondent no. 250

```
Interest",xlab=" ",xlim=c(-15,15),ylim=c(0,.35))
> plot(density(out$betadraw[250,8,500:1000]),main="Low Annual
Fee",xlab=" ",xlim=c(-15,15),ylim=c(0,.35))
> plot(density(out$betadraw[250,10,500:1000]),main="Out-of-
State",xlab=" ",xlim=c(-15,15),ylim=c(0,.35))
> plot(density(out$betadraw[250,12,500:1000]),main="High
Rebate",xlab=" ",xlim=c(-15,15),ylim=c(0,.35))
> plot(density(out$betadraw[250,14,500:1000]),main="Long Grace
Period",xlab=" ",xlim=c(-15,15),ylim=c(0,.35))
> mtext("Part-Worth Distributions for Respondent
250",side=3,outer=TRUE,cex=1.3)
>
```

Point estimates of the model parameters can be easily generated from the saved draws using the `apply` command. The mean and standard deviations of the coefficient matrix Δ are:

```
> t(matrix(apply(out$Deltadraw[500:1000,],2,mean),ncol=14))
             [,1]          [,2]          [,3]        [,4]
[1,]   2.51252961 -0.0130934197  0.0109217374  0.10601398
[2,]   4.88334510 -0.0253067253  0.0205866845  0.32386912
[3,]   3.12238931  0.0017723979  0.0254455424 -0.35367756
[4,]   0.06106729  0.0047622654  0.0005271013 -0.24818139
[5,]  -0.39090470  0.0216154219  0.0140768695 -0.22417571
[6,]  -0.29720158  0.0192823095  0.0159921716 -0.24252345
[7,]   2.14173758 -0.0038188098  0.0024588301  0.66838042
[8,]   4.15796844 -0.0095154517  0.0039886913  1.30239133
[9,]  -0.39701125  0.0007499859  0.0029704428  0.12386156
[10,] -3.75767707 -0.0029890846  0.0129988106 -0.05409753
[11,]  1.42600619 -0.0079048966  0.0027301444  0.23053783
[12,]  2.45602178 -0.0142630568  0.0208663278  0.37872870
[13,]  1.11570025 -0.0095763934 -0.0029792910  0.36773564
[14,]  3.39939098 -0.0201861790  0.0192602035  0.29644011

> t(matrix(apply(out$Deltadraw[500:1000,],2,sd),ncol=14))
           [,1]        [,2]        [,3]      [,4]
[1,] 0.11498124 0.008107402 0.004534418 0.1943663
[2,] 0.17897434 0.012265078 0.007142939 0.3418811
[3,] 0.20341931 0.013359792 0.007444453 0.4252213
[4,] 0.09873792 0.006967416 0.003527673 0.1590828
[5,] 0.12626135 0.009705119 0.005814927 0.2495430
[6,] 0.17520820 0.012854631 0.008517397 0.3767015
[7,] 0.12045459 0.008761116 0.005717579 0.2454403
[8,] 0.19746405 0.013435594 0.008607683 0.3741644
[9,] 0.09928150 0.007453370 0.004176626 0.2140552
[10,] 0.23588829 0.018475045 0.009567594 0.4383213
[11,] 0.09206959 0.007181072 0.004703831 0.2367981
[12,] 0.16244095 0.011654455 0.008262366 0.3555747
[13,] 0.10998549 0.008830440 0.005254209 0.2575200
[14,] 0.13678767 0.009390693 0.005692094 0.2537742
>
```

Table A.4 Posterior mean of Δ

Attribute levels	Intercept	Age	Income	Gender
Medium Fixed Interest	2.513	−0.013	0.011	0.106
Low Fixed Interest	4.883	−0.025	0.021	0.324
Medium Variable Interest	3.122	0.002	0.025	−0.354
Reward Program 2	0.061	0.005	0.001	−0.248
Reward Program 3	−0.391	0.022	0.014	−0.224
Reward Program 4	−0.297	0.019	0.016	−0.243
Medium Annual Fee	2.142	−0.004	0.002	0.668
Low Annual Fee	4.158	−0.010	0.004	1.302
Bank B	−0.397	0.001	0.003	0.124
Out-of-State Bank	−3.758	−0.003	0.013	−0.054
Medium Rebate	1.426	−0.008	0.003	0.231
High Rebate	2.456	−0.014	0.021	0.379
High Credit Line	1.116	−0.010	−0.003	0.368
Long Grace Period	3.399	−0.020	0.019	0.296

Similar statistics are readily computed for the elements of V_β and any of the individual-level parameters $\{\beta_h\}$. Estimates of the posterior mean of Δ and V_β are summarized in Tables A.4 and A.5.

Table A.5 Posterior mean of V_β

Attribute levels														
Medium Fixed Interest	2.8	4.5	4.3	0.0	0.2	0.5	1.1	2.0	0.3	1.2	0.7	1.4	0.9	1.4
Low Fixed Interest	4.5	8.8	8.2	0.0	0.4	0.9	1.8	3.4	0.5	2.1	1.2	2.5	1.2	2.4
Medium Variable Interest	4.3	8.2	10.2	−0.1	0.5	1.2	1.9	3.7	0.1	1.5	1.5	2.9	2.1	3.0
Reward Program 2	0.0	0.0	−0.1	1.2	0.2	−0.2	−0.3	−0.5	0.0	0.5	−0.3	−0.5	−0.4	−0.5
Reward Program 3	0.2	0.4	0.5	0.2	2.0	1.8	−0.2	−0.7	0.5	1.4	−0.4	−0.9	0.4	−1.0
Reward Program 4	0.5	0.9	1.2	−0.2	1.8	3.0	−0.3	−0.8	0.2	0.3	−0.6	−1.2	0.3	−1.1
Medium Annual Fee	1.1	1.8	1.9	−0.3	−0.2	−0.3	4.1	6.9	1.4	4.4	1.0	2.2	2.3	2.6
Low Annual Fee	2.0	3.4	3.7	−0.5	−0.7	−0.8	6.9	13.5	2.7	8.1	2.1	4.7	4.3	5.1
Bank B	0.3	0.5	0.1	0.0	0.5	0.2	1.4	2.7	3.5	5.6	1.2	2.5	2.0	0.9
Out-of-State Bank	1.2	2.1	1.5	0.5	1.4	0.3	4.4	8.1	5.6	15.9	2.3	4.8	3.9	2.0
Medium Rebate	0.7	1.2	1.5	−0.3	−0.4	−0.6	1.0	2.1	1.2	2.3	2.4	4.0	2.0	2.1
High Rebate	1.4	2.5	2.9	−0.5	−0.9	−1.2	2.2	4.7	2.5	4.8	4.0	8.5	3.6	4.5
High Credit Line	0.9	1.2	2.1	−0.4	0.4	0.3	2.3	4.3	2.0	3.9	2.0	3.6	6.3	2.6
Long Grace Period	1.4	2.4	3.0	−0.5	−1.0	−1.1	2.6	5.1	0.9	2.0	2.1	4.5	2.6	5.0

Appendix B
A Guide to Installation and Use of *bayesm*

An R package is a collection of functions, data sets and documentation which can be easily installed and updated from within R. The package resides on the CRAN network of world-wide mirror sites so that its availability is assured. *bayesm* is an R package that implements the methods in this book. The tables at the end of this appendix serve as a quick reference to the most important *bayesm* functions (however, please install the latest version of *bayesm* to see the full menu of functions available).

B.1 INSTALLING bayesm

bayesm can be installed from a CRAN mirror site by selecting 'Packages' on the menu bar in the R software package. Users must first set the CRAN mirror, then install *bayesm* directly from CRAN by simply clicking on '*bayesm*' in the pop-up window displayed (see Appendix A for details). You can also install *bayesm* by using the R command

```
install.packages("bayesm")
```

The `install` command downloads the content of the package and builds a directory tree in the `library` subdirectory of the R installation. If you upgrade your version of R, you must reinstall *bayesm*.

B.2 USING bayesm

Once *bayesm* has been installed, you must load the package for each R session. This is done with the command `library(bayesm)`. In order to avoid giving this command each time you invoke R, it is convenient to include the command in the `.Rprofile` file. `.Rprofile` is executed each time R is invoked. `.Rprofile` should be located in directory pointed to by the environmental variable, HOME. Environmental variables are set by right-clicking on the 'My Computer' icon on your desktop, selecting 'Properties', and then selecting the 'Advanced' tab.

Bayesian Statistics and Marketing P. E. Rossi, G. M. Allenby and R. McCulloch

Sample content of .Rprofile is given below:

```
# Example of .Rprofile
     .First <- function() {
        library(bayesm)
     }
     .Last <- function()  cat("\n   Goodbye!\n\n")
```

Once loaded, you can use any of the more than 40 functions in *bayesm* just by inserting the function call in your source code. For all except the most trivial problems, it is best to edit a source file with your code and then 'source' this file into R. See Appendix A for more details.

The following tips are useful in using *bayesm*:

1. If you do not know much about R, read Appendix A. *bayesm* has many functions defined in it. The 'turn-key' or 'end-user' functions start with the letter `r`, e.g. `rmnpGibbs` is the Gibbs sampler for the multinomial probit model.
2. Check the examples. Each function has an example file. To view the example for a function, use the R command `?`, e.g. `?rmnpGibbs`. The example will be listed at the bottom of the displayed help text. You can also find the examples in the R program directory tree, e.g. C:\Program Files\R\rwXXXX\library\bayesm\R-ex. You may have to unzip these files.
3. The best way to work with *bayesm* functions is to copy the examples into a .R file and then edit the file to read in your own data and run the function. At first, use as many defaults as possible (for example, priors) to make sure that the function is working properly on your example.

B.3 OBTAINING HELP ON bayesm

Once *bayesm* has been loaded, you can obtain help in various ways.

To see a list of all the functions in *bayesm*, use the command `library(help= bayesm)`. This command will produce only brief summaries of each function in *bayesm*.

To learn more, use the `help()` or `?` commands as for any function in R. This will provide the standard information on usage, arguments and output from the function. The most useful aspect of these files is the examples section. For example, `?rmnlIndepMetrop` produces:

```
rmnlIndepMetrop     package:bayesm     R Documentation

MCMC Algorithm for Multinomial Logit Model
Description:
     'rmnIndepMetrop' implements Independence Metropolis for the
        MNL.

Usage:
     rmnlIndepMetrop(Data, Prior, Mcmc)
```

```
Arguments:
    Data: list(p,X,y)
   Prior: list(A,betabar)  optional
    Mcmc: list(R,keep,nu)

Details:
     Model:   Pr(y=j) = exp(x_j'beta)/sum_k{e {x_k'beta}}.

     Prior:   beta ~ N(betabar,A^ {-1})

     list arguments contain:
   'p' number of alternatives
   'X' nobs*m x nvar matrix
   'y' nobs vector of multinomial outcomes (1,..., m)
   'A' nvar x nvar pds prior prec matrix (def: .01I)
   'betabar' nvar x 1 prior mean (def: 0)
   'R' number of MCMC draws
   'keep' MCMC thinning parm: keep every keepth draw (def: 1)
   'nu' degrees of freedom parameter for independence t density
        (def: 6)

Value:
     a list containing:
betadraw: R/keep x nvar array of beta draws
 acceptr: acceptance rate of Metropolis draws

See Also:
     'rhierMnlRwMixture'

Examples:
     ##
     if(nchar(Sys.getenv("LONG_TEST")) != 0) # set env var
        LONG_TEST to run
     {

     set.seed(66)
     n=200; p=3; beta=c(1,-1,1.5,.5)
     simout=simmnl(m,n,beta)
     A=diag(c(rep(.01,length(beta)))); betabar=rep(0,length(beta))

     R=2000
     Data=list(y=simout$y,X=simout$X,p=p);
     Mcmc=list(R=R,keep=1) ; Prior=list(A=A,betabar=betabar)
     out=rmnlIndepMetrop(Data=Data,Prior=Prior,Mcmc=Mcmc)
     cat(" Betadraws ",fill=TRUE)
     mat=apply(out$betadraw,2,quantile,probs=c(.01,.05,.5,
                                                        .95,.99))
     mat=rbind(beta,mat); rownames(mat)[1]="beta"; print(mat)
     }
```

The examples section could be clipped out and used as a template for running your own data.

To list all functions related to a specific topic, use the command `help.search("topic")`. For example, `help.search("bayes")` will produce:

```
Help files with alias or concept or title matching 'bayes' using
fuzzy matching:
breg(bayesm)               Posterior Draws from a Univariate
                           Regression with Unit Error Variance
eMixMargDen(bayesm)        Compute Marginal Densities of A Normal
                           Mixture Averaged over MCMC Draws
logMargDenNR(bayesm)       Compute Log Marginal Density Using
                           Newton-Raftery Approx
rbprobitGibbs(bayesm)      Gibbs Sampler (Albert and Chib) for Binary
                           Probit
rhierLinearModel(bayesm)
                           Gibbs Sampler for Hierarchical Linear
                              Model
rhierMnlRwMixture(bayesm)
                           MCMC Algorithm for Hierarchical
                              Multinomial
                           Logit with Mixture of Normals
                              Heterogeneity
rivGibbs(bayesm)           Gibbs Sampler for Linear "IV" Model
rmnlIndepMetrop(bayesm)
                           MCMC Algorithm for Multinomial Logit Model
rmnpGibbs(bayesm)          Gibbs Sampler for Multinomial Probit
rmultireg(bayesm)          Draw from the Posterior of a Multivariate
                           Regression
rmvpGibbs(bayesm)          Gibbs Sampler for Multivariate Probit
rnmixGibbs(bayesm)         Gibbs Sampler for Normal Mixtures
rscaleUsage(bayesm)        MCMC Algorithm for Multivariate Ordinal
                           Data with Scale Usage Heterogeneity.
runireg(bayesm)            Draw from Posterior for Univariate
                           Regression
runiregGibbs(bayesm)       Gibbs Sampler for Univariate Regression
```

Note that this list may contain functions from other packages. To restrict search to only *bayesm*, use the command, `help.search("topic",package="bayesm")`

The manual for *bayesm* is available in the help menu in R. Use 'Help' item from the menu bar and click on 'Html help', then 'Packages', and then '*bayesm*'. This will give you a linked HTML document with all of the defined functions.

A very nice-looking manual for *bayesm* can be found from this html menu under the top link 'browse directory' in the HTML page for *bayesm*. This is a PDF file created from a LaTeX version of the documentation. This same file can be found by going to the CRAN website and clicking on the 'packages' menu and then clicking on *bayesm*.

The file `bayesm.pdf` is the manual for the most recent version of the package. Google "CRAN" to find a link to the CRAN website.

B.4 TIPS ON USING MCMC METHODS

All of the important functions in *bayesm* are implementations of various MCMC methods. The following tips are useful:

1. If you are unfamiliar with MCMC methods, read Chapter 3. Try some of our test examples before trying your own data.
2. The 'output' of an MCMC method is a set of draws of the parameters. You must decide how many draws to make and also how to analyze the draws produced. Unlike most classical methods, the MCMC methods provide an estimate of the entire posterior distribution, not just a few moments. Summarize the distribution by using histograms or quantiles. Resist the temptation to simply report the posterior mean and posterior standard deviation. For nonnormal distributions, these moments have little meaning!
3. Most of the MCMC methods implemented in *bayesm* run very fast so it is possible to make tens of thousands of draws even for relatively large data sets in less than 1/2 hour. Use this power where possible. Only the hierarchical models, `rhierLinearModel`, `rhierBinLogit`, `rmnlRwMixture`, and `rscaleUsage` will take appreciably longer (more than 1/2 hour, in some cases several hours, will be required).
4. If you are having problems with using too much memory, set `keep` in the `Mcmc` parameter list to more than 1.

B.5 EXTENDING AND ADAPTING OUR CODE

We hope that many researchers will find our functions useful. Some will want to adapt and extend our code. To find the R and C source code, go to the CRAN website, click on the 'packages' menu link. Zipped files with source for the package are available. When this file is 'unzipped' it will create a directory structure. Look at the directories 'R' for source for the functions and 'src' for the C/C++ code.

B.6 UPDATING bayesm

From time to time, we will release new versions of *bayesm*. To obtain the latest release, use the `update.packages()` function or choose 'Packages' from the menu bar and then 'Update packages'. If you would like to be informed about releases of R packages, consider subscribing to the R-help or R-packages email lists. See the mailing lists link on the R homepage.

Table B.1 `bayesm` MCMC functions

Name	Description	Model	Priors	Required arguments			Output
				Data	Prior	Mcmc	
`rbprobitGibbs`	Gibbs sampler for binary probit	$z = X\beta + \varepsilon, \varepsilon \sim N(0, 1)$ $y = 1$ if $z > 0$	$\beta \sim N(\bar{\beta}, A^{-1})$	X, y		R (no. of MCMC draws)	betadraw
`rhierBinLogit`	Hybrid sampler for hierarchical binary logit – designed for conjoint data. Note: x variables are expressed as a difference between alternatives. See Appendix A.	$y_{hi} = 1$ with $\text{pr} = \frac{\exp(x_{hi}'\beta_h)}{1 + \exp(x_{hi}'\beta_h)}$ $h = 1, \ldots, H$; $i = 1, \ldots, I_h$ H panel units $\beta_h \sim N(\Delta' z_h, V_\beta)$	$V_\beta \sim \text{IW}(\nu, V)$ $\text{vec}(\Delta) \sim N(\text{vec}(\bar{\Delta}), V_\beta \otimes A^{-1})$	lgtdata, lgtdata[[h]]$X is X matrix for unit h lgtdata[[h]]$y is dep. var. for unit h		R	betadraw Deltadraw Vbetadraw llike reject

Table B.1 (*continued*)

`rhierLinearModel`	Gibbs sampler for linear hierarchical model	$y_i = X_i\beta_i + \varepsilon_i$, $\varepsilon_i \sim N(0, \tau_i)$ $\beta_i \sim N(\Delta' z_i, V_\beta)$	$\tau_i \sim \nu_\varepsilon \text{ssq}_i/\chi^2_{\nu_\varepsilon}$ $V_\beta \sim \text{IW}(\nu, V)$ $\text{vec}(\Delta) \sim N(\text{vec}(\bar{\Delta}), V_\beta \otimes A^{-1})$	regdata, regdata[[i]]$X is X matrix for unit i regdata[[i]]$y is dep. var. for unit i		R	betadraw taudraw Deltadraw Vbetadraw
`rhierMnlRwMixture`	Hybrid sampler for hierarchical logit with mixture of normals prior	$y_i \sim$ Multinomial $(P_r(X_i, \beta_i))$, $i = 1, \ldots,$ no. of units $\beta_i \sim \Delta' z_i + u_i$ $u_i \sim N(\mu_{\text{ind}_i}, \Sigma_{\text{ind}_i})$ $\text{ind}_i \sim \text{Multinomial}(pvec)$ $pvec$ is ncomp -vector	pvec $\sim$ Dirichlet(a) $\text{vec}(\Delta) \sim N(\bar{\delta}, A_\delta^{-1})$ $\mu_j \sim N(\bar{\mu}, \Sigma_j \otimes a_\mu^{-1})$ $\Sigma_j \sim \text{IW}(\nu, V)$	p, lgtdata p is no. of choices lgtdata[[i]]$y is vector of multinomial choices for unit i lgtdata[[i]]$X is X matrix (use createX to make) for ith unit.	ncomp (no. of normal comps)	R	Deltadraw betadraw probdraw compdraw a list of lists of draws of normal comps
`rivGibbs`	Gibbs sampler for linear instrumental variables model	$x_i = z_i'\delta + \varepsilon_1$ $y_i = \beta x_i + w_i'\gamma + \varepsilon_2$ $\varepsilon \sim N(0, \Sigma)$	$\delta \sim N(\mu_\delta, A_\delta^{-1})$ $\text{vec}(\beta, \gamma) \sim N(\mu_{\beta\delta}, A_{\beta\delta}^{-1})$ $\Sigma \sim \text{IW}(\nu, V)$	z, w, x, y z, w are matrices of instruments, exog. vars x, y are nobs -vectors		R	deltadraw betadraw gammadraw Sigmadraw

Table B.1 (*continued*)

`rmnlIndepMetrop`	Independence Metropolis sampler for Multinomial logit model	$y \sim$ Multinomial$(P_r(X, \beta))$	$\beta \sim N(\bar{\beta}, A^{-1})$	p, y, X p is no. of choices X is nobs$*m$ (use createX) y is nobs -vector of multinomial indicat.		R	betadraw accept
`rmnpGibbs`	Gibbs sampler for multinomial probit	$w_i = X_i\beta + \varepsilon, \varepsilon \sim N(0, \Sigma)$ $y_i = j$, if $w_{ij} > \max(w_{i,-j}, 0)$ $j = 1, \ldots, p-1$ else $y_i = p$	$\beta \sim N(\bar{\beta}, A^{-1})$ $\Sigma \sim \text{IW}(\nu, V)$	p, y, X p is no. of choices y is nobs -vector of multinomial indicat. X is nobs $*(p-1) \times k$ use createX(DIFF=TRUE) to make X		R	betadraw sigmadraw normalize beta and Sigma draws to identified quantities!
`rmvpGibbs`	Gibbs sampler for the multivariate probit	$w_i = X_i\beta + \varepsilon_i, \varepsilon_i \sim N(0, \Sigma)$ $y_{ij} = 1$ if $w_{ij} > 0$, else 0 $j = 1, \ldots, p$	$\beta \sim N(\bar{\beta}, A^{-1})$ $\Sigma \sim \text{IW}(\nu, V)$	p, y, X p is no. of choices y is nobs -vector of multinomial indicat. X is nobs$*p \times k$		R	betadraw sigmadraw normalize to R and divide beta draws by std dev. for ident!

Table B.1 *(continued)*

`rnmixGibbs`	Gibbs sampler for mixture of normals	$y_i \sim \mathrm{N}(\mu_{\mathrm{ind}_i}, \Sigma_{\mathrm{ind}_i})$ $\mathrm{ind}_i \sim$ Multinomial(pvec) pvec is ncomp -vector	pvec $\sim$ Dirichlet(a) $\mu_j \sim$ $\mathrm{N}(\bar{\mu}, \Sigma_j \otimes a^{-1})$ $\Sigma_j \sim \mathrm{IW}(\nu, V)$	y is $n \times k$ matrix of k-dim. obs	ncomp	R	probdraw zdraw (indicators) compdraw a list of lists of draws of normal comps
`rscaleUsage`	Gibbs sampler for multivariate ordinal data with scale usage heterogeneity. Designed for survey questions on k-point scale	$x_{ij} = d$ if $c_{d-1} < y_{ij} < c_d$ $d = 1, \ldots, k$ $c_d = a + bd + ed^2$ $y_i = \mu + \tau_i \iota + \sigma_i z_i$ $z_i \sim N(0, \Sigma)$	$\mathrm{vec}(\tau_i, \log(\sigma_i)) \sim$ $N(\phi, \Lambda)$ $\phi = \begin{pmatrix} 0 \\ \lambda_{22} \end{pmatrix}$ $\mu \sim N(\bar{\mu}, A_\mu^{-1})$ $\Sigma \sim$ $\mathrm{IW}(\nu, V)$, $\Lambda \sim$ $\mathrm{IW}(\nu_\Lambda, V_\Lambda)$ $e \sim$ unif. on grid	k, x x is nobs $\times$ nvar matrix of responses to nvar survey questions, each one is on k-point scale			Sigmadraw mudraw taudraw sigmadraw Lambdadraw edraw use defaults!
`runiregGibbs`	Gibbs sampler for univariate regression model	$y = X\beta + \varepsilon$, $\varepsilon \sim$ $\mathrm{N}(0, \sigma^2 I_{nobs})$	$\beta \sim N(\bar{\beta}, A^{-1})$ $\sigma^2 \sim \nu ssq/\chi^2_\nu$	y, X		R	betadraw sigmasqdraw

Table B.2 Key *bayesm* utilities

Name	Description	Arguments	Output	Notes
`condMom`	Moments of ith element of multivariate normal given others	x vector of conditioning values (ith not used) mu mean sigi inverse of covariance matrix i index of element to compute cond dist for	list(cmean,cvar) conditional mean and variance	
`createX`	Creates X matrix for MNL and probit routines	p number of choice alternatives na number of alternative specific vars in Xa nd number of non-alternative specific vars (demos) Xa n*p×na matrix of alternative specific vars Xd n × nd matrix of non-alternative specific vars INT logical flag for intercepts DIFF logical flag to diff. with respect to base alternative base number of base alternative	X matrix	Use with rmnlIndepMetrop, rhierMnlRwMixture, rmnpGibbs, rmvpGibbs, llmnl, llmnp
`ghkvec`	Computes GHK approximation to integral of normal over half plane defined by vector of truncation points	L lower Cholesky root of Σ matrix trunpt vector of truncation points (see notes) above vector of indicators for above(1) or below(0) r number of draws in GHK	approx. to integral. vector of length= length(trunpt) /length(above)	Allows for same density and same side of axes but different truncation points to produce a vector
`llmnl`	Log-likelihood for MNL logit model	beta is k-vector of coefficients y is n-vector of multinomial outcomes from p alternatives X is n*p xk design matrix (use createX)	value of log-likelihood	

Table B.2 (*continued*)

`llmnp`	Log-likelihood for MNP model	beta is k-vector of coefficients Sigma is covariance matrix of errors X $n^*(p-1) \times k$ array (use createX with DIFF on) y n-vector of multinomial outcomes (p alternative) r is number of draws to use in GHK	value of log-likelihood	
`logMargDenNR`	Computes NR approx. to marginal density of data ('marginal likelihood')	ll vector of log-likelihood values	value of NR approx.	
`nmat`	Coverts covariance matrix stored as a vector to correlation matrix stored as a vector	vec k $\times$ k covariance matrix stored as vector	vector of correlation matrix	
`numEff`	Computes numerical std error and relative num. efficiency	x vector of draws m number of lags for computation of autocorrelations	list(stderr,f)	f is multiple for ratio of variances. sqrt(f) is relevant quantity
`rmultireg`	Draws from posterior of multivariate regression with conjugate prior	Y $n \times m$ matrix of observations on m dep. vars X $n \times k$ matrix of indep. vars Bbar prior mean ($k \times m$) A $k \times k$ prior precision matrix nu d.f. for IW prior V IW location parameters	list (B, Σ) draw from posterior	Priors $\mathrm{vec}(B)\|\Sigma \sim$ $N(\mathrm{vec}(\bar{\beta}), \Sigma \otimes A^{-1})$ $\Sigma \sim \mathrm{IW}(\nu, V)$

Table B.2 (*continued*)

`rtrun`	Draws for truncated univ. normal	mu mean, sigma std dev., *a* lower bound, *b* upper bound	draw (possibly vector) from truncated normal	Vector inputs result in vector outputs
`rwishart`	Draws from IW and Wishart	nu d.f. parameter, V pds location parameters	W Wishart draw IW inverted Wishart draw C upper triangle root of W CI inverse of C	If you want to draw from IW(nu, V), use rwishart(nu, V^{-1})!!!

bayesm MCMC functions

Name	Description	Model	Priors	Required arguments			Output
				Data	Prior	Mcmc	
`rbprobitGibbs`	Gibbs sampler for binary probit	$z = X\beta + \varepsilon, \varepsilon \sim N(0, 1)$ $y = 1$ if $z > 0$	$\beta \sim N(\bar{\beta}, A^{-1})$	X, y		R (no. of MCMC draws)	betadraw
`rhierBinLogit`	Hybrid sampler for hierarchical binary logit – designed for conjoint data. Note: x variables are expressed as a difference between alternatives. See Appendix A.	$y_{hi} = 1$ with $\text{pr} = \dfrac{\exp(x_{hi}'\beta_h)}{1 + \exp(x_{hi}'\beta_h)}$ $h = 1, \ldots, H \quad ; i = 1, \ldots, I_h$ H panel units $\beta_h \sim N(\Delta' z_h, V_\beta)$	$V_\beta \sim \text{IW}(\nu, V)$ $\text{vec}(\Delta) \sim$ $N(\text{vec}(\bar{\Delta}), V_\beta \otimes A^{-1})$	lgtdata lgtdata[[h]]$X is X matrix for unit h lgtdata[[h]]$y is dep. var. for unit h		R	betadraw Deltadraw Vbetadraw llike reject
`rhierLinearModel`	Gibbs sampler for linear hierarchical model	$y_i = X_i\beta_i + \varepsilon_i, \varepsilon_i \sim N(0, \tau_i)$ $\beta_i \sim N(\Delta' z_i, V_\beta)$	$\tau_i \sim \nu_\varepsilon \text{ssq}_i / \chi^2_{\nu_\varepsilon}$ $V_\beta \sim \text{IW}(\nu, V)$ $\text{vec}(\Delta) \sim$ $N(\text{vec}(\bar{\Delta}), V_\beta \otimes A^{-1})$	regdata regdata[[i]]$X is X matrix for unit i regdata[[i]]$y is dep. var. for unit i		R	betadraw taudraw Deltadraw Vbetadraw
`rhierMnlRwMixture`	Hybrid sampler for hierarchical logit with mixture of normals prior	$y_i \sim \text{Multinomial}\ (P_r(X_i, \beta_i))$, $i = 1, \ldots$, no. of units $\beta_i \sim \Delta' z_i + u_i$ $u_i \sim N(\mu_{\text{ind}_i}, \Sigma_{\text{ind}_i})$ $\text{ind}_i \sim \text{Multinomial}(pvec)$ $pvec$ is ncomp -vector	$\text{pvec} \sim \text{Dirichlet}(a)$ $\text{vec}(\Delta) \sim N(\bar{\delta}, A_\delta^{-1})$ $\mu_j \sim N(\bar{\mu}, \Sigma_j \otimes a_\mu^{-1})$ $\Sigma_j \sim \text{IW}(\nu, V)$	p, lgtdata p is no. of choices lgtdata[[i]]$y is vector of multinomial choices for unit i lgtdata[[i]]$X is X matrix (use createX to make) for ith unit.	ncomp (no. of normal comps)	R	Deltadraw betadraw probdraw compdraw a list of lists of draws of normal comps
`rivGibbs`	Gibbs sampler for linear instrumental variables model	$x_i = z_i'\delta + \varepsilon_1$ $y_i = \beta x_i + w_i'\gamma + \varepsilon_2$ $\varepsilon \sim N(0, \Sigma)$	$\delta \sim N(\mu_\delta, A_\delta^{-1})$ $\text{vec}(\beta, \gamma) \sim$ $N(\mu_{\beta\delta}, A_{\beta\delta}^{-1})$ $\Sigma \sim \text{IW}(\nu, V)$	z, w, x, y z, w are matrices of intstruments, exog. vars x, y are nobs -vectors		R	deltadraw betadraw gammadraw Sigmadraw

`rmnlIndepMetrop`	Independence Metropolis sampler for Multinomial logit model	$y \sim \text{Multinomial}(P_r(X, \beta))$	$\beta \sim N(\bar{\beta}, A^{-1})$	p, y, X p is no. of choices X is nobs $*m$ (use `createX`) y is nobs -vector of multinomial indicat.		R	betadraw
`rmnpGibbs`	Gibbs sampler for multinomial probit	$w_i = X_i\beta + \varepsilon, \varepsilon \sim N(0, \Sigma)$ $y_i = j$, if $w_{ij} > \max(w_{i,-j}, 0)$ $j = 1, \ldots, p-1$ else $y_i = p$	$\beta \sim N(\bar{\beta}, A^{-1})$ $\Sigma \sim \text{IW}(\nu, V)$	p, y, X p is no. of choices y is nobs -vector of multinomial indicat. X is nobs $*(p-1) \times k$ use createX(DIFF=TRUE) to make X		R	betadraw sigmadraw normalize beta and Sigma draws to identified quantities!
`rmvpGibbs`	Gibbs sampler for the multivariate probit	$w_i = X_i\beta + \varepsilon_i, \varepsilon_i \sim N(0, \Sigma)$ $y_{ij} = 1$ if $w_{ij} > 0$, else 0 $j = 1, \ldots, p$	$\beta \sim N(\bar{\beta}, A^{-1})$ $\Sigma \sim \text{IW}(\nu, V)$	p, y, X p is no. of choices y is nobs -vector of multinomial indicat. X is nobs $*p \times k$		R	betadraw sigmadraw normalize to R and divide beta draws by std dev. for ident!
`rnmixGibbs`	Gibbs sampler for mixture of normals	$y_i \sim \text{N}(\mu_{\text{ind}_i}, \Sigma_{\text{ind}_i})$ $\text{ind}_i \sim \text{Multinomial(pvec)}$ pvec is ncomp -vector	$\text{pvec} \sim \text{Dirichlet}(\boldsymbol{a})$ $\mu_j \sim \text{N}(\bar{\mu}, \Sigma_j \otimes \boldsymbol{a}^{-1})$ $\Sigma_j \sim \text{IW}(\nu, V)$	y is $\boldsymbol{n} \times \boldsymbol{k}$ matrix of $\boldsymbol{k}$-dim. obs	ncomp	R	probdraw zdraw (indicators) compdraw a list of lists of draws of normal comps
`rscaleUsage`	Gibbs sampler for multivariate ordinal data with scale usage heterogeneity. Designed for survey questions on $\boldsymbol{k}$-point scale	$x_{ij} = \boldsymbol{d}$ if $c_{d-1} < y_{ij} < c_d$ $\boldsymbol{d} = 1, \ldots, \boldsymbol{k}$ $c_d = a + bd + ed^2$ $y_i = \mu + \tau_i\iota + \sigma_i z_i$ $z_i \sim N(0, \Sigma)$	$\text{vec}(\tau_i, \log(\sigma_i)) \sim N(\phi, \Lambda)$ $\phi = \begin{pmatrix} 0 \\ \lambda_{22} \end{pmatrix}$ $\mu \sim N(\bar{\mu}, A_\mu^{-1})$ $\Sigma \sim \text{IW}(\nu, V), \Lambda \sim \text{IW}(\nu_\Lambda, V_\Lambda)$ $e \sim$ unif. on grid	$\boldsymbol{k}, \boldsymbol{x}$ $\boldsymbol{x}$ is nobs $\times$ nvar matrix of responses to nvar survey questions, each one is on $\boldsymbol{k}$-point scale			Sigmadraw mudraw taudraw sigmadraw Lambdadraw edraw use defaults!
`runiregGibbs`	Gibbs sampler for univariate regression model	$y = X\beta + \varepsilon, \varepsilon \sim \text{N}(0, \sigma^2 I_{nobs})$	$\beta \sim N(\bar{\beta}, A^{-1})$ $\sigma^2 \sim \nu ssq/\chi^2_\nu$	y, X		R	betadraw sigmasqdraw

Key bayesm utilities

Name	Description	Arguments	Output	Notes
`condMom`	Moments of ith element of multivariate normal given others	x vector of conditioning values (ith not used) mu mean sigi inverse of covariance matrix i index of element to compute cond dist for	list(cmean,cvar) conditional mean and variance	
`createX`	Creates X matrix for MNL and probit routines	p number of choice alternatives na number of alternative specific vars in Xa nd number of non-alternative specific vars (demos) Xa n*p×na matrix of alternative specific vars Xd n × nd matrix of non-alternative specific vars INT logical flag for intercepts DIFF logical flag to diff. with respect to base alternative base number of base alternative	X matrix	Use with rmnlIndepMetrop, rhierMnlRwMixture, rmnpGibbs, rmvpGibbs, llmnl, llmnp
`ghkvec`	Computes GHK approximation to integral of normal over half plane defined by vector of truncation points	L lower Cholesky root of Σ matrix trunpt vector of truncation points (see notes) above vector of indicators for above(1) or below(0) r number of draws in GHK	approx. to integral. vector of length= length(trunpt) /length(above)	Allows for same density and same side of axes but different truncation points to produce a vector
`llmnl`	Log-likelhood for MNL logit model	beta is k-vector of coefficients y is n-vector of multinomial outcomes from p alternatives X is n*p xk design matrix (use createX)	value of log-likelihood	
`llmnp`	Log-likelihood for MNP model	beta is k-vector of coefficients Sigma is covariance matrix of errors X $n^*(p-1) \times k$ array (use createX with DIFF on) y n-vector of multinomial outcomes (p alternative) r is number of draws to use in GHK	value of log-likelihood	

logMargDenNR	Computes NR approx. to marginal density of data ('marginal likelihood')	ll vector of log-likelihood values	value of NR approx.	
nmat	Coverts covariance matrix stored as a vector to correlation matrix stored as a vector	vec $k \times k$ covariance matrix stored as vector	vector of correlation matrix	
numEff	Computes numerical std error and relative num. efficiency	x vector of draws m number of lags for computation of autocorrelations	list(stderr,f)	f is multiple for ratio of variances. sqrt(f) is relevant quantity
rmultireg	Draws from posterior of multivariate regression with conjugate prior	Y $n \times m$ matrix of observations on m dep. vars X $n \times k$ matrix of indep. vars Bbar prior mean ($k \times m$) A $k \times k$ prior precision matrix nu d.f. for IW prior V IW location parameters	list (B, Σ) draw from posterior	Priors $\text{vec}(B)\|\Sigma \sim$ $N(\text{vec}(\bar{\beta}), \Sigma \otimes A^{-1})$ $\Sigma \sim \text{IW}(\nu, V)$
rtrun	Draws for truncated univ. normal	mu mean, sigma std dev., a lower bound, b upper bound	draw (possibly vector) from truncated normal	Vector inputs result in vector outputs
rwishart	Draws from IW and Wishart	nu d.f. parameter, V pds location parameters	W Wishart draw IW inverted Wishart draw C upper triangle root of W CI inverse of C	If you want to draw from IW(nu, V), use rwishart(nu, V^{-1})!!!

References

Ainslie, A. and Rossi, P.E. (1998) Similarities in choice behavior across product categories. *Marketing Science*, **17**, 91–106.

Albert, J. and Chib, S. (1993) Bayesian analysis of binary and polychotomous response data. *Journal of the American Statistical Association*, **88**, 669–679.

Allenby, G.M.(1990) Hypothesis testing with scanner data: the advantage of Bayesian methods. *Journal of Marketing Research*, **27**, 379–389.

Allenby, G.M. and Ginter, J.L. (1995) Using extremes to design products and segment markets. *Journal of Marketing Research*, **32**, 392–403.

Allenby, G.M. and Lenk, P.J. (1994) Modeling household purchase behavior with logistic normal regression. *Journal of American Statistical Association*, **89**, 1218–1231.

Allenby, G.M. and Lenk, P.J. (1995) Reassessing brand loyalty, price sensitivity, and merchandising effects on consumer brand choice. *Journal of Business & Economic Statistics*, **13**, 281–289.

Allenby, G.M. and Rossi, P.E. (1991) Quality perceptions and asymmetric switching between brands. *Marketing Science*, **10**, 185–205.

Allenby, G.M. and Rossi, P.E. (1993) A Bayesian approach to estimating household parameters. *Journal of Marketing Research*, **30**, 171–182.

Allenby, G.M., Arora, N. and Ginter, J.L. (1998) On the heterogeneity of demand. *Journal of Marketing Research*, **35**, 384–389.

Allenby, G.M. and Rossi, P.E. (1999) Marketing models of consumer heterogeneity. *Journal of Econometrics*, **89**, 57–78.

Allenby, G.M., Leone, R.P. and Jen, L. (1999) A dynamic model of purchase timing with application to direct marketing. *Journal of the American Statistical Association*, **94**, 365–374.

Allenby, G.M., Shively, T.S., Yang, S. and Garratt, M.J. (2004) A choice model for packaged goods: Dealing with discrete quantities and quantity discounts. *Marketing Science*, **23**, 95–108.

Ansari, A., Jedidi, K. and Jagpal, S. (2000) A hierarchical Bayesian methodology for treating heterogeneity in structural equation models, *Marketing Science*, **19**, 328–347.

Barnard, J., Meng, X. and McCulloch, R.E. (2000) Modelling covariance matrices in terms of standard deviations and correlations, with application to shrinkage. *Statistica Sinica*, **10**, 1281–1312.

Berger, J.O. (1985) *Statistical Decision Theory and Bayesian Analysis*. New York: Springer-Verlag.

Berger, J.O. and Wolpert, R.L. (1984) *The Likelihood Principle*, IMS Lecture Notes – Monograph Series. Hayward, CA: Institute of Mathematical Statistics.

Bernardo, J. and Smith, A. (1994) *Bayesian Theory*. Chichester: John Wiley & Sons, Ltd.

Berry, S., Levinsohn, J. and Pakes, A. (1995) Automobile prices in market equilibrium. *Econometrica*, **63**, 841–890.

Bayesian Statistics and Marketing P. E. Rossi, G. M. Allenby and R. McCulloch

Blattberg, R.C. and George, E.I. (1991) Shrinkage estimation of price and promotional elasticities: Seemingly unrelated equations. *Journal of the American Statistical Association*, **86**, 304–315.

Boatwright, P., McCulloch, R. and Rossi, P.E. (1999) Account-level modeling for trade promotion: An application of a constrained parameter hierarchical model. *Journal of the American Statistical Association*, **94**, 1063–1073.

Bronnenberg, B. and Mahajan, V.J. (2001) Unobserved retailer behavior in multimarket data: Joint spatial dependence in market shares and promotion variables. *Marketing Science*, **20**, 284–299.

Casella, G. and Berger, R.L. (2002) *Statistical Inference*. Pacific Grove, CA: Duxbury.

Chamberlain, G. (1980) Analysis of covariance with qualitative data. *Review of Economic Studies*, **47**, 225–238.

Chamberlain, G. (1984) Panel data. In Z. Griliches and M.D. Intriligator (eds), *Handbook of Econometrics, Volume 2*. Amsterdam: North-Holland.

Chang, K., Siddarth, S. and Weinberg C.B. (1999) The impact of heterogeneity in purchase timing and price responsiveness on estimates of sticker shock effects. *Marketing Science*, **18**, 178–192.

Chib, S. (1992) Bayes inference in the tobit censored regression model. *Journal of Econometrics*, **51**, 79–99.

Chib, S. (1995) Marginal likelihood from Gibbs output. *Journal of the American Statistical Association*, **90**, 1313–1321.

Chib, S. and Greenberg, E. (1995a) Understanding the Metropolis–Hastings algorithm. *American Statistician*, **49**, 327–335.

Chib, S. and Greenberg, E. (1995b) Hierarchical analysis of SUR models with extensions to correlated serial errors and time varying parameter models. *Journal of Econometrics*, **68**, 339–360.

Chib, S. and Greenberg, E. (1998) Analysis of multivariate probit models. *Biometrika*, **85**, 347–361.

Chib, S. and Jeliazkov, I. (2001) Marginal likelihood from the Metropolis–Hastings output. *Journal of the American Statistical Association*, **96**, 270–281.

Cressie, A.C.N. (1991) *Statistics for Spatial Data*. New York: John Wiley & Sons, Inc.

Deaton, A. and Muellbauer, J. (1980) *Economics and Consumer Behavior*. Cambridge: Cambridge University Press.

Dellaportas, P. and Papageorgiou, I. (2004) Multivariate mixtures of normals with unknown number of components. Working paper, Athens University of Economics and Business.

DiCiccio, T.J., Kass, R., Raftery, A. and Wasserman L. (1997) Computing Bayes factors by combining simulation and asymptotic approximations. *Journal of the American Statistical Association*, **92**, 903–915.

Diebolt, J. and Robert, C.P. (1994) Estimation of finite mixture distributions through Bayesian sampling. *Journal of the Royal Statistical Society, Series B*, **56**, 363–375.

Edwards, Y. and Allenby, G. (2003) Multivariate analysis of multiple response data. *Journal of Marketing Research*, **40**, 321–334.

Efron, B. and Morris, C. (1975) Data Analysis Using Stein's estimator and its generalizations. *Journal of the American Statistical Association*, **70**, 311–319.

Frühwirth-Schnatter, S. (2001) Markov chain Monte Carlo estimation of classical and dynamic switching and mixture models. *Journal of the American Statistical Association*, **96**, 194–209.

Frühwirth-Schnatter, S. (2004) Estimating Marginal Likelihoods for Mixture and Markov Switching Models Using Bridge Sampling Techniques. *Econometrics Journal*, 7, 143–167.

Frühwirth-Schnatter, S., Tüchler, R. and Otter, T. (2004) Bayesian analysis of the heterogeneity model. *Journal of Business & Economic Statistics*, **22**, 2–15.

Gelfand, A. and Dey, D. (1994) Bayesian Model Choice: Asymptotics and exact calculations. *Journal of the Royal Statistical Society, Series B*, **56**, 501–514.

Gelman, A. and Rubin, D.B. (1992) Inference from iterative simulation using multiple sequences. *Statistical Science*, 7, 457–511.

Gelman, A., Roberts, G.O., and Gilks, W.R. (1996) Efficient Metropolis jumping rules. In J.M. Bernardo, J.O. Berger, A.P. Dawid and A.F.M. Smith (eds), *Bayesian Statistics 5*, pp. 599–608. New York: Oxford University Press.

Gelman, A., Carlin, J.B., Stern, H.S. and Rubin, D.B. (2004) *Bayesian Data Analysis* (2nd edn). Boca Raton, FL: Chapman & Hall/CRC.

Gentzkow, M. (2005) Valuing new goods in a model with complementarity: Online newspapers. Working paper, University of Chicago.

Geweke, J. (1989) Bayesian inference in econometric models using Monte Carlo integration. *Econometrica*, **57**, 1317–1339.

Geweke, J. (1996) Bayesian reduced rank regression in econometrics. *Journal of Econometrics*, **75**, 121–146.

Geweke, J. (2004) Getting it right: Joint distribution tests of posterior simulators, *Journal of the American Statistical Association*, **99**, 799–804.

Geweke, J. (2005) *Contemporary Bayesian Econometrics and Statistics*. New York: John Wiley & Sons.

Gilbride, T.J. and Allenby, G.M. (2004) A choice model with conjunctive, disjunctive, and compensatory screening rules. *Marketing Science*, **23**, 391–406.

Gilks, W., Best, N. and Tan, K. (1995) Adaptive rejection Metropolis sampling within Gibbs sampling. *Applied Statistics*, **44**, 455–472.

Gilks, W.R. (1997) Discussion of the paper by Richardson and Green. *Journal of the Royal Statistical Society, Series B*, **59**, 731–792.

Golub, G. and Van Loan, C. (1989) *Matrix Computations*. Baltimore, MD: Johns Hopkins University Press.

Hajivassiliou, V., McFadden, D. and Ruud, P. (1996) Simulation of multivariate normal rectangle probabilities and their derivatives. *Journal of Econometrics*, **72**, 85–134.

Hastings, W.K. (1970) Monte Carlo sampling methods using Markov chains and their applications. *Biometrika*, **57**, 97–109.

Heckman, J. and Singer, B. (1984) A method for minimizing the impact of distributional assumptions in econometric models for duration data. *Econometrica*, **52**, 271–320.

Hobert, J. and Casella, G. (1996) The effect of improper priors on Gibbs sampling in hierarchical models. *Journal of the American Statistical Association*, **91**, 1461–1473.

Hobert, J.P., Robert, C.P. and Goutis, C. (1997) Connectedness conditions for the convergence of the Gibbs sampler. *Statistics and Probability Letters*, **33**, 235–240.

Imai, K. and Van Dyk, D.A. (2005) A Bayesian analysis of the multinomial probit model using marginal data augmentation. *Journal of Econometrics*, **124**, 311–334.

Jacquier, E. and Polson, N.G. (2002) Odds ratios in MCMC: Application to stochastic volatility modeling. Working paper, Graduate School of Business, University of Chicago.

Jarner, S.F. and Tweedie, R.L. (2001) Necessary conditions for geometric and polynomial ergodicity of random walk-type Markov chains. Working paper.

Johnson V. and Albert, J. (1999) *Ordinal Data Modeling*. New York: Springer-Verlag.

Keane, M. (1994) A computationally practical simulation estimator for panel data. *Econometrica*, **62**, 95–116.

Kim, J., Allenby, G.M. and Rossi P.E. (2002) Modeling consumer demand for variety, *Marketing Science*, **21**, 29–250.

Koop, G. (2003) *Bayesian Econometrics*. Chichester: John Wiley & Sons, Ltd.

Lancaster, T. (2004) *An Introduction to Modern Bayesian Econometrics*, London: Blackwell Publishing.

Lenk, P. and DeSarbo, W. (2000) Bayesian inference for finite mixtures of generalized linear models with random effects. *Psychometrika*, **65**, 93–119.

Lenk, P., DeSarbo, W., Green, P. and Young, M. (1996) Hierarchical Bayes conjoint analysis: recovery of part-worth heterogeneity from reduced experimental designs. *Marketing Science*, **15**, 173–191.

LeSage, J.P. (2000) Bayesian estimation of limited dependent variable spatial autoregressive models. *Geographical Analysis*, **32**, 19–35.

Liu, J. (2001) *Monte Carlo Strategies in Scientific Computing*. New York: Springer-Verlag.

Magnus, J.R. and Neudecker, H. (1988) *Matrix Differential Calculus with Applications in Statistics and Econometrics*. Chichester: John Wiley & Sons, Ltd.

Manchanda, P., Ansari, A. and Gupta, S. (1999) The 'shopping basket': A model for multicategory purchase incidence decisions, *Marketing Science*, **18**, 95–114.

Manchanda, P., Chintagunta, P.K. and Rossi, P. (2004) Response modeling with non-random marketing mix variables. *Journal of Marketing Research*, **XLI**, 467–478.

Marsaglia, G. (1999) Random numbers for C. End at last? USENET post. http://www.mathematik.uni-bielefeld.de/sillke/ALGORITHMS/random/marsaglia-inline-c (accessed May 30, 2005).

Marsaglia, G. and Tsang, W.W. (2000a) The ziggurat method for generating random variables. *Journal of Statistical Software*, **5**.

Marsaglia, G. and Tsang, W.W. (2000b) A simple method for generating gamma variables. *ACM Transactions on Mathematical Software*, **26**, 363–372.

Matsumoto, M. and Nishimura, T. (1998) Mersenne twister: A 623-dimensionally equidistributed uniform pseudo-random number generator. *ACM Transactions on Modeling and Computer Simulation*, **8**, 3–30.

McCulloch, R. and Rossi, P.E. (1994) An exact likelihood analysis of the multinomial probit model with fully identified parameters. *Journal of Econometrics*, **64**, 207–240.

McCulloch, R., Polson, N. and Rossi, P.E. (2000) A Bayesian analysis of the multinomial probit model with fully identified parameters *Journal of Econometrics*, **99**, 173–193.

Meng, X. and Van Dyk, D. (1999) Seeking efficient data augmentation schemes via conditional and marginal augmentation. *Biometrika*, **86**, 301–320.

Meng, X. and Wong, W.H. (1996) Simulating ratios of normalizing constants via a simple identity: a theoretical exploration. *Statistica Sinica*, **6**, 831–860.

Montgomery, A.L. (1997) Creating micro-marketing pricing strategies using supermarket scanner data. *Marketing Science*, **16**, 315–337.

Montgomery, A.L. and Bradlow, E.T. (1999) Why analyst overconfidence about the functional form of demand models can lead to overpricing. *Marketing Science*, **18**, 569–583.

Morris, C. (1983) Parametric empirical Bayes inference: Theory and applications. *Journal of the American Statistical Association*, **78**, 47–65.

Muirhead, R. (1982) *Aspects of Multivariate Statistical Theory*. New York: John Wiley & Sons, Inc.

Neelamegham, R. and Chintagunta. P. (1999) A Bayesian model to forecast new product performance in domestic and international markets. *Marketing Science*, **18**, 115–136.

Newton, M. and Raftery, A. (1994) Approximate Bayesian inference by the weighted likelihood bootstrap, *Journal of the Royal Statistical Society, Series B*, **56**, 3–48.

Nobile, A. (1998) A hybrid Markov chain for the Bayesian analysis of the multinomial probit model. *Statistics and Computing*, **8**, 229–242.

Otter, T., Frühwirth-Schnatter, S. and Tüchler, R. (2003) Unobserved preference changes in conjoint analysis. Working paper, University of Vienna.

Richardson, S. and Green, P.J. (1997) On Bayesian analysis of mixtures with an unknown number of components. *Journal of the Royal Statistical Society, Series B*, **59**, 731–792.

Robert, C. and Casella, G. (2004) *Monte Carlo Statistical Methods* (2nd edn). New York: Springer-Verlag.

Roberts, G. and Rosenthal, J. (2001) Optimal scaling for various Metropolis-Hastings algorithms. *Statistical Science*, **16**, 351–367.

Rossi, P.E., McCulloch, R.E. and Allenby, G.M. (1996) The value of purchase history data in target marketing, *Marketing Science*, **15**, 321–340.

Rossi, P.E., Gilula, Z. and Allenby, G.M. (2001) Overcoming scale usage heterogeneity: A Bayesian hierarchical approach. *Journal of the American Statistical Association*, **96**, 20–31.

Sawtooth Software (2001) *Proceedings of the 2001 Sawtooth Software Conference*. Sequim, WA: Sawtooth Software.

Schwarz, G. (1978) Estimating the dimension of a model. *Annals of Statistics*, **6**, 461–464.

Seetharaman, P.B., Ainslie, A. and Chintagunta, P. (1999) Investigating household state dependence effects across categories. *Journal of Marketing Research*, **36**, 488–500.

Smith, T. and LeSage, J. (2000) A Bayesian probit model with spatial dependencies. Working paper, University of Pennsylvania.

Steenburgh, T.J., Ainslie, A. and Engebretson, P.H. (2003) Massively categorical variables: Revealing the information in zipcodes. *Marketing Science*, **22**, 40–57.

Stephens, M. (2000) Dealing with label switching in mixture models. *Journal of the Royal Statistical Society, Series B*, **62**, 795–809.

Tanner, T. and Wong, W. (1987) The calculation of posterior distributions by data augmentation. *Journal of the American Statistical Association*, **82**, 528–549.

Ter Hofstede, F., Wedel, M. and Steenkamp, J.E.M. (2002) Identifying spatial segments in international markets. *Marketing Science*, **21**, 160–178.

Tierney, L. (1994) Markov chains for exploring posterior distributions. *Annals of Statistics*, **22**, 1701–1728.

Van Dyk, D. and Meng, X. (2001) The art of data augmentation. *Journal of Computational and Graphical Statistics*, **10**, 1–50.

Venables, W.N. and Ripley, B.D. (2000) *S Programming*. New York: Springer-Verlag.

Venables, W.N., Smith, D.M. and R Development Core Team (2001) *An Introduction to R*. Bristol: Network Theory Limited.

Vilas-Boas, J. and Winer, R.S. (1999) Endogeneity in brand choice models. *Management Science*, **45**, 1324–1238.

Yang, S. and Allenby, G.M. (2000) A model for observation, structural and household heterogeneity in panel data, *Marketing Letters*, **11**(2), 137–149.

Yang, S. and Allenby, G.M. (2003) Modeling interdependent consumer preferences. *Journal of Marketing Research*, **40**, 282–294.

Yang, S., Allenby, G.M. and Fennell, G. (2002) Modeling variation in brand preference: The roles of objective environment and motivating conditions. *Marketing Science*, **21**, 14–31.

Yang, S., Chen, Y. and Allenby,G.M. (2003) Bayesian analysis of simultaneous demand and supply. *Quantitative Marketing and Economics*, **1**, 251–276.

Zellner, A. (1971) *An Introduction to Bayesian Inference in Econometrics*. New York: John Wiley & Sons, Inc.

Zellner, A. (1986) On assessing prior distributions and Bayesian regression analysis with g-prior distributions. In P. Goel and A. Zellner (eds), *Bayesian Inference and Decision Techniques: Essays in Honour of Bruno de Finetti*. Amsterdam: North-Holland.

Zellner, A. and Rossi, P.E. (1984) Bayesian analysis of dichotomous quantal response models. *Journal of Econometrics*, **25**, 365–394.

Index

Page numbers followed by *f* indicate figures; page numbers followed by *t* indicate tables.

WILEY SERIES IN PROBABILITY AND STATISTICS

ESTABLISHED BY WALTER A. SHEWHART AND SAMUEL S. WILKS

The ***Wiley Series in Probability and Statistics*** is well established and authoritative. It covers many topics of current research interest in both pure and applied statistics and probability theory. Written by leading statisticians and institutions, the titles span both state-of-the-art developments in the field and classical methods.

Reflecting the wide range of current research in statistics, the series encompasses applied, methodological and theoretical statistics, ranging from applications and new techniques made possible by advances in computerized practice to rigorous treatment of theoretical approaches.

This series provides essential and invaluable reading for all statisticians, whether in academia, industry, government, or research.

ABRAHAM and LEDOLTER · Statistical Methods for Forecasting
AGRESTI · Analysis of Ordinal Categorical Data
AGRESTI · An Introduction to Categorical Data Analysis
AGRESTI · Categorical Data Analysis, *Second Edition*
ALTMAN, GILL, and McDONALD · Numerical Issues in Statistical Computing for the Social Scientist
AMARATUNGA and CABRERA · Exploration and Analysis of DNA Microarray and Protein Array Data
ANDĚL · Mathematics of Chance
ANDERSON · An Introduction to Multivariate Statistical Analysis, *Third Edition*
*ANDERSON · The Statistical Analysis of Time Series
ANDERSON, AUQUIER, HAUCK, OAKES, VANDAELE, and WEISBERG · Statistical Methods for Comparative Studies
ANDERSON and LOYNES · The Teaching of Practical Statistics
ARMITAGE and DAVID (editors) · Advances in Biometry
ARNOLD, BALAKRISHNAN, and NAGARAJA · Records
*ARTHANARI and DODGE · Mathematical Programming in Statistics
*BAILEY · The Elements of Stochastic Processes with Applications to the Natural Sciences
BALAKRISHNAN and KOUTRAS · Runs and Scans with Applications
BARNETT · Comparative Statistical Inference, *Third Edition*
BARNETT · Environmental Statistics: Methods & Applications
BARNETT and LEWIS · Outliers in Statistical Data, *Third Edition*
BARTOSZYNSKI and NIEWIADOMSKA-BUGAJ · Probability and Statistical Inference
BASILEVSKY · Statistical Factor Analysis and Related Methods: Theory and Applications
BASU and RIGDON · Statistical Methods for the Reliability of Repairable Systems
BATES and WATTS · Nonlinear Regression Analysis and Its Applications
BECHHOFER, SANTNER, and GOLDSMAN · Design and Analysis of Experiments for Statistical Selection, Screening, and Multiple Comparisons
BELSLEY · Conditioning Diagnostics: Collinearity and Weak Data in Regression

*Now available in a lower priced paperback edition in the Wiley Classics Library.

BELSLEY, KUH, and WELSCH · Regression Diagnostics: Identifying Influential Data and Sources of Collinearity
BENDAT and PIERSOL · Random Data: Analysis and Measurement Procedures, *Third Edition*
BERNARDO and SMITH · Bayesian Theory
BERRY, CHALONER, and GEWEKE · Bayesian Analysis in Statistics and Econometrics: Essays in Honor of Arnold Zellner
BHAT and MILLER · Elements of Applied Stochastic Processes, *Third Edition*
BHATTACHARYA and JOHNSON · Statistical Concepts and Methods
BHATTACHARYA and WAYMIRE · Stochastic Processes with Applications
BILLINGSLEY · Convergence of Probability Measures, *Second Edition*
BILLINGSLEY · Probability and Measure, *Third Edition*
BIRKES and DODGE · Alternative Methods of Regression
BLISCHKE and MURTHY (editors) · Case Studies in Reliability and Maintenance
BLISCHKE and MURTHY · Reliability: Modeling, Prediction, and Optimization
BLOOMFIELD · Fourier Analysis of Time Series: An Introduction, *Second Edition*
BOLLEN · Structural Equations with Latent Variables
BOROVKOV · Ergodicity and Stability of Stochastic Processes
BOULEAU · Numerical Methods for Stochastic Processes
BOX · Bayesian Inference in Statistical Analysis
BOX · R. A. Fisher, the Life of a Scientist
BOX and DRAPER · Empirical Model-Building and Response Surfaces
*BOX and DRAPER · Evolutionary Operation: A Statistical Method for Process Improvement
BOX, HUNTER, and HUNTER · Statistics for Experimenters: An Introduction to Design, Data Analysis, and Model Building
BOX and LUCEÑO · Statistical Control by Monitoring and Feedback Adjustment
BRANDIMARTE · Numerical Methods in Finance: A MATLAB-Based Introduction
BROWN and HOLLANDER · Statistics: A Biomedical Introduction
BRUNNER, DOMHOF, and LANGER · Nonparametric Analysis of Longitudinal Data in Factorial Experiments
BUCKLEW · Large Deviation Techniques in Decision, Simulation, and Estimation
CAIROLI and DALANG · Sequential Stochastic Optimization
CHAN · Time Series: Applications to Finance
CHATTERJEE and HADI · Sensitivity Analysis in Linear Regression
CHATTERJEE and PRICE · Regression Analysis by Example, *Third Edition*
CHERNICK · Bootstrap Methods: A Practitioner's Guide
CHERNICK and FRIIS · Introductory Biostatistics for the Health Sciences
CHILÈS and DELFINER · Geostatistics: Modeling Spatial Uncertainty
CHOW and LIU · Design and Analysis of Clinical Trials: Concepts and Methodologies, *Second Edition*
CLARKE and DISNEY · Probability and Random Processes: A First Course with Applications, *Second Edition*
*COCHRAN and COX · Experimental Designs, *Second Edition*
CONGDON · Applied Bayesian Modelling
CONGDON · Bayesian Statistical Modelling
CONGDON · Bayesian Models for Categorical Data
CONOVER · Practical Nonparametric Statistics, *Second Edition*
COOK · Regression Graphics
COOK and WEISBERG · Applied Regression Including Computing and Graphics
COOK and WEISBERG · An Introduction to Regression Graphics
CORNELL · Experiments with Mixtures, Designs, Models, and the Analysis of Mixture Data, *Third Edition*

*Now available in a lower priced paperback edition in the Wiley Classics Library.

COVER and THOMAS · Elements of Information Theory
COX · A Handbook of Introductory Statistical Methods
*COX · Planning of Experiments
CRESSIE · Statistics for Spatial Data, *Revised Edition*
CSÖRGÖ and HORVÁTH · Limit Theorems in Change Point Analysis
DANIEL · Applications of Statistics to Industrial Experimentation
DANIEL · Biostatistics: A Foundation for Analysis in the Health Sciences, *Sixth Edition*
*DANIEL · Fitting Equations to Data: Computer Analysis of Multifactor Data, *Second Edition*
DASU and JOHNSON · Exploratory Data Mining and Data Cleaning
DAVID and NAGARAJA · Order Statistics, *Third Edition*
*DEGROOT, FIENBERG, and KADANE · Statistics and the Law
DEL CASTILLO · Statistical Process Adjustment for Quality Control
DENISON, HOLMES, MALLICK, and SMITH · Bayesian Methods for Nonlinear Classification and Regression
DETTE and STUDDEN · The Theory of Canonical Moments with Applications in Statistics, Probability, and Analysis
DEY and MUKERJEE · Fractional Factorial Plans
DILLON and GOLDSTEIN · Multivariate Analysis: Methods and Applications
DODGE · Alternative Methods of Regression
*DODGE and ROMIG · Sampling Inspection Tables, *Second Edition*
*DOOB · Stochastic Processes
DOWDY and WEARDEN, and CHILKO · Statistics for Research, *Third Edition*
DRAPER and SMITH · Applied Regression Analysis, *Third Edition*
DRYDEN and MARDIA · Statistical Shape Analysis
DUDEWICZ and MISHRA · Modern Mathematical Statistics
DUNN and CLARK · Applied Statistics: Analysis of Variance and Regression, *Second Edition*
DUNN and CLARK · Basic Statistics: A Primer for the Biomedical Sciences, *Third Edition*
DUPUIS and ELLIS · A Weak Convergence Approach to the Theory of Large Deviations
EDLER and KITSOS (editors) · Recent Advances in Quantitative Methods in Cancer and Human Health Risk Assessment
*ELANDT-JOHNSON and JOHNSON · Survival Models and Data Analysis
ENDERS · Applied Econometric Time Series
ETHIER and KURTZ · Markov Processes: Characterization and Convergence
EVANS, HASTINGS, and PEACOCK · Statistical Distributions, *Third Edition*
FELLER · An Introduction to Probability Theory and Its Applications, Volume I, *Third Edition,* Revised; Volume II, *Second Edition*
FISHER and VAN BELLE · Biostatistics: A Methodology for the Health Sciences
*FLEISS · The Design and Analysis of Clinical Experiments
FLEISS · Statistical Methods for Rates and Proportions, *Second Edition*
FLEMING and HARRINGTON · Counting Processes and Survival Analysis
FULLER · Introduction to Statistical Time Series, *Second Edition*
FULLER · Measurement Error Models
GALLANT · Nonlinear Statistical Models
GELMAN and MENG (editors): Applied Bayesian Modeling and Casual Inference from Incomplete-Data Perspectives
GHOSH, MUKHOPADHYAY, and SEN · Sequential Estimation
GIESBRECHT and GUMPERTZ · Planning, Construction, and Statistical Analysis of Comparative Experiments
GIFI · Nonlinear Multivariate Analysis
GLASSERMAN and YAO · Monotone Structure in Discrete-Event Systems
GNANADESIKAN · Methods for Statistical Data Analysis of Multivariate Observations, *Second Edition*

*Now available in a lower priced paperback edition in the Wiley Classics Library.

GOLDSTEIN and LEWIS · Assessment: Problems, Development, and Statistical Issues
GREENWOOD and NIKULIN · A Guide to Chi-Squared Testing
GROSS and HARRIS · Fundamentals of Queueing Theory, *Third Edition*
HAHN and MEEKER · Statistical Intervals: A Guide for Practitioners
*HAHN and SHAPIRO · Statistical Models in Engineering
HALD · A History of Probability and Statistics and Their Applications before 1750
HALD · A History of Mathematical Statistics from 1750 to 1930
HAMPEL · Robust Statistics: The Approach Based on Influence Functions
HANNAN and DEISTLER · The Statistical Theory of Linear Systems
HEIBERGER · Computation for the Analysis of Designed Experiments
HEDAYAT and SINHA · Design and Inference in Finite Population Sampling
HELLER · MACSYMA for Statisticians
HINKELMAN and KEMPTHORNE: · Design and Analysis of Experiments, Volume 1: Introduction to Experimental Design
HOAGLIN, MOSTELLER, and TUKEY · Exploratory Approach to Analysis of Variance
HOAGLIN, MOSTELLER, and TUKEY · Exploring Data Tables, Trends and Shapes
*HOAGLIN, MOSTELLER, and TUKEY · Understanding Robust and Exploratory Data Analysis
HOCHBERG and TAMHANE · Multiple Comparison Procedures
HOCKING · Methods and Applications of Linear Models: Regression and the Analysis of Variance, *Second Edition*
HOEL · Introduction to Mathematical Statistics, *Fifth Edition*
HOGG and KLUGMAN · Loss Distributions
HOLLANDER and WOLFE · Nonparametric Statistical Methods, *Second Edition*
HOSMER and LEMESHOW · Applied Logistic Regression, *Second Edition*
HOSMER and LEMESHOW · Applied Survival Analysis: Regression Modeling of Time to Event Data
HUBER · Robust Statistics
HUBERTY · Applied Discriminant Analysis
HUNT and KENNEDY · Financial Derivatives in Theory and Practice, *Revised Edition*
HUSKOVÁ, BERAN, and DUPAC · Collected Works of Jaroslav Hájek—with Commentary
IMAN and CONOVER · A Modern Approach to Statistics
JACKSON · A User's Guide to Principle Components
JOHN · Statistical Methods in Engineering and Quality Assurance
JOHNSON · Multivariate Statistical Simulation
JOHNSON and BALAKRISHNAN · Advances in the Theory and Practice of Statistics: A Volume in Honor of Samuel Kotz
JUDGE, GRIFFITHS, HILL, LÜTKEPOHL, and LEE · The Theory and Practice of Econometrics, *Second Edition*
JOHNSON and KOTZ · Distributions in Statistics
JOHNSON and KOTZ (editors) · Leading Personalities in Statistical Sciences: From the Seventeenth Century to the Present
JOHNSON, KOTZ, and BALAKRISHNAN · Continuous Univariate Distributions, Volume 1, *Second Edition*
JOHNSON, KOTZ, and BALAKRISHNAN · Continuous Univariate Distributions, Volume 2, *Second Edition*
JOHNSON, KOTZ, and BALAKRISHNAN · Discrete Multivariate Distributions
JOHNSON, KOTZ, and KEMP · Univariate Discrete Distributions, *Second Edition*
JUREČKOVÁ and SEN · Robust Statistical Procedures: Asymptotics and Interrelations
JUREK and MASON · Operator-Limit Distributions in Probability Theory
KADANE · Bayesian Methods and Ethics in a Clinical Trial Design

*Now available in a lower priced paperback edition in the Wiley Classics Library.

KADANE and SCHUM · A Probabilistic Analysis of the Sacco and Vanzetti Evidence
KALBFLEISCH and PRENTICE · The Statistical Analysis of Failure Time Data, *Second Edition*
KARIYA and KURATA · Generalized Least Squares
KASS and VOS · Geometrical Foundations of Asymptotic Inference
KAUFMAN and ROUSSEEUW · Finding Groups in Data: An Introduction to Cluster Analysis
KEDEM and FOKIANOS · Regression Models for Time Series Analysis
KENDALL, BARDEN, CARNE, and LE · Shape and Shape Theory
KHURI · Advanced Calculus with Applications in Statistics, *Second Edition*
KHURI, MATHEW, and SINHA · Statistical Tests for Mixed Linear Models
KLEIBER and KOTZ · Statistical Size Distributions in Economics and Actuarial Sciences
KLUGMAN, PANJER, and WILLMOT · Loss Models: From Data to Decisions
KLUGMAN, PANJER, and WILLMOT · Solutions Manual to Accompany Loss Models: From Data to Decisions
KOTZ, BALAKRISHNAN, and JOHNSON · Continuous Multivariate Distributions, Volume 1, *Second Edition*
KOTZ and JOHNSON (editors) · Encyclopedia of Statistical Sciences: Volumes 1 to 9 with Index
KOTZ and JOHNSON (editors) · Encyclopedia of Statistical Sciences: Supplement Volume
KOTZ, READ, and BANKS (editors) · Encyclopedia of Statistical Sciences: Update Volume 1
KOTZ, READ, and BANKS (editors) · Encyclopedia of Statistical Sciences: Update Volume 2
KOVALENKO, KUZNETZOV, and PEGG · Mathematical Theory of Reliability of Time-Dependent Systems with Practical Applications
LACHIN · Biostatistical Methods: The Assessment of Relative Risks
LAD · Operational Subjective Statistical Methods: A Mathematical, Philosophical, and Historical Introduction
LAMPERTI · Probability: A Survey of the Mathematical Theory, *Second Edition*
LANGE, RYAN, BILLARD, BRILLINGER, CONQUEST, and GREENHOUSE · Case Studies in Biometry
LARSON · Introduction to Probability Theory and Statistical Inference, *Third Edition*
LAWLESS · Statistical Models and Methods for Lifetime Data, *Second Edition*
*LAWSON · Statistical Methods in Spatial Epidemiology
LE · Applied Categorical Data Analysis
LE · Applied Survival Analysis
LEE and WANG · Statistical Methods for Survival Data Analysis, *Third Edition*
LePAGE and BILLARD · Exploring the Limits of Bootstrap
LEYLAND and GOLDSTEIN (editors) · Multilevel Modelling of Health Statistics
LIAO · Statistical Group Comparison
LINDVALL · Lectures on the Coupling Method
LINHART and ZUCCHINI · Model Selection
LITTLE and RUBIN · Statistical Analysis with Missing Data, *Second Edition*
LLOYD · The Statistical Analysis of Categorical Data
MAGNUS and NEUDECKER · Matrix Differential Calculus with Applications in Statistics and Econometrics, *Revised Edition*
MALLER and ZHOU · Survival Analysis with Long Term Survivors
MALLOWS · Design, Data, and Analysis by Some Friends of Cuthbert Daniel
MANN, SCHAFER, and SINGPURWALLA · Methods for Statistical Analysis of Reliability and Life Data
MANTON, WOODBURY, and TOLLEY · Statistical Applications Using Fuzzy Sets
MARCHETTE · Random Graphs for Statistical Pattern Recognition
MARDIA and JUPP · Directional Statistics

*Now available in a lower priced paperback edition in the Wiley Classics Library.

MASON, GUNST, and HESS · Statistical Design and Analysis of Experiments with Applications to Engineering and Science, *Second Edition*
McCULLOCH and SEARLE · Generalized, Linear, and Mixed Models
McFADDEN · Management of Data in Clinical Trials
McLACHLAN · Discriminant Analysis and Statistical Pattern Recognition
McLACHLAN and KRISHNAN · The EM Algorithm and Extensions
McLACHLAN and PEEL · Finite Mixture Models
McNEIL · Epidemiological Research Methods
MEEKER and ESCOBAR · Statistical Methods for Reliability Data
MEERSCHAERT and SCHEFFLER · Limit Distributions for Sums of Independent Random Vectors: Heavy Tails in Theory and Practice
*MILLER · Survival Analysis, *Second Edition*
MONTGOMERY, PECK, and VINING · Introduction to Linear Regression Analysis, *Third Edition*
MORGENTHALER and TUKEY · Configural Polysampling: A Route to Practical Robustness
MUIRHEAD · Aspects of Multivariate Statistical Theory
MURRAY · X-STAT 2.0 Statistical Experimentation, Design Data Analysis, and Nonlinear Optimization
MURTHY, XIE, and JIANG · Weibull Models
MYERS and MONTGOMERY · Response Surface Methodology: Process and Product Optimization Using Designed Experiments, *Second Edition*
MYERS, MONTGOMERY, and VINING · Generalized Linear Models. With Applications in Engineering and the Sciences
NELSON · Accelerated Testing, Statistical Models, Test Plans, and Data Analyses
NELSON · Applied Life Data Analysis
NEWMAN · Biostatistical Methods in Epidemiology
OCHI · Applied Probability and Stochastic Processes in Engineering and Physical Sciences
OKABE, BOOTS, SUGIHARA, and CHIU · Spatial Tesselations: Concepts and Applications of Voronoi Diagrams, *Second Edition*
OLIVER and SMITH · Influence Diagrams, Belief Nets and Decision Analysis
PALTA · Quantitative Methods in Population Health: Extensions of Ordinary Regressions
PANKRATZ · Forecasting with Dynamic Regression Models
PANKRATZ · Forecasting with Univariate Box–Jenkins Models: Concepts and Cases
*PARZEN · Modern Probability Theory and Its Applications
PEÑA, TIAO, and TSAY · A Course in Time Series Analysis
PIANTADOSI · Clinical Trials: A Methodologic Perspective
PORT · Theoretical Probability for Applications
POURAHMADI · Foundations of Time Series Analysis and Prediction Theory
PRESS · Bayesian Statistics: Principles, Models, and Applications
PRESS · Subjective and Objective Bayesian Statistics, *Second Edition*
PRESS and TANUR · The Subjectivity of Scientists and the Bayesian Approach
PUKELSHEIM · Optimal Experimental Design
PURI, VILAPLANA, and WERTZ · New Perspectives in Theoretical and Applied Statistics
PUTERMAN · Markov Decision Processes: Discrete Stochastic Dynamic Programming
*RAO · Linear Statistical Inference and Its Applications, *Second Edition*
RAUSAND and HØYLAND · System Reliability Theory: Models, Statistical Methods and Applications, *Second Edition*
RENCHER · Linear Models in Statistics
RENCHER · Methods of Multivariate Analysis, *Second Edition*
RENCHER · Multivariate Statistical Inference with Applications

*Now available in a lower priced paperback edition in the Wiley Classics Library.

RIPLEY · Spatial Statistics
RIPLEY · Stochastic Simulation
ROBINSON · Practical Strategies for Experimenting
ROHATGI and SALEH · An Introduction to Probability and Statistics, *Second Edition*
ROLSKI, SCHMIDLI, SCHMIDT, and TEUGELS · Stochastic Processes for Insurance and Finance
ROSENBERGER and LACHIN · Randomization in Clinical Trials: Theory and Practice
ROSS · Introduction to Probability and Statistics for Engineers and Scientists
ROSSI, ALLENBY, and McCULLOCH · Bayesian Statistics and Marketing
ROUSSEEUW and LEROY · Robust Regression and Outlier Detection
RUBIN · Multiple Imputation for Nonresponse in Surveys
RUBINSTEIN · Simulation and the Monte Carlo Method
RUBINSTEIN and MELAMED · Modern Simulation and Modeling
RYAN · Modern Regression Methods
RYAN · Statistical Methods for Quality Improvement, *Second Edition*
SALTELLI, CHAN, and SCOTT (editors) · Sensitivity Analysis
*SCHEFFÉ · The Analysis of Variance
SCHIMEK · Smoothing and Regression: Approaches, Computation, and Application
SCHOTT · Matrix Analysis for Statistics
SCHOUTENS · Lévy Processes in Finance: Pricing Financial Derivatives
SCHUSS · Theory and Applications of Stochastic Differential Equations
SCOTT · Multivariate Density Estimation: Theory, Practice, and Visualization
*SEARLE · Linear Models
SEARLE · Linear Models for Unbalanced Data
SEARLE · Matrix Algebra Useful for Statistics
SEARLE, CASELLA, and McCULLOCH · Variance Components
SEARLE and WILLETT · Matrix Algebra for Applied Economics
SEBER · Multivariate Observations
SEBER and LEE · Linear Regression Analysis, *Second Edition*
SEBER and WILD · Nonlinear Regression
SENNOTT · Stochastic Dynamic Programming and the Control of Queueing Systems
*SERFLING · Approximation Theorems of Mathematical Statistics
SHAFER and VOVK · Probability and Finance: It's Only a Game!
SMALL and McLEISH · Hilbert Space Methods in Probability and Statistical Inference
SRIVASTAVA · Methods of Multivariate Statistics
STAPLETON · Linear Statistical Models
STAUDTE and SHEATHER · Robust Estimation and Testing
STOYAN, KENDALL, and MECKE · Stochastic Geometry and Its Applications, *Second Edition*
STOYAN and STOYAN · Fractals, Random Shapes and Point Fields: Methods of Geometrical Statistics
STYAN · The Collected Papers of T. W. Anderson: 1943–1985
SUTTON, ABRAMS, JONES, SHELDON, and SONG · Methods for Meta-analysis in Medical Research
TANAKA · Time Series Analysis: Nonstationary and Noninvertible Distribution Theory
THOMPSON · Empirical Model Building
THOMPSON · Sampling, *Second Edition*
THOMPSON · Simulation: A Modeler's Approach
THOMPSON and SEBER · Adaptive Sampling
THOMPSON, WILLIAMS, and FINDLAY · Models for Investors in Real World Markets
TIAO, BISGAARD, HILL, PEÑA, and STIGLER (editors) · Box on Quality and Discovery: with Design, Control, and Robustness

*Now available in a lower priced paperback edition in the Wiley Classics Library.

TIERNEY · LISP-STAT: An Object-Oriented Environment for Statistical Computing and Dynamic Graphics
TSAY · Analysis of Financial Time Series
UPTON and FINGLETON · Spatial Data Analysis by Example, Volume II: Categorical and Directional Data
VAN BELLE · Statistical Rules of Thumb
VESTRUP · The Theory of Measures and Integration
VIDAKOVIC · Statistical Modeling by Wavelets
WEISBERG · Applied Linear Regression, *Second Edition*
WELSH · Aspects of Statistical Inference
WESTFALL and YOUNG · Resampling-Based Multiple Testing: Examples and Methods for p-Value Adjustment
WHITTAKER · Graphical Models in Applied Multivariate Statistics
WINKER · Optimization Heuristics in Economics: Applications of Threshold Accepting
WONNACOTT and WONNACOTT · Econometrics, *Second Edition*
WOODING · Planning Pharmaceutical Clinical Trials: Basic Statistical Principles
WOOLSON and CLARKE · Statistical Methods for the Analysis of Biomedical Data, *Second Edition*
WU and HAMADA · Experiments: Planning, Analysis, and Parameter Design Optimization
YANG · The Construction Theory of Denumerable Markov Processes
*ZELLNER · An Introduction to Bayesian Inference in Econometrics
ZELTERMAN · Discrete Distributions: Applications in the Health Sciences
ZHOU, OBUCHOWSKI, and McCLISH · Statistical Methods in Diagnostic Medicine

*Now available in a lower priced paperback edition in the Wiley Classics Library.

Printed and bound in the UK by
CPI Antony Rowe, Eastbourne